Student Interactive Workbook

Biology
The Unity and Diversity of Life

TWELFTH EDITION

Cecie Starr

Ralph Taggart

Christine Evers

Lisa Starr

Prepared by

Dave Cox
Lincoln Land Community College

Marty Zahn
Thomas Nelson Community College

Karen Francl
Redford University

Richard W. Cheney, Jr.
Christopher Newport University

Kathleen Hecht
Nassau Community College

Michael Windelspecht
Appalachian State University

Cyndi Maurstad
San Jacinto College – South

Jane B. Taylor
Northern Virginia Community College

Wenda Ribeiro
Thomas Nelson Community College

John D. Jackson
North Hennepin Community College

BROOKS/COLE
CENGAGE Learning

Australia • Brazil • Japan • Korea • Mexico • Singapore • Spain • United Kingdom • United States

For product information and technology assistance, contact us at
**Cengage Learning Customer & Sales Support,
1-800-354-9706**

For permission to use material from this text or product, submit all requests online at **www.cengage.com/permissions**
Further permissions questions can be emailed to
permissionrequest@cengage.com

ISBN-13: 978-0-495-55807-1
ISBN-10: 0-495-55807-9

Brooks/Cole
10 Davis Drive
Belmont, CA 94002-3098
USA

Cengage Learning is a leading provider of customized learning solutions with office locations around the globe, including Singapore, the United Kingdom, Australia, Mexico, Brazil, and Japan. Locate your local office at: **www.cengage.com/international**

Cengage Learning products are represented in Canada by Nelson Education, Ltd.

To learn more about Brooks/Cole, visit **www.cengage.com/brookscole**

Purchase any of our products at your local college store or at our preferred online store **www.ichapters.com**

Printed in the United States of America
1 2 3 4 5 6 7 12 11 10 09 08

CONTENTS

PREFACE

Tell me and I will forget, show me and I might remember, involve me and I will understand.
—Chinese Proverb

The proverb outlines three levels of learning in increasing order of effectiveness. The writer of the proverb understood that humans learn most efficiently when they involve themselves in the material to be learned. This student workbook is like a tutor; when properly used it increases the efficiency of your study periods. The interactive exercises actively involve you in the most important terms and central ideas of your text. Specific tasks ask you to recall key concepts and test your understanding of the facts. As you complete the exercises, you will discover items that you need to reexamine or clarify. Your performance on these tasks provides an estimate of your next test score based on specific material. Most importantly though, this biology student workbook, together with the text, helps you make informed decisions about matters that affect your own well-being and that of your environment. In the years to come, human survival on planet Earth will demand administrative and managerial decisions based on an informed background in biology.

HOW TO USE THIS STUDENT WORKBOOK

Following this preface, you will find an outline that will show you how the student workbook is organized and will help you use it efficiently. Each chapter in the student workbook begins with a title, chapter introduction, and list of focal points. The Interactive Exercises follow, wherein each chapter is divided into sections of one or more of the main (1st level) headings labeled 1.1, 1.2, and so on. *For easy reference to an answer or definition, each question and term in this unique student workbook is accompanied by the appropriate text page number(s).* The Interactive Exercises begin with a list of Selected Words (other than boldfaced terms) chosen by the authors as those most likely to enhance understanding of the material in the chapter. This is followed by a list of Boldfaced, Page-Referenced Terms that appear in the text. These terms are essential to understanding each student workbook section of a particular chapter. Space is provided by each term for you to formulate a definition in your own words. Next is the Interactive Exercises which is a series of exercises designed to enhance your understanding of the main topics in the chapter. These exercises include questions types like matching, fill-in-the-blank, true/false, and labeling.

A Self-Test immediately follows the Interactive Exercises. This test is composed primarily of multiple-choice questions. Any wrong answers in the Self-Test indicate portions of the text you need to reexamine. A series of Chapter Objectives / Review Questions follow each Self-Test. These serve as a quick review of the tasks that you should be able to accomplish if you have understood the assigned reading in the text. The final part of each chapter is Integrating and Applying Key Concepts. It invites you to try your hand at applying major concepts to situations in which there is not necessarily a single correct answer. Therefore, answers are not provided in the answer section at the end of the student workbook. Your text generally will provide enough clues to get you started on an answer, but this part is intended to stimulate your thinking and provoke group discussions.

A person's mind, once stretched by a new idea, can never return to its original dimension.
—Oliver Wendell Holmes

STRUCTURE OF THIS STUDENT WORKBOOK

The following outline shows how each chapter in this student workbook is organized.

Chapter Number ────────▶

32

Chapter Title ────────▶

ANIMAL TISSUES AND ORGAN SYSTEMS

Introduction ────────▶

This includes a brief overview to the subject matter of the chapter.

Chapter 32 starts with a description of how cells are organized and connected to form organs, organ systems, and multicellular animals. The four types of animal tissues are described: epithelial, connective, muscle, and nerve. Vertebrate organ systems are introduced along with a more detailed description of how tissues are integrated into a specific organ system—the integumentary system.

Focal Points ────────▶

- Figure 32.2 [p.540] illustrates how cells are attached to each other to form coherent tissues.
- Figure 32.4 [p.541] shows various types of epithelium.
- Figure 32.5 and 32.6 [pp.542-543] illustrate connective tissues.
- Figure 32.8 [p.544] has images of muscle tissues.
- Figure 32.9 [p.545] shows a typical neuron.
- Figure 32.11 and 32.12 [pp.546-547] describe anatomical terms and outline the major human organ systems.
- Figure 32.13 [p.568] diagrams human skin structure.

Interactive Exercises ────────▶

The Interactive Exercises are divided into numbered sections by titles of main headings and page references. Each section begins with a list of author-selected words that appear in the chapter. This is followed by a list of important boldfaced, page-referenced terms from each section of the chapter. Each section ends with interactive exercises that vary in type and require constant interaction with the important chapter information

Self-Test ────────▶

This is a set of questions designed to provide a quick assessment of how well you understand the information from the chapter.

Chapter Objectives / Review Questions ────────▶

This section provides a list of concepts that you need to master before proceeding to the next chapter. Page numbers from the text are provided should you be unable to answer the questions.

Integrating and Applying Key Concepts ────────▶

These represent "big-picture" or "real-life" applications of the concepts presented in the chapter.

Answers to Interactive Exercises and Self-Test ────────▶

Answers for all the Interactive Exercises and the Self-Test can be found at the end of the student workbook. The answers are arranged by chapter number and main heading.

CREDITS AND ACKNOWLEDGMENTS

This page constitutes an extension of the copyright page. We have made every effort to tract the ownership of all copyrighted material and to secure permission from copyright holders. In the event of any question arising as to the use of any material, we will be pleased to make the necessary corrections in future printings. Thanks are due to the following authors, publishers, and agents for permission to use the material indicated.

Chapter 2
p. 16 (17): Lisa Starr with PDV ID:IBNA; H.R. Drew, R>M. Wing, T. Takano, C. Broka, S. Tanaka, K. Itakura, R.E. Dickerson; Structure of a B-DNA Dodecamer. Conformation and Dynamics, PNAS.
p. 17 (2): Lisa Starr with PDB ID:IBNA; H.R. Drew, R.M. Wing, T. Takano, C., Broka, S. Tanaka, K. Itakura, R.E> Dickerson; Structure of B-DNA Dodecamer. Conformation and Dynamics, PNAS.

Chapter 3
p. 32 (1): PDB files fro Klotho Biochemical Compounds Declarative Database.

Chapter 4
p. 42 (19-40): Lisa Starr

Chapter 22
p. 219 (1-6): D.J. Patterson / Seaphot Limited: Plant Earth Pictures.

Chapter 23
p. 233 (1-7): Lisa Starr
p. 226 (photo): A.&E. Boomford, Ardea, London; Lee Casebere

Chapter 24
p. 248 (28): Carolina Biological Supply Company

Chapter 28
p. 292 (1-7): Carolina Biological Supply Company
p. 295 (photo, right, top): C.E. Jeffree, et al., *Planta*, 172(1):20-37, 1987. Reprinted by permission of C.E. Jeffree and Springer-Verlag.
p. 295 (photo, right, bottom): Jeremy Burgess / SPL / Photo Researchers, Inc.
p. 296 (8-10): Chuck Brown

Chapter 29
p. 305 (13-15): Micrograph Chuck Brown

Chapter 30
p. 312 (1-4): © Gary Head
p. 317: Patricia Schulz

Chapter 36
p. 393: Yokochi and J. Rohen, Photographic Anatomy of the Human Body, 2nd Ed., Igaku-Shoin, Ltd., 1979

Chapter 47
p. 521: D. Robert Franz; art After Paul Hertz.

1

INVITATION TO BIOLOGY

INTRODUCTION

Chapter 1 introduces you to the study of biology, presenting terminology that help scientists interpret patterns in nature. This chapter discusses the levels of organization in nature and current classification systems (taxonomy), and introduces ideas that help explain the diversity of organisms today. Finally, it introduces the concepts of scientific theory and critical thinking, emphasizing the steps of the scientific method.

FOCAL POINTS

- Nature is comprised of many levels of organization, beginning at the level of atoms and extending to the level of the biosphere [pp.4-5, Figure 1.2].
- Energy flows through a system while nutrients cycle through a system [p.6, Figure 1.3]
- Organisms grow and reproduce, passing along inherited traits encoded in their DNA [p.7]
- Taxonomy classifies organisms based on their shared traits [pp.8-9]
- Mutations in DNA, combined with the process of natural selection, help explain why life is so diverse [p.10]
- Scientific investigation involves asking specific questions in nature, creating a hypothesis about a single variable, making observations, and drawing conclusions [pp.12-13, Figure 1.3]

INTERACTIVE EXERCISES

1.1. LIFE'S LEVELS OF ORGANIZATION [pp.4-5]

Selected Words: molecules of life [p.4]

Boldfaced, Page-Referenced Terms

[p.4] nature_____

_____

[p.5] emergent property _____

_____

Matching [pp.4-5]

Match each of the following levels of biological organization with its correct definition.

1. _____ molecule
2. _____ cell
3. _____ community
4. _____ ecosystem
5. _____ organ system
6. _____ organ
7. _____ population
8. _____ biosphere
9. _____ tissue
10. _____ atom
11. _____ multicelled organism

A. Interaction of two or more tissues to perform a common task

B. All the regions of Earth that hold organisms

C. The smallest unit of life that can survive and reproduce on its own

D. Interaction of organs physically or chemically to perform a common task

E. Two or more atoms bonded together

F. All populations of all species occupying a given area

G. The smallest unit that retains an element's properties

H. The interaction of a community and the physical environment

I. Group of individuals of the same species occupying a specific area

J. Organized arrays of cells that interact for a specific task

K. Individual composed of different types of cells

Sequence [pp.4-5]

Arrange the following levels of organization in nature in the correct hierarchical order. Write the letter of the least inclusive level next to 12. Write he letter of the most inclusive level next to 22.

12. _____
13. _____
14. _____
15. _____
16. _____
17. _____
18. _____
19. _____
20. _____
21. _____
22. _____

A. community
B. tissue
C. cell
D. organ
E. organ system
F. atom
G. biosphere
H. molecule
I. population
J. multicelled organism
K. ecosystem

1.2. OVERVIEW OF LIFE'S UNITY [pp.6-7]

Selected Words: receptor stimulation [p.6], internal environment [p.7]

Boldfaced, Page-Referenced Terms

[p.6] energy _____

_____

[p.6] nutrient _____

_____

[p.6] producer _____

_____

[p.6] consumer _____

_____

[p.6] photosynthesis _____

_____

[p.6] receptor _____

_____

[p.7] homeostasis _____

_____

[p.7] DNA _____

_____

[p.7] trait _____

_____

[p.7] inheritance _____

_____

[p.7] reproduction _____

_____

[p.7] development _____

_____

Fill-in-the-Blanks [p.6]

Energy refers to the capacity to do (1) _____. Organisms spend a great deal of time acquiring energy

and (2) _____, which are essential atoms or molecules that an organism cannot make for itself. If an organism

is able to acquire its energy and raw materials from the environment, and subsequently, make their own food, they

are called (3) _____. Plants are capable of making their own food through the process of (4) _____, in

which they obtain energy from sunlight and convert it to a usable form. On the other hand, (5) _____ cannot

make their own food and must acquire it by eating plants and other organisms.

Matching [p.7]

6. _____ homeostasis

7. _____ reproduction

8. _____ development

9. _____ inheritance

A. The actual transmission of traits (through DNA) from parents to offspring

B. Maintenance of a constant internal environment

C. The orderly transformation of the first cell of a new individual into an adult

D. The mechanism by which parents transmit DNA to offspring

1.3. OVERVIEW OF LIFE'S DIVERSITY [pp.8-9]

Selected Words: phylum [p.8], kingdom [p.8], domain [p.8]

Boldfaced, Page-Referenced Terms

[p.8] species _____

_____

[p.8] genus _____

_____

[p.8] bacteria _____

_____

[p.8] archaea _____

_____

[p.8] eukaryote _____

_____

[p.8] protist _____

_____

[p.8] fungi _____

_____

[p.9] plant _____

_____

[p.9] animal _____

_____

Choice [pp.8-9]

For each description in questions 1-7, choose the correct domains.

    a. Bacteria   b. Archaea   c. Eukarya

1. _____ Single-celled, lack nucleus; most ancient lineage
2. _____ Includes plants, animals, and fungi
3. _____ Includes all multicelled organisms
4. _____ Single-celled, lack nucleus; most closely related to more derived multicelled organisms
5. _____ Includes protists
6. _____ Often located in extreme environments, like hydrothermal vents and frozen desert rocks
7. _____ Contains a nucleus in each cell

1.4. AN EVOLUTIONARY VIEW OF DIVERSITY [p.10]

Selected Words: survival of the fittest [p.10], selective agents [p.10], artificial selection [p.10]

Boldfaced, Page-Referenced Terms

[p.10] mutation _____

_____

[p.10] adaptive trait _____

_____

[p.10] natural selection _____

_____

[p.10] evolution _____

_____

Dichotomous Choice [p.10]

Circle one of the two possible answers given between parentheses in each statement.

1. The differential survival and reproduction of individuals in a population that differ in the details of their inheritable traits is called (evolution / natural selection).

2. Human manipulation of breeding of pigeons is an effort to select for particular traits is an example of (natural / artificial) selection.

3. Individuals of a population show variation in their shared, heritable traits. Such variation arises through (mutations in / sharing of) DNA.

1.5. CRITICAL THINKING IN SCIENCE [p.11]

1.6. HOW SCIENCE WORKS [pp.12-13]

Selected Words: if-then process [p.12], scientific method [p.12]

Boldfaced, Page-Referenced Terms

[p.11] critical thinking _____

_____

[p.11] science _____

_____

[p.12] hypothesis _____

_____

[p.12] prediction _____

_____

[p.12] model _____

_____

[p.12] scientific theory _____

_____

[p.13] experiment _____

_____

[p.13] variable _____

_____

[p.13] experimental group _____

_____

[p.13] control group _____

_____

Sequence [pp.12-13]

Arrange the following steps of the scientific method in correct chronological order. Write the letter of the first step next to 1, the letter of the second step next to 2, and so on.

1. _____

2. _____

3. _____

4. _____

5. _____

6. _____

7. _____

A. Develop a hypothesis

B. Repeat the tests or devise new ones

C. Devise ways to test the accuracy of predictions drawn from the hypothesis (use of observations, models, and experiments)

D. Make a prediction, using the hypothesis as a guide; the "if-then" process

E. If tests do not provide expected results, check to see what might have gone wrong

F. Objectively analyze and report the results from tests and the conclusions drawn

G. Observe some aspect of nature and research what others have found out about it

Matching [pp.11-13]

Using Table 1.3 from your text as a guide, match each of the following aspects of a sunflower experiment with its correct description.

8. _____ hypothesis

9. _____ control group

10. _____ report

11. _____ question

12. _____ observation

13. _____ experimental group

14. _____ assessment

15. _____ prediction

A. Fertilizer may help my sunflowers grow taller

B. Does fertilizer help my sunflowers grow taller

C. Ten plants are not treated with fertilizer

D. Submit the results and conclusions of my sunflower fertilizer experiments to the scientific community for review and publication

E. If fertilizer helps my sunflowers grow, then plants treated with fertilizer will grow taller than plants not treated with fertilizer

F. Complie results from sunflower fertilizer experiments and draw conclusions fro them

G. Ten plants are treated with fertilizer

H. I noticed that many people buy fertilizer for their sunflowers

1.7. THE POWER OF EXPERIMENTAL TESTS [pp.14-15]

1.8. SAMPLING ERROR IN EXPERIMENTS [pp16]

Selected Words: single-variable experiment [p.15]

Boldfaced, Page-Referenced Terms

[p.16] sampling error _____

_____

Short Answer [pp.14-15]

1. Because the participants in the Olestra experiment were not selected randomly, what kinds of biases may have been introduced into the experiment? _____

    _____

    _____

2. Why do researchers try to design single-variable experiments, rather than examining multiple variables all at once? _____

    _____

    _____

True/False [pp.14-16]

If the statement is true, write a "T" in the blank. If the statement is false, make it correct by changing the underlined word and writing the correct word in the blank.

3. _____ Natural processes are often influenced by <u>many</u> interacting variables.

4. _____ Researchers unravel cause and effect in complex natural processes by studying the effects of <u>two</u> variables at a time.

5. _____ Sampling error is a difference between results from a subset and results from the whole. It happens most often when sample sizes are <u>small</u>.

SELF-TEST

_____ 1. Two or more atoms bonded together are
called a _____. [p.4]
a. molecule
b. organism
c. population
d. tissue

_____ 2. All populations of all species occupying a
given area are called a _____. [p.5]
a. biosphere
b. organ system
c. community
d. multicelled organism

_____ 3. An example of an organism that uses
photosynthesis to obtain energy from
sunlight is a _____. [p.6]
a. mouse
b. bird
c. tree
d. protist

_____ 4. An example of a prediction is: [p.12]
a. If my car will not start, then the battery
must be dead.
b. Why won't my car start?
c. If I recharge the battery, will my car
start?
d. My car is too old to start.

_____ 5. The only domain that includes
multicellular organisms with a nucleus is
_____. [p.8]
a. Archaea
b. Eubacteria
c. Bacteria
d. Eukarya

_____ 6. The orderly transformation of the first cell
of a new individual into an adult is called
_____. [p.7]
a. development
b. reproduction
c. inheritance
d. evolution

_____ 7. An example of a control group would be:
[p.13]
a. A group that is composed only of males
b. A group that does not receive the
treatment of the experimental variable
c. A group that starts the experiment later
than other groups
d. A group that receives two treatment
variables instead of three

_____ 8. In the scientific method, which of the
following should be the first step taken?
[p.12]
a. A prediction
b. An observation of nature
c. A hypothesis
d. Development of an experiment

_____ 9. Energy may be defined as _____. [p.6]
a. maintenance of a constant internal
environment
b. the ability to withstand photosynthesis
c. the capacity to do work
d. the ability to pass nutrients from
parents   to offspring

_____ 10. Different species of Galapagos Island
finches have different beak types to obtain
different kinds of food. One species
removes tree bark with sharp beak to forage
for insect larvae and pupae, whereas
another species has a large, powerful beak
capable of crushing and eating large, heavy,
coated seeds. _____ selection is
responsible for the evolution of beaks in
these finches. [p.10]
a. Adaptive
b. Natural
c. Artificial
d. Discriminate

_____ 11. Which of these statements is true? [p.6]
a. Nutrients cycle within a system, while
energy flows through a system.
b. Both nutrients and energy flow through
a system.
c. Both nutrients and energy cycle within
a system.
d. Nutrients flow through a system, while
energy cycles within a system.

CHAPTER OBJECTIVES/REVIEW QUESTIONS

This section lists general and detailed chapter objectives that can be used as review questions. You can make maximum use of these items by writing answers on a separate sheet of paper. Fill in answers where blanks are provided. To check for accuracy, compare your answers with information given in the chapter or glossary.

1. Diagram the levels of organization in nature, beginning with the *atom* and ending with the *biosphere*. [pp.4-5]
2. Distinguish between the terms *energy* and *nutrient*. [p.6]
3. Describe how producers are different from consumers [p.6]
4. Explain why homeostasis is important for living organisms. [p.7]
5. Describe the differences among inheritance, reproduction, and development. [p.7]
6. Distinguish among the three domains: Eukarya, Bacteria, and Archaea. [p.8]
7. Distinguish between natural and artificial selection and give an example of each. [p.10]
8. Describe how adaptive traits are related to evolution. [p.10]
9. Define *scientific theory* and give an example of one. [p.12]
10. List and explain the general steps used in the scientific method. [p.12]
11. Distinguish between a control group and experimental group in a scientific test. [p.13]
12. Define *sampling error* and give an example of one. [p.16]

INTEGRATING AND APPLYING KEY CONCEPTS

1. Humans have the ability to maintain body temperature very close to 37°C.
   a. What conditions would tend to make body temperature drop?
   b. What measures do you think your body takes to raise body temperature when it drops?
   c. What conditions would cause body temperature to rise?
   d. What measures do you think your body takes to lower body temperature when it rises?
2. If a theory has not been proven false in 100 years of testing, can we conclude that it is a scientific truth? Why or why not?
3. What would happen if scientists failed to use control groups in all of their experiments? How might this change our interpretation of their results?
4. Can we use the scientific method to answer the following question: "Is it wrong for me to be angry when squirrels take birdseed out of my birdfeeder?" Justify your answer.
5. Re-read the potato chip experiment described on page 14 of your text. Name two ways that you might improve the quality of the experiment if you were to complete it again.

2

LIFE'S CHEMICAL BASIS

INTRODUCTION

This chapter looks at the basic chemistry needed to understand biology. Topics covered include atoms and their structure, chemical bonding, properties of water, and acids, bases, and buffers. Many topics discussed here will reappear throughout the book, so it is important that you understand these concepts before moving on.

FOCAL POINTS

- Figure 2.1 [p.20] looks at the chemicals that make up humans, and the percentages of the more common ones in the human body.
- Using Figure 2.3 [p.22] you can determine the types of bonding an atom might undergo.
- Figure 2.5 [p.24] demonstrates how to form shell models which reveal electron vacancies in atoms.
- Figure 2.6 [p.25] demonstrates how ions are formed.
- Figure 2.8 [p.27] looks at the various kinds of bonds found in biological molecules.
- Figure 2.9 [p.27] shows how hydrogen bonds are formed and how, while weak individually, collectively they can stabilize the structure of large molecules.
- Section 2.6 [pp.30-31] and Figure 2.13 [p.30] describe acids, bases, and buffers.
- Table 2.2 [p.32] summarizes the most important concepts of the chapter.

INTERACTIVE EXERCISES

Impacts, Issues: What Are You Worth? [p.20]

2.1. START WITH ATOMS [p.22]

2.2. PUTTING RADIOISOTOPES TO USE [p.23]

Selected Words: atomic number [p.22], mass number [p.22], isotope [p.22], PET [p.23]

Boldfaced, Page-Referenced Terms

[p.22] element _____

_____

[p.22] atom _____

_____

[p.22] proton _____

_____

[p.22] electron _____

_____

[p.22] neutron _____

_____

[p.22] periodic table of the elements _____

_____

[p.23] radioisotope _____

_____

[p.23] radioactive decay _____

_____

[p.23] tracer _____

_____

Matching [pp.22-23]

Choose the most appropriate answer for each term.

1. _____ atoms
2. _____ protons
3. _____ nucleus
4. _____ PET
5. _____ neutrons
6. _____ electrons
7. _____ atomic number
8. _____ mass number
9. _____ elements
10. _____ isotopes
11. _____ radioisotopes
12. _____ tracers
13. _____ Henri Becquereal
14. _____ Dmetry Mendeleev
15. _____ Melvin Calvin

A. Compounds that have a radioisotope attached to allow determination of the pathway or destination of a substance

B. Subatomic particles with a negative charge

C. Positively charged subatomic particles within the nucleus

D. Positron-emission tomography; obtains images of particular body tissues

E. Atoms of a given element that differ in the number of neutrons

F. The number of protons in an atom

G. A form of isotope that contains an unstable nucleus that emits energy and particles in an attempt to stabilize its structure

H. Creator of the Periodic Table of the Elements

I. The number of protons and neutrons in the nucleus of one atom

J. Fundamental forms of matter that occupy space, have mass, and cannot be broken down into something else

K. Smallest units that retain the properties of a given element

L. Subatomic particles within the nucleus carrying no charge

M. Discovered Uranium radioisotope

N. Used radioisotopes to identify specific reaction steps in photosynthesis

O. The atoms central core

True/False [pp.22-23]

If the statement is true, write a T in the blank. If the statement is false, rewrite it to make it correct.

16. _____ Ions are formed when an atom gains or loses neutrons

17. _____ The mass number of each element refers to the total number of protons and neutrons in its nucleus

18. _____ An inert gas reacts easily with other elements

19. _____ An atom of hydrogen consists of one proton and one electron

20. _____ Radioactive decay is dependent on external forces such as temperature and pressure

2.3. WHY ELECTRONS MATTER [pp.24-25]

Selected Words: electron orbital [p.24], lowest energy levels [p.24], shell model [p.24]

Boldfaced, Page-Referenced Terms

[p.25] chemical bond _____

_____

[p.25] molecule _____

_____

[p.25] compounds _____

_____

[p.25] mixture _____

_____

[p.25] electronegativity _____

_____

[p.24] ion _____

_____

Matching [pp.24-25]

Choose the most appropriate answer for each term.

1. _____ mixture
2. _____ shell model
3. _____ lowest energy level
4. _____ orbitals
5. _____ compounds
6. _____ molecule
7. _____ higher energy levels
8. _____ electronegativity
9. _____ ion

A. Regions of space around an atom's nucleus where electrons are likely to be at any one instant

B. Results when two or more atoms bond together

C. Two or more elements that may combine in various proportions

D. A graphic representation of the distribution of electrons in their energy levels

E. Types of molecules composed of two or more different elements in proportions that never vary

F. Energy of electrons farther from the nucleus than the first orbital

G. Energy of electrons in the orbital closest to the nucleus

H. An atom with a different number of electrons and protons

I. Measure of an atom's ability to pull electrons from other atoms

Complete the Table

10. Referring to the Periodic Table in the Appendix, enter the information for each of the following elements: *atomic number, atomic mass, number of protons, number of neutrons, and number of electrons*

a. sodium (Na) _____

_____

b. fluorine (F) _____

_____

c. carbon (C) _____

_____

d. hydrogen (H) _____

_____

e. oxygen (O) _____

_____

f. helium (He) _____

_____

g. chlorine (Cl) _____

_____

Fill-in-the-Blanks

An (11) _____ is uncharged only when it has as many electrons as protons. An (11) with different

numbers of electrons and protons is called an (12) _____. An (11) is formed when an atom gains or loses (13)

_____. A (14) _____ is an attractive force that arises between two atoms when their electrons interact.

Water is an example of a (15) _____. When you dissolve sugar in water, you form a (16) _____.

Identification [pp.24-25]

17. The following example demonstrates a shell model for the element helium. In Figures a–f, the number of protons and the number of neutrons are shown within the nucleus. For each energy level, indicate the number of electrons present, using the form 2e, 3e, and so on. In the space under each diagram, indicate the name of the element.

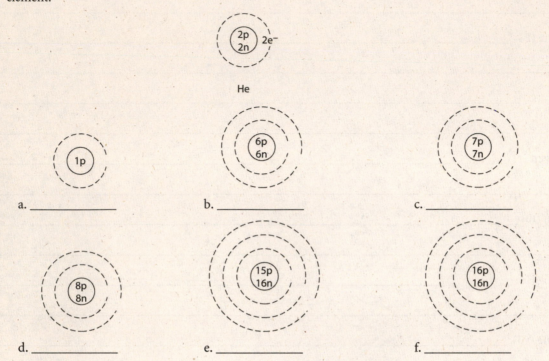

He

a. _____   b. _____   c. _____

d. _____   e. _____   f. _____

2.4. WHAT HAPPENS WHEN ATOMS INTERACT? [pp.26-27]

Selected Words: structural formula [p.26], chemical formula [p. 26], single covalent bond [p.26], double covalent bond [p.26], triple covalent bond [p.26], nonpolar covalent bond [p.27], polar covalent bond [p.27], hydrogen bond [p. 27]

Boldfaced, Page-Referenced Terms

[p.26] ion _____

_____

[p.26] ionic bond _____

_____

[p.26] covalent bond_____

_____

[p.27] hydrogen bond_____

_____

Short Answer [p.26]

1. When $_{11}$Na combines with $_{17}$Cl (the subscripts give the atomic number), the equation is written

    NA + Cl→NACl. Write a balanced equation for each of the following.

    a. $_{11}$Na and $_{16}$S _____

    b. $_{12}$Mg and $_{17}$Cl _____

    c. $_{19}$K and $_9$F _____

2. Referring to the following example of hydrogen gas, complete the two diagrams below it by placing electrons (as dots) in the outer shells to identify the covalent bonding that forms oxygen gas and water molecules.

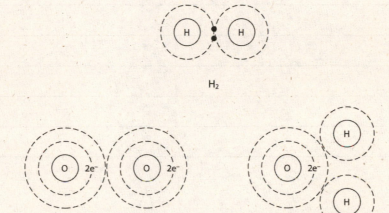

Short Answer [p.27]

3. Distinguish between a nonpolar covalent bond and a polar covalent bond. Cite an example of each.

    _____

    _____

    _____

4. Explain the importance of hydrogen bonds in establishing the structure of DNA.

    _____

    _____

    _____

2.5. WATER'S LIFE-GIVING PROPERTIES [pp.28-29]

Selected Words: no net charge [p.28], polarity [p.28], solvent [p.28], dissolved [p.28]

Boldfaced, Page-Referenced Terms

[p.28] hydrophilic_____

_____

[p.28] hydrophobic_____

_____

[p.29] temperature _____

_____

[p.29] evaporation _____

_____

[p.28] solute _____

_____

[p.29] cohesion_____

_____

Fill-in-the-Blanks

The (1) _____ of water molecules allows them to hydrogen-bond with each other. Water molecules hydrogen-bond with polar molecules, which are (2) _____ (water-loving) substances. Polarity causes water to repel oil and other nonpolar substances, which are (3) _____ (water-dreading). (4) _____ is a measure of the molecular motion of a given substance. Liquid water changes its temperature more slowly than air because of the great amount of heat required to break the high number of (5) _____ bonds between water molecules; this property helps (6) _____ temperature in aquatic habitats and cells. (7) _____ occurs when large energy inputs increase molecular motion to the point where hydrogen bonds stay broken, releasing individual molecules from the water surface. Below 0°C, water molecules become locked in the lattice-like bonding pattern of (8) _____, which is less dense than water. Water is an excellent (9) _____, in which ions and polar molecules readily dissolve. Substances dissolved in water are known as (10) _____. Collective hydrogen bonding creates a high tension on surface water molecules, resulting in (11) _____, the property of water that explains how long, narrow water columns rise to the tops of tall trees.

2.6. ACIDS AND BASES [pp.30-31]

Selected Words: donate H^+ [p.30], accept H^+ [p.30], neutral solutions [p.30], alkaline [p.30], acid stomach [p.30], acid rain [p.31], coma [p.31], respiratory acidosis [p.31], tetany [p.31], alkalosis [p.31]

Boldfaced, Page-Referenced Terms

[p.30] pH scale _____

_____

[p.30] acids _____

_____

[p.30] bases _____

_____

[p.31] salts _____

_____

[p.31] buffer systems _____

_____

Matching [pp.30-31]

Choose the most appropriate answer for each term.

1. _____ acids
2. _____ pH scale
3. _____ alkaline
4. _____ H^+
5. _____ bases
6. _____ OH
7. _____ buffer system
8. _____ salt

A. Hydroxide ion
B. Substances that accept H^+ when dissolved in water
C. Used to represent H^+ concentration in fluids
D. Partnership between a weak acid and the base that forms when it dissolves in water; counteracts slight pH shifts
E. Hydrogen ion or proton
F. Substances that donate H^+ when dissolved in water
G. A term used to describe a basic solution
H. A compound that dissolves easily in water and releases ions other than H^+ and OH^-

Ordering [p.30]

9. Refer to Fig 2.13 and order the following materials from most acidic to least acidic. Include the pH of each (round the pH to the nearest whole number):

    ammonia, beer, black coffee, blood, hair remover, lemon juice, pure water, seawater

    _____

    _____

    _____

    _____

    _____

    _____

    _____

    _____

Short Answer [p. 30]

10. If pH changes from 3 to 5, what happens to the hydrogen ion concentration? _____

    _____

    _____

_____

SELF-TEST

___ 1. Each element has a unique _____ which refers to the number of protons present in its atoms. [p.22]
   a. isotope
   b. mass number
   c. atomic number
   d. radioisotope

___ 2. An ionic bond is one in which _____. [p.26]
   a. electrons are shared equally
   b. electrically neutral atoms have a mutual attraction
   c. two charged atoms have a mutual attraction due to a transfer of electrons
   d. electrons are shared unequally

___ 3. A covalent bond is one in which _____. [p.26]
   a. electrons are shared
   b. electrically neutral atoms have a mutual attraction
   c. two charged atoms have a mutual attraction due to a transfer of electrons
   d. electrons are lost

___ 4. If neon has an atomic number of 10 and an atomic mass of 20, it has _____ neutron(s) in its nucleus. [p.22]
   a. one
   b. two
   c. five
   d. ten
   e. twenty

___ 5. Substances that are nonpolar and repelled
     by water are _____. [p.28]
     a. hydrolyzed
     b. nonpolar
     c. hydrophilic
     d. hydrophobic

___ 6. A nonpolar covalent bond implies that
     _____. [p.27]
     a. one negative atom bonds with a
        hydrogen atom
     b. it is a double bond
     c. there is no difference in charge at
        the ends (the two poles) of the bond
     d. atoms of different elements do not
        exert the same pull on shared
        electrons

___ 7. This type of bond contributes to the 3-D
     shape of large molecules. [p.27]
     a. hydrogen
     b. ionic
     c. covalent
     d. inert
     e. single

___ 8. A solution with a pH of 2 is_____
     times more acidic than one with a pH of 5.
     [p.30]
     a. 3
     b. 10
     c. 30
     d. 100
     e. 1,000

___ 9. A control that minimizes unsuitable pH
     shifts is a(n) _____. [p.30]
     a. hydrophilic compound
     b. salt
     c. base
     d. acid
     e. buffer

___ 10. Which of the following properties of water
     protects aquatic organisms during a long,
     cold winter? [p.29]
     a. Cohesion
     b. Solvent properties
     c. Temperature-stabilizing
     d. None of the above

CHAPTER OBJECTIVES/REVIEW QUESTIONS

1. Define *element*. [p.22]
2. What are the four most abundant elements in living organisms? [p.20]
3. List and describe the three types of subatomic particles. [p.22]
4. Distinguish between the atomic mass and atomic number of an element and tell what subatomic particles contribute to the value of each. [p.22]
5. Explain why helium, neon, and argon are known as inert elements. [p.22]
6. What information may be obtained about an atom from a shell model? [pp.24-25]
7. Explain the difference between molecules and compounds. [p.25]
8. How does a mixture differ from a compound? [p.25]
9. Explain why H2 is an example of a nonpolar covalent bond and why H2O has two polar covalent bonds. [p.27]
10. Explain the biological importance of hydrogen bonds. [p.27]
11. What is meant by the temperature of an object at the molecular level? [p.29]
12. What occurs at the molecular level during the process of evaporation? [p.29]
13. Describe what happens when a substance is dissolved in water. [p.28]
14. Explain the difference between acids and bases in terms of their relationship to hydrogen ions (H$^+$). [p.30]
15. Define *buffer system*; cite an example and describe how buffers operate. [p.31]

INTEGRATING AND APPLYING KEY CONCEPTS

1. Explain what would happen if water were a nonpolar molecule instead of a polar molecule. Would water be a good solvent for the same kinds of substances? Would the nonpolar molecule's specific heat likely be higher or lower than that of water? Would surface tension be affected? How about the cohesive nature? Would the ability to form hydrogen bonds change? Is it likely that the nonpolar molecules could form unbroken columns of liquid? What implications would that hold for trees?

2. What would be the implications for life on Earth if water did not have temperature-stabilizing effects? What would be different within individual organisms? What would change in ecosystems such as ponds and rivers? What would change on a global (biosphere) level? Why then do scientists search for evidence of water on other planets?

3. Interrelate the following terms in a concept map: atom, atomic number, electron, element, mass number, negative charge, neutron, nucleus, positive charge, and proton.

4. Why are buffers important to biological molecules? Do all buffers maintain the same pH? Why is this important?

3

MOLECULES OF LIFE

INTRODUCTION

Often students question why a course in the biological sciences needs to start with a discussion of chemistry. Yet, chemistry is the foundation of life. Your body is powered by chemical reactions that convert food into energy and raw materials for the building of cells and tissues. In this chapter you will examine the basic principles of organic chemistry, which focuses on the element carbon. You will then examine the biologically important organic compounds, namely the carbohydrates, fats, proteins, and nucleic acids.

FOCAL POINTS

- Figure 3.4 [p.38] identifies the common functional groups found in biological molecules.
- Figure 3.6 [p.39] illustrates condensation and hydrolytic reactions, which are the cornerstone of biochemistry. The principles are the same for all of the biologically important molecules.
- Figure 3.8 [p.41] shows the difference in structure between starch, cellulose, and glycogen, all polymers of glucose. The different structures are due to different monomer bonding patterns.
- Figure 3.10 [p.42] shows the formation of a triglyceride.
- Figure 3.13 [p.43] shows the structure of a phospholipid.
- Figure 3.15 [p.44] shows the generalized structure of an amino acid.
- Figure 3.17 [p.45] shows the different levels of protein structure and how they contribute to a three-dimensional molecule. Proteins are the working molecules of the cell. Their function is based on their shape.
- Figure 3.21 [p.48] shows the structure of the four DNA nucleotides.
- Table 3.2 [p.50] provides an excellent summary of the chapter.

INTERACTIVE EXERCISES

Impacts, Issues: Fear of Frying [p.34]

3.1. ORGANIC MOLECULES [pp.36-37]

3.2. FROM STRUCTURE TO FUNCTION [pp.38-40]

Selected Words: "organic" substances [p.36], "inorganic" substances [p.36], hydroxyl groups [p.38], carbonyl groups [p.38], carboxyl groups [p.38], phosphate groups [p.38], sulfhydryl groups [p.38], functional-group transfer [p.39], electron transfer [p.39], rearrangement [p.39], condensation [p.39], cleavage [p.39]

Boldfaced, Page-Referenced Terms

[p.36] organic compounds _____

_____

[p.38] functional groups _____

_____

[p.39] monomers _____

_____

[p.39] polymers_____

_____

[p.38] alcohols _____

_____

[p.39] condensation reaction _____

_____

[p.39] hydrolysis_____

_____

Fill-in-the-Blanks [pp.36-38]

The molecules of life are (1) _____ compounds, which are defined as containing the element (2) _____ and at least one (3) _____ atom. The term is a holdover from a time when chemists thought "organic" substances were the ones made naturally in living (4) _____ only, as opposed to (5) _____ substances that formed abiotically. The term persists, although scientists now synthesize organic compounds in (6) _____ and have reason to believe that organic compounds were present on Earth before organisms were. Carbon's importance to life starts with its versatile (7) _____ behavior. Each carbon atom can covalently bond with as many as (8) _____ other atoms. The (9) _____ consist only of hydrogen atoms covalently bonded to (10) _____. Examples are (11) _____ and other fossil fuels. Each organic compound has one or more (12) _____ groups, which are particular atoms or clusters of atoms covalently bonded to (13) _____. The hydroxyl functional group has a (14) _____ characteristic, which allows it to form (15) _____ bonds and dissolve quickly.

True/False [pp.36-40]

If the statement is true, write a T in the blank. If the statement is false, rewrite it to make it correct.

16. _____ Enzymes are a class of proteins that make reactions proceed faster than they would otherwise.

17. _____ All functional groups are polar.

18. _____ Hydrocarbons are the simplest organic molecule.

19. _____ Structural details give clues to how molecules function.

20. _____ Carbon only makes up a small percentage of living organisms.

Matching [pp.38-39]

Match each of the following functional groups with its correct description.

21. _____ sulfhydryl group

22. _____ hydroxyl group

23. _____ carbonyl group

24. _____ phosphate group

25. _____ carboxyl group

26. _____ amine group

A. Stabilizes the structure of proteins

B. Found within the structure of ATP and DNA

C. The functional group of the alcohols

D. Used in the building of fats and carbohydrates

E. A key component of the amino acids and fatty acids

F. Contains nitrogen and is important in some nucleotide base

Labeling [p.38]

For each of the following molecules, shade in all of the atoms associated with the indicated functional group. Use Figure 3.4 as a reference.

27. methyl

28. carbonyl

29. carboxyl

30. carbonyl

31. hydroxyl

32. amino

33. phosphate

Matching [p.39]

Choose the most appropriate answer for each term.

34. metabolism

35. condensation reaction

36. monomers

37. hydrolysis

38. polymers

39. functional-group transfer

40. cleavage

41. rearrangement

42. electron transfer

A. Activity by which cells acquire and use energy for cellular process

B. A type of reaction that splits molecules using water

C. The individual subunits of organic molecules

D. Any reaction that splits a molecule into two smaller molecules

E. The type of chemical reaction that moves electrons between molecules

F. The movement of functional groups between molecules

G. The formation of a covalent bond by the removal of –OH and H functional groups, forming water

H. Long chains of subunits, sometimes consisting of millions of individual subunits

I. A change in the internal bond structure of a molecule

Identification [pp.39, 44]

43. Study the structural formulas of the two adjacent amino acids. Identify the enzyme action causing formation of a covalent bond and a water molecule (through a condensation reaction) by circling an H atom from one amino acid and an –OH group from the other amino acid. Also circle the covalent bond that formed the dipeptide.

amino acid   amino acid   →   dipeptide

3.3. CARBOHYDRATES [pp.40-41]

Selected Words: "saccharide" [p.40], monosaccharide [p.40], oligosaccharide [p.40], disaccharides [p.40], "complex" carbohydrates [p.40], polysaccharides [p.40], sucrose [p.40], glucose [p.40], cellulose [p.41], glycogen [p.41], starch [p.41], chitin [p.41]

Identification [p.39]

1. In the diagram, identify condensation reaction sites between the two glucose molecules by circling the components of the water molecule whose removal allows a covalent bond to form between the glucose molecules (Figure 3.7, p.40). Note that the reverse reaction is hydrolysis and that both condensation and hydrolysis reactions require enzymes in order to proceed effificiently.

Choice [p.40]

Choose the class of carbohydrates (a–c) associated with the terms in items 2–9.

    a. oligosaccharides   b. polysaccharides   c. monosaccharides

2. _____ "complex" carbohydrates

3. _____ chitin

4. _____ disaccharides

5. _____ ribose and deoxyribose

6. _____ lactose, sucrose, and maltose

7. _____ glucose and fructose

8. _____ starch and glycogen

9. _____ cellulose

Fill In [p.41]

    (10) _____ is a structural carbohydrate associated with plant cell walls. The chains of (11) _____ used to form this molecule are stabilized by (12) _____ bonds. (13) _____ is also a polymer of glucose and is the primary energy source for plants. (14) _____ is a third class of glucose polymer which provides back up energy stores for animals.

Matching [pp.40-41]

Match each of the following carbohydrates with its correct function.

15. _____ sucrose

16. _____ chitin

17. _____ glucose

18. _____ ribose

19. _____ cellulose

20. _____ amylose

21. _____ glycogen

A. Instant energy source for most organisms; precursor of many organic molecules; serves as building block for larger carbohydrates

B. A form of starch produced by plants to store excess sugars

C. Most plentiful sugar in nature, formed from glucose and fructose

D. Animal starch that is stored in liver and muscle tissue of mammals

E. Main structural material in some external skeletons and other hard body parts of some animals and fungi

F. Structural material of plant cell walls

G. Five-carbon sugar occurring in DNA and RNA

3.4. GREASY, OILY—MUST BE LIPIDS [pp.42-43]

Selected Words: unsaturated [p.42], saturated [p.42], "vegetable oils" [p.42], neutral fats [p.42], glycerol [p.42], adipose tissue [p.42], trans fats [p.42], cholesterol [p.43]

Boldfaced, Page-Referenced Terms

[p.42] lipids_____

_____

[p.42] fats _____

_____

[p.42] fatty acid_____

_____

[p.42] triglycerides _____

_____

[p.43] phospholipids_____

_____

[p.43] waxes _____

_____

[p.43] steroids_____

_____

Labeling [p.42]

1. In the appropriate blanks, label the molecules shown as *saturated* or *unsaturated*. For the unsaturated molecules, circle the regions that make them unsaturated.

a. _____

b. _____

c. _____

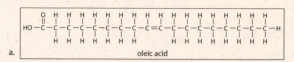

a.
   oleic acid

b.
   stearic acid

c.
   linolenic acid

Choice [pp.42-43]

For questions 2-17, choose from the answers below. Some answers may be used more than once.

   a. triglycerides   b. phospholipids   c. waxes   d. steroids

2. _____ The sex hormones are formed from this class

3. _____ Richest source of body energy

4. _____ The lipid found in honeycombs

5. _____ Cholesterol belongs to this class

6. _____ This class may have either saturated or unsaturated structure

7. _____ This molecule has a phosphate functional group in place of a fatty acid chain

8. _____ This is the only class of lipids that lack fatty acid tails

9. _____ The primary component of cell membranes

10. _____ All possess a rigid backbone of four fused carbon rings

11. _____ Found in the cuticles of plants

12. _____ Precursors of vitamin D and bile salts

13. _____ This class is made from three fatty acids combined with a unit of glycerol

14. _____ Vegetable oil belongs to this class

15. _____ Used by vertebrates for insulation

16. _____ Furnishes protection and lubrication for hair, skin, and feathers

17. _____ The neutral fats belong to this class

3.5. PROTEINS—DIVERSITY IN STRUCTURE AND FUNCTION [pp.44-45]

3.6. WHY IS PROTEIN STRUCTURE SO IMPORTANT? [pp.46-47]

Selected Words: peptide bond [p.43], primary structure [p.44], secondary structure [p.44], "domain" [p.44], tertiary structure [pp.44-45], quaternary structure [p.45], sickle-cell anemia [p.46]

Boldfaced, Page-Referenced Terms

[p.44] amino acid _____

_____

[p.44] polypeptide _____

_____

[p.44] protein _____

_____

[p.46] heme _____

_____

[p.46] denaturation _____

_____

Labeling [p.44]

1. In the following model of an amino acid, label the R group, amino group, and carboxyl group.

a. _____

b. _____

c. _____

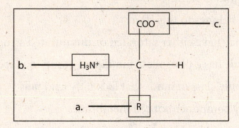

Matching [pp.44-47]

Choose the most appropriate statement for each term.

2. _____ amino acid
3. _____ peptide bond
4. _____ polypeptide chain
5. _____ primary structure
6. _____ secondary structure
7. _____ tertiary structure
8. _____ domain
9. _____ quaternary structure
10. _____ mutation
11. _____ amino acid substitution
12. _____ denaturation

A. Coils or twists in amino acids caused by hydrogen bonds

B. Three or more amino acids joined in a linear chain

C. An event that can alter protein structure enough to block or enhance its function

D. The type of covalent bond linking one amino acid to another

E. Globular proteins and hemoglobin are examples of this level of protein structure

F. The unwinding of protein structure causing a change in shape

G. The lowest level of protein structure consisting of a linear, unique sequence of amino acids

H. A small organic compound having an amino group, an acid group, a hydrogen atom, and an R group

I. The level of organization determined by interacting domain

J. A mutation that results in replacement of the correct amino acid with an incorrect amino acid

K. A structurally stable unit of a polypeptide chain

Fill-in-the-Blanks [pp.45-47]

What is the take-home lesson? A protein's (13) _____ dictates its function. Hemoglobin, hormones, (14) and _____ transporters—such proteins help us survive. Twists and folds in their (15) _____ chains form anchors, or (16) _____-spanning barrels, or jaws that grip enemy agents in the body. (17) _____ can alter the chains enough to block or enhance an anchoring, transport, or defensive function. Sometimes the consequences are awful. Yet changes in sequences and functional (18) _____ also give rise to variation in (19) _____—the raw material for (20) _____.

For questions 21-27, choose from the answers below. Answers may be used more than once. Refer to Figure 3.17.

    a. primary protein structure   b. secondary protein structure

    c. tertiary protein structure   d. quartenary protein structure

21. _____ Each protein's unique sequence of amino acids

22. _____ A structurally stable unit known as a "domain" comprises this level of protein structure

23. _____ Involves polypeptide chains forming sheets and/or helices

24. _____ Held in place by hydrogen bonds

25. _____ Two or more polypeptide chains associated as one molecule

26. _____ Results in a functional protein

27. _____ Leads to the formation of fibrous proteins that are involved in structure and organization of cells and tissues

3.7. NUCLEIC ACIDS [pp.48-49]

Selected Words: adenine [p.48], thymine [p.48], guanine [p.48], cytosine [p.48]

Boldfaced, Page-Referenced Terms

[p.48] ATP _____

_____

[p.48] nucleic acids _____

_____

[p.48] DNA_____

_____

[p.48] RNA _____

_____

Labeling [p.48]

1. In the following diagram of a nucleotide, label the phosphate groups, nitrogenous base, and five-carbon sugar subunits.

    a. _____

    b. _____

    c. _____

Matching [p.48]

Choose the most appropriate answer for each term.

2. _____ adenosine triphosphate

3. _____ nucleic acids

4. _____ double helix

5. _____ RNAs

6. _____ DNA

A. A molecule that can transfer phosphate groups, making molecules reactive

B. Single-stranded molecules made of ribonucleotides

C. Chains of nucleotides in which the sugar of one is bonded at the phosphate group of the other

D. The signature molecule of life, made from deoxyribonucleotides

E. Two nucleotide chains twisted together to form a DNA molecule

SELF-TEST

Choice

For questions 1-10, choose the class of organic molecule to which the item belongs. Some answers may be used more than once.

a. lipids [pp.42-43]   b. nucleic acids [pp.48-49]   c. proteins [pp.44-47]   d. carbohydrates [pp.40-41]

1. _____ glycoproteins

2. _____ phospholipids

3. _____ glycogen

4. _____ adenosine triphosphate

5. _____ sucrose and maltose

6. _____ triglycerides

7. _____ DNA and RNA

8. _____ cholesterol

9. _____ glycogen and starch

10. _____ waxes

Multiple Choice

__ 11. Amino acids are linked by _____bonds to form the primary structure of a protein. [p.44-45]
   a. disulfide
   b. hydrogen
   c. ionic
   d. peptide

__ 12. Proteins _____. [pp.44-45]
   a. are weapons against disease-causing bacteria and other invaders
   b. are composed of amino acid subunits
   c. may act as hormones
   d. may function as enzymes
   e. all of the above

___ 13. Which of the following does not belong to the lipid class of organic molecules? [p.42]
   a. Sterols
   b. Waxes
   c. Phospholipids
   d. Glycoproteins
   e. Triglycerides

___ 14. DNA _____. [p.48]
   a. is one of the adenosine phosphates
   b. is one of the nucleotide coenzymes
   c. is double-stranded
   d. is composed of monosaccharides

___ 15. Denaturation is a change in _____ shape. [p.46]
   a. lipid
   b. carbohydrate
   c. protein
   d. nucleic acid

___ 16. Carbon is a part of so many different substances because _____. [p.36]
   a. carbon generally forms two covalent bonds with a variety of other atoms
   b. a carbon atom generally forms four covalent bonds with a variety of atoms
   c. carbon ionizes easily
   d. carbon is a polar compound

___ 17. Which of the following levels of protein structure is not correctly linked to its description? [p.44-45]
   a. Primary—the linear sequence of amino acids
   b. Secondary—coiling of a polypeptide due to the action of hydrogen bonds
   c. Tertiary—interactions between the domains of a protein
   d. Quaternary—chemical interactions between multiple polypeptide chains
   e. All of the above are correct

___ 18. _____ are molecules used by cells as structural materials, as energy transport molecules, or as storage forms of energy. [pp.40-41]
   a. Lipids
   b. Nucleic acids
   c. Carbohydrates
   d. Proteins

___ 19. Hydrolysis could be correctly described as the _____. [p.39]
   a. heating of a compound in order to drive off its excess water and concentrate its volume
   b. breaking of a long-chain compound into its subunits by adding water molecules to its structure between the subunits
   c. linking of two or more molecules by the removal of one or more water molecules
   d. constant removal of hydrogen atoms from the surface of a carbohydrate
   e. prime example of a condensation class of reactions

CHAPTER OBJECTIVES/REVIEW QUESTIONS

1. Distinguish between organic and inorganic compounds. [p.36]
2. Explain why carbon is important in the study of biological chemistry. [p.36]
3. Explain the role of a functional group. [p.38]
4. Be able to identify select functional groups. [p.38]
5. Describe the importance of enzymes to chemical reactions. [p.39]
6. Define the following: *functional-group transfer, electron transfer, rearrangement, condensation,* and *cleavage.* [p.39]
7. Define and distinguish between *condensation reactions* and *hydrolysis*; cite a general example. [p.39]
8. Give the general characteristics of a carbohydrate. [p.40]
9. Name and generally define the three classes of carbohydrates. Give an example of each. [pp.40-41]
10. For the complex carbohydrates, indicate which is associated with animals, which with plants, and which with fungi. [pp.40-41]
11. Give the general characteristics of lipids and list their general functions. [p.42]
12. Distinguish a saturated fatty acid from an unsaturated fatty acid. [p.42]
13. Describe the structure of triglycerides. [p.42]
14. Describe the structure of phospholipids and give their biological importance. [p.43]
15. Describe how waxes are used by organisms. [p.43]
16. Give the general characteristics of steroids and describe how their structure differs from the other classes of lipids. [p.43]
17. List the functions of steroids. [p.43]
18. Describe the structure of an amino acid and how amino acids form peptide bonds. [p.44]
19. Describe how the primary, secondary, tertiary, and quaternary structure of proteins results in complex three-dimensional structures. [pp.44-45]
21. Define *denaturation.* [p.46]
22. List the three parts of every nucleotide. [p.48]
23. Distinguish between DNA and RNA. [p.48]

INTEGRATING AND APPLYING KEY CONCEPTS

1. Humans can obtain energy from many different food sources. Do you think this ability is an advantage or a disadvantage in terms of long-term survival? Why?
2. Modern diets often focus on proteins as a primary energy source. What are the preferred energy molecules of the cell? What is the overall role of proteins? Do you think that the metabolic pathways have evolved for long-term protein use for energy?
3. Condensation reactions are sometimes called dehydration synthesis reactions. Explain why this term is also appropriate.
4. In the early 20th century, scientists thought that proteins might be the hereditary material. What characteristics of proteins would have supported this idea? Why wouldn't carbohydrates or lipids be good candidates for hereditary material?
5. A common statement in nutrition is "you are what you eat." Explain why this statement is both true and false.

4

CELL STRUCTURE AND FUNCTION

INTRODUCTION

For biologists, the cell is where it all begins—it is considered to be the fundamental unit of life. In this chapter you will begin to explore some of the basic principles of cell structure. As you will see, cells are intricate, efficient, biological machines. Even the bacteria, which some call "simple cells," are complex in their interactions with the environment. Your primary goal for this chapter is to understand the terminology used to describe cellular structures and their functions. Future chapters of the textbook will explore in greater detail how cells operate in their environment. Understanding the interrelated structure and function of cellular organelles will provide you with the foundation needed later on in the textbook.

FOCAL POINTS

- Figure 4.5 animated [p.56] demonstrates the relationship between cell surface area and cell volume.
- Figure 4.6 animated [p.57] illustrates the basic phospholipid structure of the cell membrane.
- Figure 4.7 animated [p.58] compares the structure of light and electron microscopes.
- Figure 4.9 [p.59] depicts relative sizes of objects and the visual tools necessary to see them.
- Figure 4.10 animated [p.60] shows the general body plan of a prokaryote.
- Table 4.1 [p.62] summarizes eukaryotic organelles and their functions.
- Figure 4.15 animated [p.63] is an excellent visual tool for studying plant and animal cellular structure and function.
- Figure 4.17 animated [p.65] shows how the structure of the nuclear envelope is related to its function.
- Figure 4.18 animated [p.66] illustrates the interrelationships between components of the endomembrane system.
- Figure 4.21 animated [p.69] reveals the internal structure of the chloroplast.
- Figure 4.22 animated [p.70] demonstrates the development of plant cell wall structure.
- Figure 4.25 animated [p.71] illustrates the different types of cellular junctions.
- Figure 4.27 animated [p.72] demonstrates the use of kinesin motor protein to transport organelles within the cell.
- Figure 4.29 animated [p.73] demonstrates the mechanism behind movement of cilia and flagella.
- Table 4.3 [p.75] is an excellent comparison of the cell components of prokaryotic and eukaryotic cells and their functions in those cells.

INTERACTIVE EXERCISES

Impacts, Issues: Food for Thought [p.52]

4.1. THE CELL THEORY [pp.54-55]

4.2. WHAT IS A CELL? [pp.56-57]

4.3. HOW DO WE SEE CELLS? [pp.58-59]

Selected Words: Escherichia coli [p.52], pathogen [p.52], micrometer [p.54], bacteria [p.54], cellulae [p.54], compound microscope [p.54], phospholipids [p.57], micrographs [p.58], phase-contrast microscopes [p.58], dyes[p.58], contrast [p.58], reflected light microscopes [p.58], fluorescence microscope [p.58], light-emitting tracer[p.58], wavelength [p.58], electron microscopes [p.58], resolve [p.59], magnetic fields [p.59], transmission electron microscopes [p.59], scanning electron microscope [p.59]

Boldfaced, Page-Referenced Terms

[p.55] cell theory _____

_____

[p.56] cell _____

_____

[p.56] eukaryotic cell _____

_____

[p.56] prokaryotic cell _____

_____

[p.56] plasma membrane _____

_____

[p.56] nucleus _____

_____

[p.56] nucleoid _____

_____

[p.56] cytoplasm _____

_____

[p.56] ribosomes _____

_____

[p.56] surface-to-volume ratio _____

_____

[p.57] lipid bilayer _____

_____

Matching [pp.56-57]

Choose the most appropriate answer for each term.

1. _____ prokaryotic cells
2. _____ plasma membrane
3. _____ cytoplasm
4. _____ ribosomes
5. _____ nucleus
6. _____ eukaryotic cells
7. _____ surface-to-volume ratio
8. _____ lipid bilayer
9. _____ nucleoid
10. _____ cell

A. An interior region of prokaryotic cells where DNA is found

B. The structural foundation of all cell membranes

C. The type of cell that lacks a nucleus

D. A physical relationship that constrains increases in cell size and shape

E. The smallest unit of life that retains all the properties of life

F. Organelles involved in building proteins

G. The thin outermost membrane of cells that separates metabolic activities from events outside the cell

H. In eukaryotic cells, the membranous sac that contains the DNA

I. The semifluid material between the plasma membrane and the region of DNA

J. A type of cell possessing internal membranes that divide the cytoplasm into compartments

Short Answer [p.55]

11. List the four basic principles of the cell theory.

_____

_____

_____

_____

_____

Matching [pp.58-59]

Choose the most appropriate statement for each term relating to microscopes.

12. _____ micrograph
13. _____ transmission electron microscope
14. _____ wavelength
15. _____ compound light microscope
16. _____ scanning electron microscope
17. _____ fluorescence microscope
18. _____ contrast

A. Glass lenses bend incoming light rays to form an enlarged image of a cell or another specimen

B. The distance from one wave's peak to the peak of the wave behind it

C. A narrow beam of electrons moves back and forth across the surface of a specimen coated with a thin metal layer

D. A photograph of an image formed with a microscope

E. Difference between light and dark sections of stained specimens; allows greater detail to be seen

F. Electrons pass through a thin section of cells to form an image of its internal details

G. A cell or molecule is the source of emitted light when a laser is focused on the specimen

Matching [pp.54-55]

Match the scientist(s) with the statement that best describes his contribution to science.

19. _____ Hans and Zacharias Janssen
20. _____ Antoni van Leeuwenhoek
21. _____ Robert Hook
22. _____ Robert Brown
23. _____ Matthias Schleiden and Theodor Schwann
24. _____ Rudolf Virchow

A. Discovered that cells divide into descendant cells

B. The first to identify a plant cell

C. Created the first compound microscope

D. First articulated the cell theory

E. First coined the term "cellulae" (cell)

F. The first to observe living microscopic "animalcules" and "beasties"

Completion [p.59]

Fill in the blanks to complete the following table. [p.59]

25. a. 1 _____ = 1/100 Meter;   b. 1 Meter = _____ cm

26. a. 1 _____ = 1/1000 Meter   b. 1 Meter = _____ mm

27. a. 1 _____ = 1/1,000,000 Meter   b. 1 Meter = _____ μm

28. a. 1 _____ = 1/1,000,000,000 Meter   b. 1 Meter = _____ nm

4.4. INTRODUCING PROKARYOTIC CELLS [pp.60-61]

4.5. MICROBIAL MOBS [p.61]

4.6. INTRODUCING EUKARYOTIC CELLS [p.62]

4.7. VISUAL SUMMARY OF EUKARYOTIC CELL COMPONENTS [p.63]

Selected Words: Domain Bacteria [p.60], Domain Archaea [p.60], plasma membrane [p.60], peptidoglycan [p.60], capsule [p.60], cytoplasm [p61], nucleoid [p.61], plasmid [p.61], microenvironment [p.61], nucleus [p.62], eu- [p.62], karyon [p.62], metabolic pathway [p.62], cytoskeleton [p.62]

Boldfaced, Page-Referenced Terms

[p.60] cell wall_____

_____

[p.60] flagella _____

_____

[p.60] pili _____

_____

[p.61] biofilms _____

_____

[p.62] organelle_____

_____

Matching [pp.60-61]

Match each of the following prokaryotic structures with the correct description.

1. _____ cell walls
2. _____ capsule
3. _____ plasma membrane
4. _____ flagella
5. _____ ribosomes
6. _____ pili
7. _____ nucleoid
8. _____ plasmid
9. _____ biofilm

A. Sites where polypeptide chains are built

B. Permeable to dissolved substances; composed of protein or peptidoglycan

C. Small circles of DNA containing a few genes

D. Selectively controls the movement of materials in and out of the cytoplasm

E. Slender cellular structures involved in mobility

F. Made of polysaccharides; allows bacteria to attach to surfaces

G. Community of prokaryotes living in a shared, sticky mass of polysaccharides and glycoproteins

H. Protein filaments that project from surface of bacterial cells; used for attachment

I. Region of bacterial cell where a single, circular DNA molecule is located

Matching [pp.62-63]

Select the function that most closely fits the eukaryotic cellular structures listed.

10. _____ lysosome
11. _____ Golgi body
12. _____ nucleus
13. _____ chloroplast
14. _____ endoplasmic reticulum (ER)
15. _____ cell wall
16. _____ mitochondrion
17. _____ cytoskeleton
18. _____ centriole

A. Protects DNA from damaging reactions in cytoplasm

B. Modifies new polypeptide chains; synthesizes lipids

C. Protects and structurally supports plant cells

D. Intracellular digestion; recycles materials

E. Produces ATP by aerobic respiration

F. Produces and organizes the microtubules

G. Finishes, sorts, and ships proteins, lipids, and enzymes

H. Photosynthetic organelle

I. Internal structural support for the cell

Labeling [p.63]

Identify each indicated part of the accompanying illustrations.

19. _____

20. _____

21. _____

22. _____

23. _____

24. _____

25. _____

26. _____

27. _____

28. _____

29. _____

30. _____

31 _____

32. _____

33. _____

34. _____

35. _____

36. _____

37. _____

38. _____

39. _____

40. _____

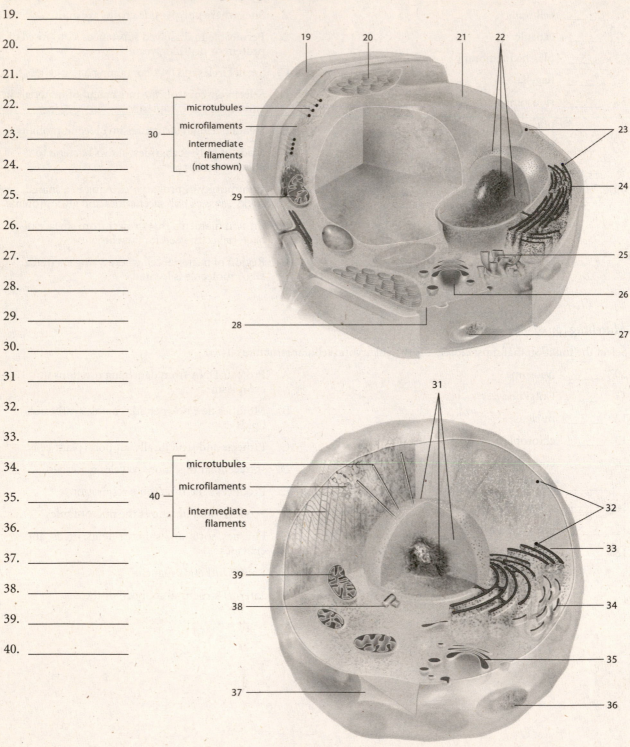

microtubules

microfilaments

intermediate
filaments
(not shown)

microtubules

microfilaments

intermediate
filaments

4.8. THE NUCLEUS [pp.64-65]

4.9. THE ENDOMEMBRANE SYSTEM [pp 66-67]

4.10. LYSOSOME MALFUNCTION [p.68]

Selected Words: nuclear membrane [p.64], membrane proteins [p.65], nuclear pores [p.65], rough ER [p.66], smooth ER [p.66], sarcoplasmic reticulum [p.66], gangliosides [p.68], Tay-Sachs disease [p.68]

Boldfaced, Page-Referenced Terms

[p.64] nuclear envelope _____

_____

[p.65] nucleoplasm _____

_____

[p. 65] nucleolus _____

_____

[p.65] chromatin_____

_____

[p.65] chromosome _____

_____

[p.66] endomembrane system _____

_____

[p.66] endoplasmic reticulum (ER) _____

_____

[p.67] vesicles _____

_____

[p.67] peroxisomes _____

_____

[p.67] vacuoles _____

_____

[p.67] Golgi body _____

_____

[p.67] lysosomes _____

_____

Short Answer [p.68]

1. Explain how a genetic mutation may cause an error in cell metabolism resulting in Tay-Sachs disease.

_____

_____

_____

Matching [pp.64-67]

Match each of the following terms with the corresponding description.

2. _____ chromosome

3. _____ nucleolus

4. _____ smooth ER

5. _____ nuclear envelope

6. _____ rough ER

7. _____ chromatin

8. _____ peroxisome

9. _____ lysosome

10. _____ Golgi body

11. _____ ribosome

12. _____ sarcoplasmic reticulum

13. _____ vesicles

14. _____ vacuoles

15. _____ nucleoplasm

A. Site of ribosome assembly

B. Finishing, sorting, and packaging of proteins and lipids

C. Viscous fluid within the nucleus

D. Contains enzymes for intracellular digestion

E. Folds polypeptides into tertiary-structured proteins

F. A type of smooth ER found in muscle cells

G. A double membrane that encloses the DNA

H. Makes lipids for the cell membrane

I. Membrane-bound intracellular trash cans

J. Synthesizes polypeptide chains

K. Small, membrane-enclosed saclike organelles

L. One DNA molecule and associated proteins

M. All the DNA and associated proteins in the nucleus

N. Contains enzymes to digest fatty acids, amino acids, alcohol, and hydrogen peroxide

Sequence [pp.66-67]

Place the numbers 1-7 in the spaces before each statement to put into proper sequence these events in the organelles of the endomembrane system.

16. _____ Rough ER folds a polypeptide chain into a protein with tertiary structure

17. _____ RNA moves through nuclear pores into the cytoplasm

18. _____ Golgi body adds the finishing touches to a protein and packages it for transport

19. _____ Using instructions from RNA, ribosomes synthesize polypeptide chains from amino acids

20. _____ DNA transcribes protein-making instructions into RNA

21. _____ A transport vesicle leaves the Golgi and merges with the plasma membrane

22. _____ A transport vesicle leaves the ER and travels to the Golgi body

4.11. OTHER ORGANELLES [pp.68-69]

Selected Words: aerobic respiration [p.68], endosymbiosis [p.68], chromoplasts [p.69], amyloplasts [p.69], grana [p.69], stroma [p.69], thylakoid membrane [p.69], ATP [p.69], NADPH [p.69]

Boldfaced, Page-Referenced Terms

[p.68] mitochondrion _____

_____

[p.69] plastids _____

_____

[p.69] chloroplasts _____

_____

[p.69] central vacuole _____

_____

Choice [pp.68-69]

For questions 1-15, choose from the following:

     a. mitochondria   b. chloroplasts   c. amyloplasts   d. central vacuole   e. chromoplasts

1. _____ Organelle specialized for photosynthesis in eukaryotic cells
2. _____ Organelle that makes ATP during aerobic respiration
3. _____ Organelle that lacks pigments and stores starch grains
4. _____ More plentiful in cells that have big demands for energy
5. _____ Organelle with an abundance of carotenoids but no chlorophylls
6. _____ Fluid pressure in this organelle keeps plant cells firm
7. _____ Organelles that convert sunlight energy into chemical energy
8. _____ Abundant in starch-storing cells of stems, potato tubers, and seeds
9. _____ Stores amino acids, sugars, ions, and toxic wastes in its fluid-filled interior
10. _____ Found in nearly all eukaryotic cells
11. _____ Function as gravity-sensing organelles
12. _____ Uses sunlight to synthesize ATP
13. _____ Builds carbohydrates from carbon dioxide and water
14. _____ Resemble bacteria in size, form, and biochemistry
15. _____ Stores pigments that give red or orange color to fruits, flowers and roots

Short Answer [p.68]

16. Define endosymbiosis. _____

_____

_____

4.12. CELL SURFACE SPECIALIZATIONS [pp.70-71]

4.13. THE DYNAMIC CYTOSKELETON [pp.72-73]

Selected Words: cell wall [p.70], pectin [p.70], cellulose [p.70], middle lamella [p.70], chitin [p.70], plasmodesmata [p.71], tight junction [p.71], adhering junction [p.71], gap junction [p.71], tubulin [p.72], actin [p.72], lamins [p.72], accessory proteins [p.72], kinesins [p.72], 9+2 array [p.73], "false feet" [p.73], dynein [p.73], basal body [p.73]

Boldfaced, Page-Referenced Terms

[p.70] primary wall _____

_____

[p.70] secondary wall _____

_____

[p.70] lignin _____

_____

[p.70] cuticle _____

_____

[p.70] extracellular matrix (ECM) _____

_____

[p.71] cell junctions _____

_____

[p.72] cytoskeleton _____

_____

[p.72] microtubules _____

_____

[p.72] microfilaments _____

_____

[p.72] cell cortex _____

_____

[p.72] intermediate filament _____

_____

[p.72] motor proteins _____

_____

[p.73] eukaryotic flagella _____

_____

[p.73] cilia _____

_____

[p.73] centriole _____

_____

[p.73] pseudopods _____

_____

Matching [pp.70-71]

Choose the most appropriate statement for each term.

1. _____ primary wall
2. _____ secondary wall
3. _____ tight junctions
4. _____ adhering junctions
5. _____ gap junctions
6. _____ plasmodesmata
7. _____ middle lamella
8. _____ pectin
9. _____ cellulose
10. _____ lignin
11. _____ cuticle
12. _____ extracellular matrix (ECM)

A. Seal abutting cells that line the outer surfaces and internal cavities of animals so no fluids can pass between them

B. The main constituent of a plant's primary cell wall

C. A strong, waterproof component of the secondary cell walls of older stems and roots

D. Layers of firm material secreted onto the primary cell wall's inner surface that strengthen the wall and maintain its shape

E. Thin, pliable structure made of polysaccharides that allows a plant cell to grow and enlarge

F. Nonliving mixture of fibrous proteins and polysaccharides surrounding the cells that secrete it

G. A sticky layer in between the primary walls of adjoining cells

H. Substance a young plant cell first secretes onto the outer surface of its plasma membrane

I. Cytoplasmic channels that cross the cell walls and join neighboring plant cells

J. A protective body covering made of cell secretions

K. Anchor cells to each other and to the extracellular matrix

L. Open channels that connect the cytoplasm of neighboring animal cells

Matching [pp.72-73]

Select the statement that best fits each component of the cytoskeleton.

13. _____ microtubules

14. _____ microfilaments

15. _____ cell cortex

16. _____ intermediate filaments

17. _____ motor protein

18. _____ kinesin

19. _____ flagella

20. _____ cilia

21. _____ centriole

22. _____ 9 + 2 array

23. _____ pseudopods

24. _____ dynein

A. Arrangement of microtubules lengthwise through a flagellum or cilium

B. A reinforcing mesh of microfilaments under the plasma membrane

C. Long hollow cylinders made of subunits of tubulin

D. A motor protein used to bend a flagellum

E. The "false feet" eukaryotic cells use in locomotion and obtaining food

F. Long slender structure used for motion; cells have one or a few of them

G. Fibers composed mainly of subunits of actin

H. A barrel-shaped organelle responsible for the growth of structures made of microtubules

I. The most stable part of the cytoskeleton; strengthen and maintain tissue structures

J. Short, slender structures used for coordinated motion; cells have many of them

K. A type of accessory protein that uses ATP to move cell parts along microtubule and microfilament tracks

L. A motor protein used to move chloroplasts toward or away from a light source

___ 1. Which of the following statements most correctly describes the relationship between cell surface area and cell volume? [p.56]
   a. As a cell expands in volume, its surface area increases at a rate faster than its diameter does
   b. Volume increases with the square of the diameter, but surface area increases only with the cube
   c. If a cell were to grow four times in diameter, its volume of cytoplasm increases 16 times and its surface area increases 64 times
   d. Volume increases with the cube of the diameter, but surface area increases only with the square

___ 2. The cellular structure in animals that is involved in the process of intracellular digestion is the _____ . [p.62]
   a. lysosome
   b. rough endoplasmic reticulum
   c. microtubules
   d. mitochondria

___ 3. Which of the following is not found in a prokaryotic cell? [pp.60-61]
   a. DNA
   b. plasma membrane
   c. cytoplasm
   d. nucleus

___ 4. The nucleolus is the site where _____. [p.65]
   a. the protein and RNA subunits of ribosomes are assembled
   b. the chromatin is formed
   c. chromosomes are bound to the inside of the nuclear envelope
   d. chromosomes duplicate themselves prior to cell division

___ 5. The _____ is free of ribosomes and curves through the cytoplasm like connecting pipe. It is the site of lipid synthesis and some detoxification. [p.66]
   a. lysosome
   b. Golgi body
   c. smooth ER
   d. rough ER

___ 6. Which of the following is not present in all cells? [p.56, 63]
   a. cell wall
   b. plasma membrane
   c. ribosomes
   d. DNA molecules

___ 7. A part of the endomembrane system the _____ finishes, packages, and sorts lipids and proteins. [p.62]
   a. endoplasmic reticulum
   b. Golgi bodies
   c. peroxisomes
   d. lysosomes

___ 8. Chloroplasts _____ [p.69]
   a. are specialists in oxygen-requiring reactions
   b. function as part of the cytoskeleton
   c. trap sunlight energy and produce organic compounds
   d. assist in carrying out cell membrane functions

___ 9. Aerobic respiration occurs in the _____. [p.68]
   a. chloroplast
   b. mitochondria
   c. peroxisome
   d. cytoplasm

___ 10. The component of the cell that is involved with cellular movement is the _____. [pp.72-73]
   a. nucleolus
   b. cell wall
   c. chromatin
   d. cytoskeleton

___ 11. The _____ controls construction of parts of the cytoskeleton in animals. [p.73]
   a. mitochondrion
   b. endoplasmic reticulum
   c. centriole
   d. ribosome

___ 12. _____ allow communication between the cytoplasm of adjacent cells. [p.71]
   a. tight junctions
   b. adhering junctions
   c. gap junctions
   d. all of the above

___ 13. The _____ creates an image by passing a
beam of electrons back and forth across the
surface of a specimen coated with a thin
layer of metal. [p.59]
a. compound light microscope
b. fluorescence microscope
c. transmission electron microscope
d. scanning electron microscope

___ 14. _____ is the sticky substance that holds
young plant cells together. [p.70]
a. pectin
b. cellulose
c. lignin
d. chitin

CHAPTER OBJECTIVES/REVIEW QUESTIONS

1. Describe the basic components of a cell. [p.56]
2. Explain how the surface-to-volume ratio of a cell determines its size. [p.56]
3. Distinguish among the different forms of microscopes including their uses and limitations. [pp.58-59]
4. Describe the basic structure of a prokaryotic cell. [pp.60–61]
5. Describe the differences between prokaryotic and eukaryotic cells. [pp.60–62]
6. Explain the advantage of complex compartmentalization (organelles) in eukaryotic cells. [p.62]
7. Briefly describe the cellular location and function of the organelles typical of most eukaryotic cells: nucleus, ribosomes, endoplasmic reticulum, Golgi body, various vesicles, mitochondria, and the cytoskeleton. [pp.62, 75]
8. List the two main functions of a nucleus. [p.64]
9. Describe the structure of a eukaryotic nucleus. [pp.64-65]
10. Distinguish between a chromosome and chromatin. [p.65]
11. List the organelles that are part of the endomembrane system and their functions. [pp.66-67]
12. Explain the variety of tasks that lysosomes and peroxisomes perform for a eukaryotic cell. [p.67]
13. Understand the role of the mitochondria. [p.68]
14. Give the general function of the following plant organelles: chloroplasts, chromoplasts, amyloplasts, and the central vacuole. [p.69]
15. Understand the purpose of a cell wall and the types of eukaryotic cells that utilize a cell wall. [p.70]
16. Explain the difference in function between gap junctions, adhering junctions, and tight junctions. [p.71]
17. Distinguish between microtubules, microfilaments, and intermediate • laments with regard to structure and function. [p.72]
18. Explain the three basic mechanisms of eukaryotic cell movement. [p.73]

INTEGRATING AND APPLYING KEY CONCEPTS

1. Which parts of a cell constitute the minimum for keeping the simplest of living cells alive?
2. How did the existence of a nucleus, compartments, and extensive internal membranes confer selective advantages on cells that developed these features?
3. Cells are frequently described as biological factories. Compare a typical eukaryotic cell to a modern automobile factory. Describe the function of the nucleus, endoplasmic reticulum, mitochondria, Golgi body, ribosomes, vesicles, lysosomes, and peroxisomes in relation to a factory.
4. Dental plaque is a biofilm that supports the growth of many types of bacteria including Streptococcus mutans, a major factor in the formation of cavities. Consider the nature of a biofilm and explain the benefits of frequent brushing and flossing of teeth.
5. Suppose that you wanted to develop a drug that inhibited the ability of a eukaryotic cell to export proteins. Explain what possible organelles you would target and the basic mechanism by which your drug would work.

5

A CLOSER LOOK AT CELL MEMBRANES

INTRODUCTION

This chapter details the structure and function of biological membranes. It looks at the lipid bilayer as a barrier and at the functions of the various integral and peripheral proteins. Diffusion and other mechanisms of transmembrane transport are discussed with special emphasis on the diffusion of water across membranes, called osmosis.

FOCAL POINTS

- Figures 5.3 and 5.4 [p.79] show the structure of the fluid mosaic model of biological membranes and have an overview of a related experiment.
- Figure 5.5 [pp.80-81] provides a detailed explanation of the function of membrane proteins.
- Figure 5.6 [p.82] illustrates the concept of selectively permeability.
- Figure 5.8 [p.83] compares passive and active forms of transport.
- Figure 5.9 [p.84] provides a detailed examination of passive transport.
- Figure 5.10 [p.85] provides a detailed examination of active transport.
- Figure 5.12 [p.86] provides a detailed overview of endocytosis and exocytosis.
- Figure 5.17 [p.89] looks at the importance of osmosis in biological systems.
- Section 5.5 [pp.86-87] takes an in-depth look at movement across the membrane with special emphasis on endocytosis and exocytosis.
- Section 5.6 [pp.88-89] looks at the importance of osmosis to biological systems.

INTERACTIVE EXERCISES

Impacts, Issues: One Bad Transporter and Cystic Fibrosis [p.76]

5.1. ORGANIZATION OF CELL MEMBRANES [pp.78-79]

5.2. MEMBRANE PROTEINS [pp.80-81]

Selected Words: cystic fibrosis [p.76], CTFR gene [p.76], mosaic [p.78], fluid [p.78], integral proteins [p.80], peripheral proteins [p.80], nonself and self [p.81]

Boldfaced, Page-Referenced Terms

[p.78] lipid bilayer _____

_____

[p.78] fluid mosaic model _____

_____

[p.80] transport proteins _____

_____

[p.80] receptor proteins _____

_____

[p.80] recognition proteins _____

_____

[p.80] adhesion proteins _____

_____

Matching [pp.78-81]

Choose the most appropriate answer for each term.

1. _____ fluid mosaic model
2. _____ phospholipid
3. _____ adhesion proteins
4. _____ transport proteins
5. _____ cholesterol
6. _____ integral proteins
7. _____ recognition proteins
8. _____ peripheral proteins
9. _____ receptor proteins
10. _____ lipid bilayer
11. _____ passive transporters
12. _____ ion-selective channels
13. _____ CTFR gene

A. The main steroid component of animal membranes

B. Proteins that allow molecules to move through the plasma membrane without expending energy

C. Transporter protein found in the plasma membrane of epithelial cells

D. A composition of phospholipids, proteins, sterols, and glycolipids

E. The general name for proteins that are physically embedded within the cell membrane

F. The primary component of the cell membrane; consists of both hydrophobic and hydrophilic regions

G. Bind extracellular substances that trigger changes in the cell's activity

H. Help cells of the same type stick together

I. A general group of proteins positioned at the surface of the membrane

J. Contain molecular gates that move small molecules

K. The double layer of phospholipids that forms the cell membrane

L. Allow materials to pass through the cell membrane using the interior of the protein; may or may not require energy

M. Act as molecular fingerprints to identify tissues or individuals

Short Answer [pp.76-78]

14. What is the cause of cystic fibrosis? _____

_____

_____

15. Explain the significance of the study illustrated in Figure 5.4B to our understanding of membrane structure. ____

_____

_____

16. With reference to the fluid mosaic model of the plasma membrane, what contributes to the mosaic nature? What contributes to the fluidity? _____

_____

_____

5.3. DIFFUSION, MEMBRANES, AND METABOLISM [pp.82-83]

5.4. PASSIVE AND ACTIVE TRANSPORT [pp.84-85]

Selected Words: net movement [p.82], "facilitated" diffusion [p.83], endocytosis [p.83], exocytosis [p.83]

Boldfaced, Page-Referenced Terms

[p.82] concentration gradient _____

_____

[p.82] diffusion _____

_____

[p.82] selective permeability _____

_____

[p.84] passive transport _____

_____

[p.85] active transport _____

_____

[p.85] calcium pump _____

_____

[p.85] sodium–potassium pump _____

_____

[P.85] cotransporter _____

_____

Fill-in-the-Blanks [pp.82-83]

If the concentration of a substance in one region differs from that in an adjoining region, it is called a(n) (1) _____. A(n) (2) _____ is a difference between the number of molecules or ions of a given substance in adjoining regions. (3) _____ is the name for the net movement of like molecules or ions down a concentration gradient; it is a factor in the movement of substances across cell membranes and through cytoplasmic fluid. The five factors that influence the rate of diffusion are (4) _____, (5) _____, (6) _____, (7) _____, and (8) _____. If a membrane has selective (9) _____, it possesses a molecular structure that permits some substances but not others to cross it in certain ways, at certain times. For example, the lipid bilayer of biological membranes is impermeable to (10) _____ and (11) _____ molecules while allowing (12) _____ and (13) _____ to pass. (14) _____ transport uses energy to move down the molecular concentration gradient without the need for energy. (15) _____ transport uses energy to pump a solute (16) _____ its concentration gradient. In the bulk movement of substances across a membrane, the process of (17) _____ moves particles into the cell by forming a vesicle from the plasma membrane. In (18) _____, a membrane-bound vesicle inside the cell fuses with the plasma membrane, allowing particles to exit the cell.

5.5. MEMBRANE TRAFFICKING [pp.86-87]

Selected Words: receptor-mediated endocytosis [p.86], bulk-phase endocytosis [p.87], membrane cycling [p.87], pseudopod [p.87]

Boldfaced, Page-Referenced Terms

[p.86] endocytosis_____

_____

[p.86] exocytosis_____

_____

[p.86] phagocytosis_____

_____

Matching [pp.86-87]

Choose the most appropriate answer for each term.

1. _____ exocytosis
2. _____ receptor-mediated endocytosis
3. _____ bulk-phase endocytosis
4. _____ phagocytosis
5. _____ membrane cycling

A. A cell engulfs microorganisms, large edible particles, and cellular debris

B. Membrane initially used for endocytic vesicles returns receptor proteins and lipids back to the plasma membrane

C. Vesicles form around small volumes of extracellular fluid of various contents

D. A cytoplasmic vesicle moves to the cell surface; its own membrane fuses with the plasma membrane while its contents are released to the environment

E. Chemical recognition and binding of specific substances; coated pits sink into the cytoplasm and close on themselves

5.6. WHICH WAY WILL WATER MOVE? [pp.88-89]

Selected Words: tonicity [p.88], hydrostatic pressure [p.89]

Boldfaced, Page-Referenced Terms

[p.88] osmosis_____

_____

[p.88] hypotonic solution _____

_____

[p.88] hypertonic solution _____

_____

[p.88] isotonic solution_____

_____

[p.88] hydrostatic pressure _____

_____

[p.88] turgor_____

_____

[p.89] osmotic pressure _____

_____

Matching [pp.88-89]

Choose the most appropriate answer for each term.

1. _____ osmosis

2. _____ tonicity

3. _____ hypotonic solution

4. _____ hypertonic solution

5. _____ isotonic solutions

6. _____ hydrostatic pressure

7. _____ osmotic pressure

8. _____ turgor

A. Refers to the relative solute concentrations of the fluids

B. Have the same solute concentrations

D. The amount of force that prevents further increase in a solution's volume

E. The fluid on one side of a membrane that contains more solutes than the fluid on the other side of the membrane

F. The diffusion of water in response to a water concentration gradient between two regions separated by a selectively permeable membrane

G. The term for hydrostatic pressure in plants

H. The fluid on one side of a membrane that contains fewer solutes than the fluid on the other side of the membrane

I. The general term for a fluid force exerted against a cell wall and/or membrane enclosing the fluid

Short Answers [pp.88-89]

Questions 9-13 refer to the following diagram, in which side A has 25 milliliters of a 3% sucrose solution and side B has 25 milliliters of a 6% sucrose solution. The membrane separating the sides is permeable to water but impermeable to sucrose.

9. In what direction will water move through the membrane? _____

_____

10. In what direction is the net movement of water?_____

_____

11. In what direction will sucrose move through the membrane?_____

_____

12. What will happen to the sucrose concentration in side A? _____

_____

13. What will happen to the fluid level in side B? _____

_____

True/False [pp.88-89]

If the statement is true, write T in the blank. If the statement is false, rewrite it to make it correct.

14. _____ A water concentration gradient is influenced by the number of solute molecules present on both sides of the membrane.

15. _____ An animal cell placed in a hypertonic solution will swell and perhaps burst.

16. _____ Multicelled organisms can counter shifts in tonicity by selectively transporting solutes across the plasma membrane.

17. _____ A plan wilts when the concentration of salt in the soil increases, causing the soil to become hypotonic.

SELF-TEST

___ 1. White blood cells use _____ to devour disease agents invading your body. [p.86]
  a. diffusion
  b. bulk flow
  c. osmosis
  d. phagocytosis

___ 2. _____ proteins bind extracellular substances, such as hormones, that trigger changes in cell activities. [p.80]
  a. Receptor
  b. Adhesion
  c. Transport
  d. Recognition

___ 3. In a lipid bilayer, the phospholipid tails point inward and form a(n) _____ region that excludes water. [p.78]
  a. acidic
  b. basic
  c. hydrophilic
  d. hydrophobic

___ 4. Which of the following is not an example of an active transport mechanism? [pp.84-85]
  a. Calcium pump
  b. Glucose transporter
  c. Sodium–potassium pump
  d. All of the above are examples of active transport

___ 5. Which of the following is not a form of passive transport? [pp.82-83, 86]
  a. Osmosis
  b. Diffusion
  c. Bulk flow
  d. Exocytosis

___ 6. O_2, CO_2, H_2O, and other small, electrically neutral molecules usually move across the cell membrane by _____. [p.82]
  a. electric gradients
  b. receptor-mediated endocytosis
  c. passive transport
  d. active transport

_ 7. Ions such as H^+, Na^+, K^+, and Ca^{++} move across cell membranes against a concentration gradient by _____. [pp.84-85]
   a. receptor-mediated endocytosis
   b. pressure gradients
   c. passive transport
   d. active transport

_ 8. The fluid mosaic model is used to describe _____. [p.78]
   a. the process by which particles are exported from the cell
   b. the structure of the cell membrane
   c. the movement of water across a membrane
   d. the action of transport proteins

_ 9. A cell is placed in a beaker containing a solution of 40 percent NaCl and 60 percent water. After a few minutes you notice that the cytoplasm of the cell is shrinking in size. The cell is _____ in relation to the contents of the beaker. [pp.88-89]
   a. isotonic
   b. hypertonic
   c. hypotonic
   d. saturated

_ 10. Which of the following types of membrane proteins would be used as a molecular identification tag? [p.80]
   a. Recognition proteins
   b. Transport proteins
   c. Adhesion proteins
   d. Receptor proteins

CHAPTER OBJECTIVES/REVIEW QUESTIONS

1. Molecules are the most abundant component of cell membranes. [p.78]
2. Describe what is meant by the term *fluid mosaic model*. [p.78]
3. Explain why the structure of a phospholipid results in the formation of a lipid bilayer. [p.78]
4. Explain the concept of selective permeability as it applies to cell membrane function. [p.82]
5. List the factors that can have an influence on the rate of diffusion. [p.83]
6. Generally distinguish passive transport from active transport. [p.83]
7. Explain why the glucose transporter is an example of passive transport. [p.83]
8. Describe how ATP is used in active transport and how it improves the passage of solutes. [pp.84-85]
10. The diffusion of water molecules in response to water concentration gradients between two regions separated by a selectively permeable membrane is known as _____. [p.88]
11. What is turgor pressure? [p.88]
12. Define *osmotic pressure*. [p.89]
13. Describe mechanisms involved in receptor-mediated endocytosis. [p.86]
14. Explain the role of endocytosis and exocytosis in membrane recycling; cite an example. [pp. 86-87]

INTEGRATING AND APPLYING KEY CONCEPTS

1. If there were no such thing as active transport, how would the lives of organisms be affected?
2. Diagram a typical biological membrane including the lipid bilayer and both integral and peripheral proteins.
3. Compare and contrast active and passive transport. Why is each important? Why can some substances use both?
4. How is a freshwater protist able to function in an osmotically hypotonic environment? What challenges are faced by organisms that live in high-salt environments?

6

GROUND RULES OF METABOLISM

INTRODUCTION

This chapter introduces energy and describes some of its properties. It also describes enzyme structure and function and the relationship between enzymes and energy. The chapter then discusses how enzyme reactions are put together in an organism's metabolism.

FOCAL POINTS

- The introductory paragraphs [p.92] look at some of the roles of alcohol dehydrogenase.
- Figure 6.3 [p.94] illustrates the concept of entropy.
- Figure 6.4 [p.95] demonstrates the one-way flow of energy into and out of living organisms.
- Figure 6.8 [p.97] demonstrates the role of activation energy in a chemical reaction.
- Figure 6.9 [p.97] illustrates the structure of ATP and the ADP/ATP cycle.
- Section 6.1 [pp.94-95] describes energy and its characteristics.
- Sections 6.2-6.4 [pp.96-97] look in detail at enzyme structure and function.
- Section 6.4 [pp.100-101] explains metabolism and metabolic pathways.

INTERACTIVE EXERCISES

Impacts, Issues: A Toast to Alcohol Dehydrogenase [p.92]

6.1. ENERGY AND THE WORLD OF LIFE [pp.94-95]

Selected Words: alcohol dehydrogenase [p.92], alcoholic hepatitis [p.92], alcoholic cirrhosis [p.92], binge drinking [p.92], chemical energy [p.94], entropy [p.94], heat [p.95]

Boldfaced, Page-Referenced Terms

[p.94] energy _____

_____

[p.94] first law of thermodynamics _____

_____

[p.94] second law of thermodynamics _____

_____

Fill-in-the-Blanks [p.92]

(1) _____ molecules move quickly from the stomach to the bloodstream. The enzyme (2) _____ is found in the liver and helps rid the body of alcohol. One possible outcome of liver cell death is (3) _____ which is characterized by inflammation and destruction of the liver tissue. The liver has many important functions; it (4) _____, (5) _____, and (6) _____. (7) _____ is defined as consuming five alcoholic drinks in a two hour period. More than 1400 students die as a result of (7) _____ every year.

Choice [pp.94-95]

In the blank preceding each item, indicate if the first law of thermodynamics (I) or the second law of thermodynamics (II) is best described.

8. _____ Apple trees absorbing energy from the sun and storing the energy in the chemical bonds of starch and sugar

9. _____ A hydroelectric plant at a waterfall, producing electricity

10. _____ A cup of hot coffee cooling over time

11. _____ The glow of an incandescent bulb following the flow of electrons through a wire

12. _____ Earth's sun continuously losing energy to its surroundings

13. _____ The movement of a gasoline-powered automobile

14. _____ Humans running the 100-meter dash following usual food intake

15. _____ The death and decay of an organism

Short Answer [pp.94-95]

16. Why is it said that the total amount of energy available for doing work in the universe is always decreasing?

_____

_____

17. Define *entropy*. _____

_____

_____

If the statement is true, write a T in the blank. If the statement is false, make it correct by changing the underlined word(s) and writing the correct word(s) in the answer blank.

18. _____ The <u>first law of thermodynamics</u> states that entropy is constantly increasing in the universe.

19. _____ A molecule of glucose has <u>less</u> entropy than the individual carbon, hydrogen, and oxygen atoms that go into making it.

20. _____ The amount of <u>low-quality</u> energy in the universe is decreasing.

21. _____ <u>No energy conversion</u> can ever be 100 percent efficient.

22. _____ In our world, energy can flow in <u>two directions</u>: from the sun, to producers, to consumers, and back again.

6.2. THE ENERGY IN THE MOLECULES OF LIFE [pp.96-97]

6.3. HOW ENZYMES MAKE SUBSTANCES REACT [pp.98-99]

Selected Words: catalysts [p.98], cofactors [p.99]

Boldfaced, Page-Referenced Terms

[p.96] reactants _____

_____

[p.96] products _____

_____

[p.96] endergonic _____

_____

[p.96] exergonic _____

_____

[p.96] free energy _____

_____

[p.96] activation energy _____

[p.97] ATP (adenosine triphosphate) _____

_____

[p.97] phosphorylation _____

_____

[p.97] ATP/ADP cycle _____

_____

[p.98] enzymes _____

_____

[p.98] substrates _____

_____

[p.98] active sites _____

_____

[p.98] transition state_____

_____

[p.98] induced-fit model _____

_____

[p.99] antioxidants _____

_____

Fill-in-the-Blanks [pp.96-97]

In most reactions, (1) _____ of the reactants differs from (1) of the products. Reactions in which the

reactants have less energy are (2) _____. The opposite is true in (3) _____. Regardless of the type of

reaction, (4) _____ is needed to start the reaction. Enzymes act as biological (5) _____ that speed up the

chemical reaction by reducing the amount of (4) needed.

Choice [p.96]

Classify each of the following reactions as *endergonic* or *exergonic*.

6. _____   Burning wood at a campfire

7. _____   The products of a chemical reaction have more energy than the reactants

8. _____   Glucose + oxygen = carbon dioxide + water plus energy

9. _____   The reactants of a chemical reaction have more energy than the product

10. _____   The reaction releases energy

Labeling [p.96]

Identify the accompanying molecule and label its parts.

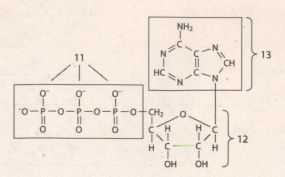

11. _____

12. _____

13. _____

14. The name of this molecule is _____

Matching

Study the sequence of reactions indicated by arrows in the accompanying diagram. Identify the numbered components of the reactions by entering the correct letter in the appropriate blank. [pp.96, 99]

15. _____

16. _____

17. _____

18. _____

19. _____

A. cofactor

B. intermediates

C. reactants

D. product

E. enzymes

Matching [pp.96-98]

Match the most appropriate letter to its number.

20. _____ exergonic reaction

21. _____ cofactors

22. _____ endergonic reaction

23. _____ reactants (substrates)

24. _____ energy carriers

25. _____ enzymes

26. _____ product

A. Mainly ATP; donate energy to reactions

B. Small molecules and metal ions that assist enzymes or serve as carriers

C. Substances able to enter into a reaction

D. Reactions that release energy

E. Proteins (usually) that catalyze reactions

F. Molecules that remain at the end of a reaction.

G. Reactions requiring a net input of energy

Identification [pp.96-97]

For the accompanying "energy hill" reaction diagram, identify the different amounts of energy.

27. _____

28. _____

29. _____

Identify the two reaction pathways.

30. _____

31. _____

Fill-in-the-Blanks

The specific substance upon which a particular enzyme acts is called its (32) _____; this substance fits into the enzyme's crevice, which is called its (33) _____. The (34) _____ model describes how a substrate contacts the site without a perfect fit. In other situations, (35) _____ molecules are squeezed out of the active site, lowering the activation energy. Enzyme induced changes in the substrate bring the (36) _____.

6.4. METABOLISM—ORGANIZED, ENZYME-MEDIATED REACTIONS [pp.100-101]

6.5. NIGHT LIGHTS [p.102]

Selected Words: biosynthetic (anabolic) pathways [p.100], degradative (catabolic) pathways [p.100], oxidized [p.101], reduced [p.101], luciferases [p.102]

Boldfaced, Page-Referenced Terms

[p.100] feedback inhibition _____

_____

[p.100] metabolic pathways _____

_____

[p.100] chemical equilibrium _____

_____

[p.100] allosteric _____

_____

[p.101] oxidation–reduction reactions _____

_____

[p.101] electron transfer chains _____

_____

[p.102] bioluminescence_____

_____

Fill-in-the-Blanks [p.100]

     (1) _____ refers to the activity by which cells acquire and use energy. A (2) _____ is a series of

enzyme-mediated reactions in which cells build, breakdown, or rearrange organic molecules. If the pathway results

in large molecules being built from small ones, it is (3) _____. If the reverse is true, the products are smaller

molecules than the reactants, it is (4) _____. Metabolic pathways may be either (5) _____, (6)

_____, or (7) _____. Enzymatic reactions may run in (8) _____ with products being converted

back to (9) _____. The (10) _____ depends on the concentration of reactants and products. (11)

_____ help cells maintain the needed concentration of thousands of different substances. These mechanisms

may increase how fast (12) _____ are made. Others may activate or (13) _____ enzymes that are already

made. An (14) _____ is a region of an enzyme that can bind regulatory molecules. Allosteric effects can cause

activation of (15) _____.

Matching [p.100]

Choose the most appropriate answer for each term.

16. _____ biosynthetic pathway
17. _____ catabolic pathway
18. _____ chemical equilibrium
19. _____ metabolic pathway
20. _____ cyclic pathway

A. A pathway in which the last step regenerates a reactant molecule used in the first step

B. An orderly series of reactions catalyzed by enzymes

C. Complex organic compounds are built from simpler molecules

D. Organic compounds are broken down to release energy

E. The rate of the forward reaction is equal to the rate of the reverse reaction

Fill-in-the-Blanks [p.101]

Oxidation–reduction refers to (21) _____ transfers. In terms of oxidation–reduction reactions, a molecule in the sequence that donates electrons is said to be (22) _____, while molecules accepting electrons are said to be (23) _____. Electrons that enter an (24) _____ are at a higher energy state than when they leave. If we think of the electron transport system as a staircase, excited electrons at the top of the staircase have the (25) _____ [choose one] (most, least) energy. Many (26) _____ deliver electrons to electron transfer chains. Energy released at certain steps is used to synthesize (27) _____.

Matching [p.101]

Match the lettered statements to the numbered items on the sketch.

28. _____
29. _____
20. _____
31. _____
32. _____

A. Represent the electron carrier molecules in an electron transport system

B. Electrons at their highest energy level

C. Released energy harnessed and used to produce ATP

D. Electrons at their lowest energy level

E. The separation of hydrogen atoms into protons and electrons

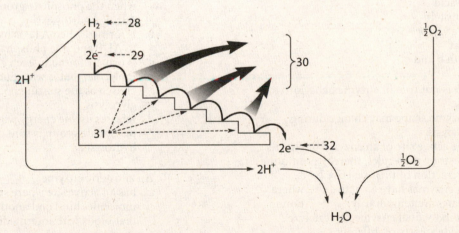

Fill-in-the-Blanks [pp.102-103]

The enzyme (33) _____ can convert chemical energy to light energy in a process called (34) _____. When genes for this process are inserted into bacteria, this glow is a visual sign of their (35) _____. This can be used to test the effectiveness of antibiotics since only (36) _____ bacteria are able to exhibit this light.

___ 1. An important principle of the second law of thermodynamics states that: [p.94]
  a. energy can be transformed into matter, and because of this, we can get something for nothing
  b. energy can only be destroyed during nuclear reactions, such as those that occur inside the sun
  c. if energy is gained by one region of the universe, another place in the universe also must gain energy in order to maintain the balance of nature
  d. energy in a system tends to become less concentrated

___ 2. What essentially does the first law of thermodynamics state? [p.94]
  a. one form of energy cannot be converted into another
  b. entropy is increasing in the universe
  c. energy cannot be created or destroyed
  d. energy cannot be converted into matter or matter into energy

___ 3. An enzyme is best described as: [p.98]
  a. an acid
  b. a protein
  c. a catalyst
  d. a fat
  e. both b and c

___ 4. Which is not true of enzyme behavior? [p.98]
  a. Enzyme shape may change during catalysis.
  b. The active site of an enzyme orients its substrate molecules, thereby promoting interaction of their reactive parts.
  c. All enzymes have an active site where substrates are temporarily bound.
  d. An individual enzyme can catalyze a wide variety of different reactions.

___ 5. When NAD+ combines with hydrogen and electrons, the NAD is _____. [p.101]
  a. reduced
  b. oxidized
  c. phosphorylated
  d. denatured

___ 6. A substance that gains electrons is _____. [p.101]
  a. oxidized
  b. a catalyst
  c. reduced
  d. a substrate

___ 7. In pathways, carbohydrates, lipids, and proteins are broken down in stepwise _____ reactions that lead to products of lower energy. [p.100]
  a. intermediate
  b. biosynthetic
  c. induced
  d. degradative

___ 8. As to major function, NAD^+, FAD, and $NADP^+$ are classified as _____. [p.99]
  a. enzymes
  b. phosphate carriers
  c. cofactors
  d. end products of metabolic pathways

___ 9. The outer phosphate bond in ATP _____. [p.97]
  a. absorbs a large amount of free energy when the phosphate group is attached during hydrolysis
  b. is formed when ATP is hydrolyzed to ADP and one phosphate group
  c. is usually found in each glucose molecule; that is why glucose is chosen as the starting point for glycolysis
  d. releases usable energy when the phosphate group is split off during hydrolysis

___ 10. An allosteric enzyme _____. [p.100]
  a. has an active site where substrate molecules bind and another site that binds with intermediate or end-product molecules
  b. is an important energy-carrying nucleotide
  c. carries out either oxidation reactions or reduction reactions but not both
  d. raises the activation energy of the chemical reaction it catalyze

CHAPTER OBJECTIVES/REVIEW QUESTIONS

1. Define energy; be able to state the first and second laws of thermodynamics. [p.94]
2. Explain how the world of life maintains a high degree of organization. [p.95]
3. Explain the functioning of the ATP/ADP cycle. [p.97]
4. A(n) _____ pathway is an orderly sequence of reactions with specific enzymes acting at each step; the pathways are linear or cyclic. [p.100]
5. Give the function of each of the following participants in metabolic pathways: substrates, intermediates, enzymes, cofactors, energy carriers, and end products. [pp.96, 98, 99]
6. What are enzymes? Explain their importance. [p.98]
7. When the "energy hill" is made smaller by enzymes so that particular reactions may proceed, it may be said that the enzyme has lowered the energy. [p.98]
8. Explain what happens to enzymes in the presence of extreme temperatures and pH. [p.99]
9. _____ are small molecules or metal ions that assist enzymes or carry atoms or electrons from one reaction site to another. [p.99]
10. Fireflies, various beetles, and some other organisms display _____ when luciferases excite the electrons of luciferins. [p.102]

INTEGRATING AND APPLYING KEY CONCEPTS

1. A piece of dry ice left sitting on a table at room temperature vaporizes. As the dry ice vaporizes into CO_2 gas, does its entropy increase or decrease? Tell why you answered as you did.
2. Why is feedback inhibition important to the well-being of organisms? What would happen if it did not exist?
3. What is bioluminescence? Why is it important to organisms?
4. Describe the various environmental factors that can affect enzyme activity.

7

WHERE IT STARTS—PHOTOSYNTHESIS

INTRODUCTION

The unique ability of photoautotrophs to harness solar energy and store it in the bonds of glucose begins the flow of energy through the biosphere. This chapter details the steps of photosynthesis from capture of solar energy by photosynthetic pigments and conversion of that light energy to chemical energy in ATP, to fixation of the carbon in CO_2 into carbohydrate. The light dependent and light independent reactions are discussed in relation to the structural design of the chloroplast. Relationships between autotrophs and heterotrophs are discussed as is the origin of today's atmosphere from the activities of ancestral photoautotrophs. Both the introduction and conclusion to this chapter address the challenges a technological society faces to supply itself with plant-derived fuels while continuing to protect the Earth, its organisms, and the environment.

FOCAL POINTS

- Figure 7.2 [p.108] shows the relationship between color and wavelength in the electromagnetic spectrum.
- Table 7.1 [p.109] summarizes various photopigments, their reflective colors, and sources.
- Figure 7.4 [p.110] (animated) compares the absorption spectra of several photopigments and illustrates Engelmann's original experiment on visible spectrum light and photosynthesis.
- Figure 7.5 [p.111] (animated) summarizes the interrelationship of chloroplast structure and function.
- Figure 7.8 [p.113] (animated) is an excellent illustration of the intricate processes involved in the noncyclic pathway of photosynthesis (light dependent reactions).
- Figure 7.9 [p.114] (animated) compares cyclic and noncyclic photophosphorylation.
- Figure 7.11 [p.115] (animated) illustrates the molecular changes and energy usage in the Calvin-Benson Cycle (light independent reactions).
- Figure 7.12 [p.116] compares the location of chloroplasts in C3 and C4 plant leaves.
- Figure 7.13 [p.117] is a good visual comparison of the differences and similarities between C3, C4, and CAM plants.

INTERACTIVE EXERCISES

Impacts, Issues: Biofuels [p.106]

7.1. SUNLIGHT AS AN ENERGY SOURCE [pp.108-109]

7.2. EXPLORING THE RAINBOW [p.110]

Selected Words: fossil fuels [p.106], biomass [p.106], biofuels [p.106], methane [p.106], ethanol [p.106], visible light [p.108], electromagnetic spectrum [p.108], nanometer (nm) [p.108], photon [p.108], prism [p.108], absorbed light [p.108], reflected light [p.108], accessory pigments [p.109], phycobilins [p.109], carotenoids [p.109], carotenes [p.109], xanthophylls [p.109], anthocyanins [p.109], retinal [p.109], absorption spectrum [p.110]

Boldfaced, Page-Referenced Terms

[p.108] wavelength _____

_____

[p.108] photosynthesis _____

_____

[p.108] pigment _____

_____

[p.109] chlorophyll *a* _____

_____

Dichotomous Choice [p.108]

Select the correct answer from the terms between parentheses in the following questions.

1. Violet light has the (highest/lowest) energy value of all visible light.

2. Infrared radiation has a (longer/shorter) wavelength than ultraviolet light.

Matching [pp.106-110]

Choose the most appropriate answer.

3. _____ chlorophyll *a*

4. _____ accessory pigments

5. _____ carotenoids

6. _____ UV light

7. _____ photons

8. _____ absorption spectrum

9. _____ phycobilins

10. _____ visible light

11. _____ pigment

12. _____ biofuels

13. _____ fossil fuels

14. _____ prism

A. Packets of electromagnetic energy that have an undulating motion through space

B. A graph that shows how efficiently different wavelengths of light are absorbed by a pigment

C. An organic molecule that absorbs specific wavelengths of light

D. Appear red, orange, and yellow; absorb violet and blue wavelengths

E. Wavelengths shorter than 380 nm; can damage DNA and other biological molecules

F. Nonrenewable organic remains of ancient plants

G. Absorbs violet, blue, and red wavelengths; the reason leaves appear green

H. Red, blue, and purple pigments; most common in deep water algae

I. Renewable source of energy; made from living plants

J. Wavelengths of light 380nm to 750 nm; drives photosynthesis

K. Absorb wavelengths of light that chlorophyll cannot absorb

L. Separates white light into its component colors by bending light waves

7.3. OVERVIEW OF PHOTOSYNTHESIS [p.111]

Selected Words: thylakoids [p.111], NADP$^+$ [p.111], NADPH [p.111], light energy [p.111], chemical energy [p.111], ATP [p.111]

Boldfaced, Page-Referenced Terms

[p.111] chloroplast _____

_____

[p.111] stroma _____

_____

[p.111] thylakoid membrane _____

_____

[p.111] photosystems _____

_____

[p.111] light-dependent reactions_____

_____

[p.111] light-independent reactions _____

_____

Completion [p.111]

Complete the following equation that summarizes the metabolic pathway of photosynthesis.

$$6H_2O + 6(1) \longrightarrow 6C(2) + (3)$$

1. _____

2. _____

3. _____

Fill-in-the-Blanks [p.111]

Supply the appropriate information to state the equation (above) for photosynthesis in words.

(4)_____ molecules of water plus six molecules of (5)_____ (in the presence of enzymes and sunlight)

yield six molecules of (6) _____ plus one molecule of (7) _____.

Fill-in-the-Blanks [p.111]

The two major sets of reactions of photosynthesis are the (8) _____ reactions and the (9) _____

reactions. The internal membranes and channels of the chloroplast form the (10) _____ membranes and are

organized into stacks. These membranes contain light-harvesting (11) _____ and (12) _____ which

convert light energy into chemical energy in the form of the molecule (13) _____. The semifluid interior area

surrounding the thylakoid membranes is known as the (14) _____ and is the area where the products of

photosynthesis are assembled.

7.4. LIGHT-DEPENDENT REACTIONS [pp.112-113]

7.5. ENERGY FLOW IN PHOTOSYNTHESIS [p.114]

Selected Words: Photosystem I [p.112], Photosystem II [p.112], p700 [p. 12], p680 [p.112], electron transport
chain [p.112], hydrogen ion gradient [p.112], ATP synthase [p.112], light-harvesting complex [p.112], noncyclic
pathway [pp.112-113], cyclic pathway [pp.112-113], cyclic photophosphorylation [p.114], cyclic
photophosphorylation [p.114], reducing agent (electron donor) [p.114]

Boldfaced, Page-Referenced Terms

[p.112] photolysis _____

_____

[p.114] photophosphorylation _____

_____

Fill-in-the-Blanks [pp.112-114]

There are two sets of light-dependent reactions, a (1) _____ pathway and a (2) _____

pathway. Both convert light energy to chemical energy in the form of (3) _____. The (4) _____

pathway is the main one in organisms containing chloroplasts. It yields (5) _____ and (6) _____ in

addition to ATP. The noncyclic pathway includes the molecules of both (7) _____ and (8) _____, while

the cyclic pathway relies on the activity of only (9) _____. Photosystem I absorbs light at a wavelength of

(10) _____ nm; Photosystem II absorbs light at a wavelength of (11) _____ nm. Absorbed light causes

(12) _____ to pop off chlorophyll *a* and enter an 13_____. Photosystem II can replace lost electrons by

stripping electrons from a molecule of water. This causes the water to split into (14) _____ ions and (15)

_____ in a process known as (16) _____. Oxygen is reactive and is released from the cell as a by-

product of photosynthesis. The energy captured from the flow of electrons through the electron transfer chain pulls in hydrogen ions from the stroma creating a (17) _____ across the thylakoid membrane. These ions flow from the thylakoid compartment back to the (18) _____ through membrane transport proteins called (19) _____ causing a phosphate group to be attached to ADP turning it into (20) _____. In the noncyclic pathway, electrons are passed from Photosystem II to (21) _____ and enter another electron transfer chain. At the end, electrons and (22) _____ are accepted by NADP$^+$ and (23) _____ is formed. If too much NADPH builds up, the noncyclic pathway stops running and cells continue to make ATP through the (24) _____ pathway.

Choice [pp.112-114]

Indicate the pathway in which each of the following occurs.

      a. cyclic pathway   b. noncyclic pathway   c. both pathways

25. _____ Uses H_2O as a reactant

26. _____ Includes Photosystem I (p700)

27. _____ Includes Photosystem II (p680)

28. _____ ATP produced

29. _____ NADPH produced

30. _____ Does not consume water nor produce NADPH and oxygen

31. _____ Causes H^+ to be shunted into the thylakoid compartments from the stroma

32. _____ Produces O_2 as a product

33. _____ Produces H^+ by breaking apart H_2O

34. _____ Uses ADP and P_i as reactants

7.6. LIGHT-INDEPENDENT REACTIONS: THE SUGAR FACTORY [p.115]

7.7. ADAPTATIONS: DIFFERENT CARBON-FIXING PATHWAYS [pp.116-117]

Selected Words: RuBP (ribulose bisphosphate) [p.115], PGA (phosphoglycerate) [p.115], PGAL (phosphoglyceraldehyde) [p.115], sucrose [p.115], starch grains [p.115], 4-carbon oxaloacetate [p.116], mesophyll cell [p.117], bundle-sheath cell [p.117], succulents [p.117], cactuses [p.117], Crassulacean Acid Metabolism [p.117]

Boldfaced, Page-Referenced Terms

[p.115] Calvin–Benson cycle_____

_____

[p.115] carbon fixation_____

_____

[p.115] rubisco_____

_____

[p.116] C3 plants_____

_____

[p.116] stomata_____

_____

[p.116] photorespiration_____

_____

[p.116] C4 plants_____

_____

[p.117] CAM plants_____

_____

Matching [pp.115-117]

Supply the full names for the following abbreviations and match it to its lettered description.

1. _____ RuBP _____
2. _____ PGAL _____
3. _____ PGA _____
4. _____ C3 Plants _____
5. _____ C4 Plants _____
6. _____ CAM Plants _____

A. Plants that fix carbon twice, first in mesophyll cell, then in bundle sheath cells

B. A 5-carbon sugar precursor that fixes inorganic carbon from carbon dioxide

C. Two 3-carbon molecules formed from an unstable 6-carbon molecule

D. Water-conserving desert plants that use an alternate carbon-fixing system at night

E. A 3-carbon sugar that has been modified by the addition of a phosphate from ATP and hydrogen and electrons from NADPH.

F. Plants that use only the Calvin-Benson cycle in mesophyll cells to fix carbon

Fill-in-the-Blanks [pp.115-117]

The light-independent reactions of the (7) _____ cycle build sugars in the (8) _____ of chloroplasts. They run on (9) _____ and (10) _____, molecules made in the light-dependent reactions. During carbon fixation, the enzyme (11) _____ is used to attach CO_2 to a five-carbon molecule (12) _____. The resulting unstable six-carbon molecule splits into two three-carbon (13) _____. A phosphate group from (14) _____ and hydrogen and electrons from (15) _____ transform each (20) into a molecule of (16) _____.Two (23) combine to form (17) _____. The ten remaining (23) combine to regenerate six (18) _____. Most of the (24) made by plants is converted to (19) _____ or (20) _____. Plants that use only the Calvin-Benson cycle to fix carbon are called (21) _____. On hot, dry days, plants close their (22) _____ to prevent excess water loss. This also prevents (23) _____ gas from entering the leaf and (24) _____ from exiting the leaf. The buildup of oxygen in the leaf causes (25) _____ to attach carbon dioxide to RuBP in a process known as (26) _____. As a result, the cell loses (27) _____ instead of fixing it. To compensate for the inefficiency caused by photorespiration, some plants called (28) _____ evolved a secondary carbon-fixation process. These plants fix carbon first in (29) _____ cells by an enzyme that does not use oxygen as rubisco does under stress conditions. The fixed carbon is transported to (30) _____ cells where it is converted to carbon dioxide and fixed for a second time in the (31) _____ cycle. A third carbon-fixing pathway is found in many desert plants such as succulents and cacti. These plants are called (32) _____ plants. These plants only open their stomata at (33) _____ when temperatures are cooler. Carbon dioxide enters the plant then and is fixed in the (34) _____ cycle into a four-carbon acid. The next day, the four-carbon acid is converted to (35) _____ that enters the Calvin-Benson cycle.

7.8. PHOTOSYNTHESIS AND THE ATMOSPHERE [p.118]

7.9 A BURNING CONCERN [p.119]

Selected Words: atmosphere [p.118], hydrogen sulfide [p.118], methane [p.118], molecular oxygen [p.118], free radicals [p.118], extinct [p.118], ozone (O^3) [p.118], ozone layer [p.118], aerobic species [p.118], aerobic respiration [p.118], biosphere [p.119], carbon cycle [p.119], coal [p.119], petroleum [p.119], natural gas [p.119], industrial revolution [p.119], carbon isotopes [p.119], global warming [p.119]

Boldfaced, Page-Referenced Terms

[p.118] autotrophs _____

_____

[p.118] heterotrophs _____

_____

[p.118] photoautotroph _____

_____

[p.118] chemoautotroph _____

_____

Fill-in-the-Blanks [pp.118-119]

"Self-nourishing" organisms are called (1) _____. Because plants use solar energy they are a kind

of (2) _____. In the process of (3) _____, they make their own food by using energy from the sun and

carbon from inorganic molecules and produce huge amounts of (4) _____ and (5) _____. (6)

_____, "nourished by others," get their energy and carbon by consuming organic molecules that have been

made by other organisms including plants. The first cells on earth were (7) _____ that obtained energy and

carbon from molecules like (8) _____ and (9) _____ gases that were plentiful in the earth's early

noxious atmosphere. Chemoautotrophs flourished for a (10) _____ years until the evolution of (11)

_____ in the first photoautotrophs. Shortly afterward, (12) _____ evolved. The (13) _____ gas

released by the splitting of water in this process began to accumulate in the atmosphere. This gas is reactive and

produced toxic (14) _____ that caused many organisms to become (15) _____. Only a few survived in

anaerobic environments. In time, oxygen-detoxifying pathways evolved. One of these was the ATP-forming

reactions of (16) _____. Oxygen molecules that rose high in the ancient atmosphere combined to form (17)

_____, molecules capable of absorbing (18) _____ radiation in sunlight. Accumulation of this gas

formed the (19) _____ that shielded, and still shields, life from damaging radiation. New (20) _____

emerged, flourished, and diversified across the face of the earth. Photosynthesis removes carbon dioxide from the

atmosphere and fixes it in organic molecules, while aerobic respiration releases organic carbon back into the

atmosphere as carbon dioxide in a balanced cycle known as the (21) _____. Since the (22) _____ in the

mid 1800's, the levels of (23) _____ in the atmosphere have been rising largely due to the burning of (24)

_____. We are adding more carbon dioxide than photoautotrophs can remove. Accumulating carbon dioxide

acts like an atmospheric blanket contributing to (25) _____.

___ 1. _____ and _____ light are best at driving photosynthesis. [p.110]
  a. orange; red
  b. red; green
  c. violet; red
  d. green; yellow

___ 2. The cyclic pathway of the light-dependent reactions functions mainly to _____. [p.114]
  a. fix CO_2
  b. make ATP
  c. produce PGAL
  d. regenerate ribulose bisphosphate

___ 3. Chemoautotrophs obtain energy by oxidizing such inorganic substances as _____. [p.118]
  a. PGA
  b. PGAL
  c. hydrogen sulfide
  d. water

___ 4. The ultimate electron and hydrogen acceptor in noncyclic photophosphorylation is _____. [p.113]
  a. $NADP^+$
  b. ADP
  c. O_2
  d. H_2O

___ 5. C4 plants have an advantage in hot, dry conditions because _____. [pp.116-117]
  a. their leaves are covered with thicker wax layers than those of C3 plants
  b. their stomata open wider than those of C3 plants, thus cooling their surfaces
  c. they have a two-step CO_2 fixation that reduces photorespiration
  d. they are also capable of carrying on photorespiration

___ 6. Light-harvesting pigments are found _____. [p.111]
  a. on the outer chloroplast membrane
  b. in the cytoplasm of the cell
  c. in the stroma
  d. in the thylakoid membrane system

___ 7. In both C4 and CAM plants, the first product of carbon fixation is _____. [p.117]
  a. malate
  b. PGA
  c. oxaloacetate
  d. glycolate

___ 8. O_2 released during photosynthesis comes from _____. [p.112]
  a. splitting CO_2
  b. splitting H_2O
  c. RuBP
  d. degradation of sugar molecules

___ 9. Plants need _____ and _____ to carry on photosynthesis. [p.111]
  a. oxygen; water
  b. oxygen; CO_2
  c. CO_2; H_2O
  d. sugar; water

___ 10. The two products of the light-dependent reactions that are required for the light-independent chemistry are _____ and _____. [pp.112-113]
  a. CO_2; H_2O
  b. O_2; NADPH
  c. O_2; ATP
  d. ATP; NADPH

___ 11. Global Warming is due in a large part to the accumulation of _____ in the atmosphere. [p.119]
  a. water
  b. oxygen
  c. carbon dioxide
  d. hydrogen sulfide

___ 12. O_2 and NADPH are produced only by _____. [pp.112-113]
  a. cyclic photophosphorylation
  b. noncyclic photophosphorylation
  c. the Calvin-Benson cycle
  d. Photosystem I

_____ 13. A pigment is an organic molecule that selectively _____ light of a specific wavelength. [p.108]
a. reflects
b. refracts
c. transmits
d. absorbs

_____ 14. The most common photosynthetic pigments in plants are _____. [p.109]
a. anthocyanins and retinal
b. phycobilins and anthocyanins
c. chlorophylls and phycobilins
d. chlorophylls and carotenoids

CHAPTER OBJECTIVES/REVIEW QUESTIONS

1. Distinguish between organisms known as autotrophs and those known as heterotrophs. [p.118]
2. Name the two major stages of photosynthesis, state in what part of the chloroplast they occur, and give an overview of the reactions they involve. [p.111]
3. The flattened channels and disk-like compartments inside the chloroplast are organized into interconnected stacks. This is the _____ membrane system that is surrounded by a semifluid interior, the _____ [p.111]
4. The most common photosynthetic pigment is _____. [p.109]. It forms the reactive center of both _____ [p.112] by absorbing light in wavelengths of 680-700 nm.
5. Describe how the pigments found on thylakoid membranes are organized into photosystems and how they relate to photon light energy. [pp.112-113]
6. Contrast the components and functioning of the cyclic and noncyclic pathways of the light-dependent reactions. [pp.112-114]
7. Two energy-carrying molecules produced in the noncyclic pathways are _____ and _____; explain why these molecules are necessary for the light-independent reactions. [pp.112-115]
8. Following evolution of the noncyclic pathway, _____ accumulated in the atmosphere and made aerobic respiration possible. [p.118]
9. Describe the Calvin–Benson cycle in terms of its reactants and products. [p.115]
10. State the various ways that glucose produced by the Calvin-Benson cycle is used by photoautotrophs. [p.115]
11. Describe the variations on carbon fixation seen in C3, C4, and CAM plants. Include locations and timing of processes and structural and chemical differences. How do these variations adapt plants to survive climatic stresses? [pp.116-117]

INTEGRATING AND APPLYING KEY CONCEPTS

1. Suppose oxygen producing pathways of photosynthesis had never evolved. What would be the effect on the lifestyle of organisms on Earth?
2. Suppose that humans acquired all the enzymes needed to carry out photosynthesis. Speculate about the attendant changes in human anatomy, physiology, and behavior that would be necessary for those enzymes to actually carry out photosynthetic reactions.
3. Consider the consequences of the development of a photosynthetic cell symbiont for human epidermal tissues.
4. Human activities, such as deforestation and polluting of waterways, have the capacity to greatly reduce the numbers of photosynthetic organisms on earth. On a chemical level, consider how the resulting reduction in consumption of photosynthetic reactants and evolution of photosynthetic products would affect the Earth and it inhabitants.

8

HOW CELLS RELEASE CHEMICAL ENERGY

INTRODUCTION

Chapter 8 looks at the various ways that cells can extract energy from food. Both aerobic and anaerobic mechanisms are covered, but a major emphasis of the chapter is aerobic respiration.

FOCAL POINTS

- The introductory section [p.122] looks at the problems that arise when mitochondria do not function properly.
- Figure 8.3 [p.125] provides an overview of aerobic respiration.
- Reaction pathway of glycolysis is presented in a detailed flow chart [p.127].
- Figure 8.5 [p.128] provides an overview of aerobic respiration in the mitochondria.
- Figure 8.6 [p.129] looks at the second stage of aerobic respiration, acetyl–CoA formation, and the Krebs cycle.
- Figure 8.7 [p.130] details the electron transfer pathway.
- Figure 8.8 [p.131] summarizes aerobic respiration.
- Figures 8.9 [p.132] looks at two of the many fermentation pathways seen in living organisms.
- Figure 8.12 [p.135] shows how various foods in your diet can enter the various stages of aerobic respiration.

INTERACTIVE EXERCISES

Impacts, Issues: When Mitochondria Spin Their Wheels [p.122]

8.1. OVERVIEW OF CARBOHYDRATE BREAKDOWN PATHWAYS [pp.124-125]

Selected Words: mitochondria [p.122], Luft's syndrome [p.122], Friedreich's ataxia [p.122], anaerobic [p.124], coenzymes [p.125], NAD+ [p.125], FAD [p.125], Krebs cycle [p.125], electron transfer phosphorylation [p.125]

Boldfaced, Page-Referenced Terms

[p.124] aerobic respiration _____

_____

[p.124] glycolysis _____

_____

[p.124] pyruvate _____

_____

[p.124] fermentation _____

_____

[p.124] aerobic _____

_____

[p.124] anaerobic-respiration _____

_____

Short Answer [p.122]

1. Why are mitochondria so crucial to normal life?_____

_____

_____

2. Compare and contrast aerobic respiration and anaerobic fermentation. _____

_____

_____

_____

Fill-in-the-Blanks [pp.124-125]

Most organisms make (3) _____ by (4) _____ which breaks down organic molecules such as

carbohydrates. The (5) _____ pathway uses oxygen while the (6) _____ pathway does not. Both of these

pathways begin with the same reaction, (7) _____. This reaction converts one molecule of (8) _____ to

two molecules of (9) _____. Aerobic respiration ends in the (10) _____, while anaerobic fermentation

ends in the (11) _____. Aerobic respiration yields about (12) _____ molecules of ATP per glucose, while

fermentation ends with a net yield of (13) _____ molecules of ATP. During aerobic respiration, glycolysis is

followed by the (14) _____. In order to begin the (14), pyruvate must be converted to (15) _____.

Hydrogen ions and electrons released by these reactions are picked up by (16) _____ and (17) _____.

The hydrogen ions and electrons are then delivered to the (18) _____, the third stage of the aerobic pathway.

Operation of the electron transfer chain allows for the formation of many molecules of (19) _____.

Completion [p.125]

20. For each of the three stages of aerobic respiration, list all of the molecules that go into the reaction and all of the molecules that come out of the reaction. Refer to Figure 8.3 in your text. _____

_____

_____

_____

_____

8.2. GLYCOLYSIS—GLUCOSE BREAKDOWN STARTS [pp.126-127]

Selected Words: energy-requiring steps [p.127], energy-generating steps [p.127], PGAL (phosphoglyceraldehyde) [p.126], NADH [p.126], net energy yield [p.126]

Boldfaced, Page-Referenced Term

[p.126] substrate-level phosphorylation _____

_____

Fill-in-the-Blanks [p.127]

Refer to the diagram of glycolysis to fill in the following blanks.

(1) [recall Ch.7] _____ organisms can synthesize and stockpile energy-rich carbohydrates and other

food molecules from inorganic raw materials for later use. When used as an energy source, (2) _____ is

partially broken down by the glycolytic pathway; at the end of this process some of its stored energy remains in two

(3) _____ molecules. Some of the energy of glucose is released during the breakdown reactions and used in

forming the energy carrier (4) _____ and the reduced coenzyme (5) _____. These reactions take place in

the cytoplasm. Glycolysis begins with two phosphate groups being transferred to (6) _____ from two (7)

_____ molecules. The addition of two phosphate groups to (6) energizes it and causes it to become unstable

and split apart, forming two molecules of (8) _____. Each (8) gains one (9) _____ group from the

cytoplasm, then (10) _____ atoms and electrons from each PGAL are transferred to NAD^+, changing this

coenzyme to NADH. At the same time, two (11) _____ molecules form by substrate-level phosphorylation; the

cell's energy investment is paid off. One (12) _____ molecule is released from each 2-PGA as a waste product.

The resulting intermediates are rather unstable; each gives up a(n) (13) _____ group to ATP. Once again, two

(14) _____ molecules have formed by (15) _____ phosphorylation. For each (16) _____ molecule

entering glycolysis, the net energy yield is two ATP molecules that the cell can use anytime to do work. The end

products of glycolysis are two molecules of (17) _____, each with a(n) (18) _____ -carbon backbone.

Sequence [p.127]

Arrange the following events of the glycolysis pathway in correct chronological sequence. Write the letter of the first step next to 19, the letter of the second step next to 20, and so on.

19. _____

20. _____

21. _____

22. _____

23. _____

24. _____

25. _____

26. _____

A. The first two ATPs form by substrate-level phosphorylation; the cell's energy debt is paid off

B. Diphosphorylated glucose (fructose-1, 6-bisphosphate) molecules split to form two PGAL; this is the first energy-releasing step

C. Two 3-carbon pyruvate molecules form as the end products of glycolysis

D. Glucose is present in the cytoplasm

E. Two more ATPs form by substrate-level phosphorylation, the cell gains ATP; net yield of ATP from glycolysis is two ATPs

F. The cell invests two ATPs; one phosphate group is attached to each end of the glucose molecule (fructose-1, 6-bisphosphate)

G. Two PGAL gain two phosphate groups from the cytoplasm

H. Hydrogen atoms and electrons from each PGAL are transferred to NAD$^+$, reducing this carrier to NADH

8.3. SECOND STAGE OF AEROBIC RESPIRATION [pp.128-129]

8.4. AEROBIC RESPIRATION'S BIG ENERGY PAYOFF [pp.130-131]

Selected Words: acetyl–CoA [p.128], oxaloacetate [p.129], citrate [p.129], NADH [p.129], FAD [p.129], FADH2 [p.129], electron transfer phosphorylation [p.130], ATP synthases [p.130], final electron acceptor [p.130]

Fill-in-the-Blanks [pp.128-131]

If sufficient oxygen is present, the end product of glycolysis enters a preparatory step, (1) _____ formation. This step converts pyruvate into (1), the molecule that enters the (2) _____ cycle. This is followed by (3) _____ phosphorylation. During these three processes, a total of (4) _____ and (5) _____ (energy-carrier molecules) are typically generated. In the preparatory conversions prior to the Krebs cycle and within the Krebs cycle, the food molecule fragments are further broken down into molecules of (6) _____. During these reactions, hydrogen atoms (7) (with their _____) are stripped from the fragments and transferred to the coenzymes (8) _____ and (9) _____.

Labeling [p.128]

In exercises 10-14, identify the structure or location in the top diagram; in exercises 15-18, identify the chemical substance involved in the lower diagram. In exercise 19, name the metabolic pathway depicted.

10. _____ of mitochondrion

11. _____ of mitochondrion

12. _____ of mitochondrion

13. _____ of mitochondrion

14. _____

15. _____

16. _____

17. _____

18. _____

19. _____

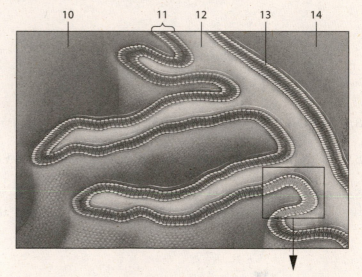

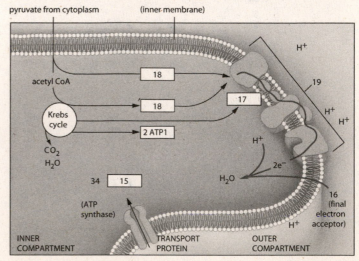

Fill-in-the-Blanks [p.130]

During (20) _____ NADH and FADH₂ deliver their electrons into an electron transfer chain. The harnessed energy from the electrons is used to (21) _____ hydrogen ions across the (22) _____ membrane of the mitochondria. A hydrogen ion gradient forms as the hydrogen ions move from the (23) _____ to the (24) _____ compartment. The hydrogen ions move from the (23) _____ through a channel protein called (25) _____. The flow of the hydrogen ions results in the production of (26) _____ molecules of ATP. Once the hydrogen ions have moved back into the inner compartment and the electrons have moved through the electron transfer chain, they are picked up by (27) _____ to form (28) _____.

8.5. ANAEROBIC ENERGY-RELEASING PATHWAYS [pp.132-133]

8.6. THE TWITCHERS [p.133]

Selected Words: ethanol [p.132], Saccharomyces cerevisiae [p.132], Lactobacillus [p.133], slow-twitch muscle fibers [p.133], fast-twitch muscle fibers [p.133]

Boldfaced, Page-Referenced Terms

[p.132] alcoholic fermentation _____

_____

[p.133] lactate fermentation _____

_____

Bacteria and single celled protists that live in a variety of anaerobic conditions are called (1) _____.
One disease caused by such a bacteria is (2) _____. Yeast and animal cells can use both (3) _____ and
(4) _____ pathways. (5) _____ is the first step in fermentation, just like in aerobic respiration. However,
in fermentation, the (6) _____ is converted to other molecules and not broken down to (7) _____ and
(8) _____. Fermentation results in the formation of a net (9) _____, no more. In (10) _____
pyruvate is converted to ethanol. Bakers use (11) _____, a yeast, to make dough rise. As the yeast ferments the
glucose, (12) _____ is released, thus the dough expands. Yeast is also used to form (13) _____ beverages.
In (14) _____ pyruvate is converted to lactic acid. The bacteria (15) _____ are used to ferment dairy
products. Skeletal muscles may be classified as (16) _____ or (17) _____ depending on how they make
ATP. (16) muscle fibers have (18) _____ mitochondria and produce ATP via (19) _____. (17) muscle
fibers have (20) _____ mitochondria and produce ATP via (21) _____.

8.7. ALTERNATIVE ENERGY SOURCES IN THE BODY [pp.134-135]

8.8. REFLECTIONS ON LIFE'S UNITY [p.136]

Selected Words: insulin [p.134], glucagon [p.134]

Fill-in-the-Blanks

After eating, increased glucose in the blood causes the pancreas to secrete (1) _____, which favors the
uptake of glucose by the liver and muscles for conversion to (2) _____ for storage. Between meals, when
glucose levels fall, the pancreas secretes (3) _____, which causes the (4) _____ to break down (2) and
release glucose. Maintaining the proper glucose concentration is very important to the (5) _____, which uses
2/3 of the circulating glucose as its only energy source.

Labeling [p.135]

Identify the process or substance indicated in the illustration.

6. _____

7. _____

8. _____

9. _____

10. _____

11. _____

12. _____

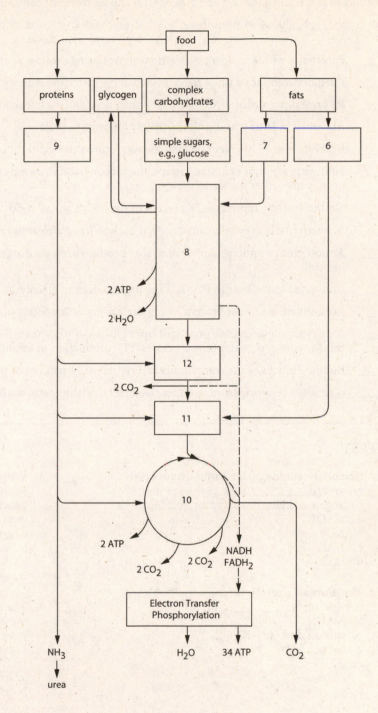

Choice [pp.134-135]

For questions 13-27, refer to the text and Figure 8.12; choose from the following:

a. glucose   b. glucose-6-phosphate   c. glycogen   d. fatty acids   e. triglycerides   f. PGAL
g. acetyl–CoA   h. amino acids   i. glycerol   j. proteins

13. _____ Fats that are broken down between meals or during exercise as alternatives to glucose

14. _____ Used between meals when free glucose supply dwindles; enters glycolysis after conversion

15. _____ Its breakdown yields much more ATP than does glucose breakdown

16. _____ Absorbed in large amounts immediately following a meal

17. _____ Represents only 1 percent or so of the total stored energy in the body

18. _____ Following removal of amino groups, the carbon backbones may be converted to fats or carbohydrates or they may enter the Krebs cycle

19. _____ On the average, represents 78 percent of the body's stored food

20. _____ Between meals liver cells can convert it back to free glucose and release it

21. _____ Amino groups undergo conversions that produce urea, a nitrogen-containing waste product excreted in urine

22. _____ Converted in the liver to PGAL, a key intermediate of glycolysis

23. _____ Accumulate inside the fat cells of adipose tissues, at strategic points under the skin

24. _____ A storage polysaccharide produced from glucose-6-phosphate following food intake that exceeds cellular energy demand (and increases ATP production to inhibit glycolysis)

25. _____ Building blocks of the compounds that represent 21 percent of the body's stored food

26. _____ A product resulting from enzymes cleaving circulating fatty acids; enters the Krebs cycle

---

SELF-TEST

___ 1. Glycolysis would quickly halt if the process ran out of _____, which serves as the hydrogen and electron acceptor. [p.126]
a. NADP$^+$
b. ADP
c. NAD$^+$
d. H$_2$O

___ 2. The ultimate electron acceptor in aerobic respiration is _____. [p.130]
a. NADH
b. carbon dioxide (CO$_2$)
c. oxygen (O$_2$)
d. ATP

___ 3. When glucose is used as an energy source, the largest amount of ATP is generated by the portion _____ of the entire respiratory process. [pp.130-131]
a. glycolytic pathway
b. acetyl–CoA formation
c. Krebs cycle
d. electron transfer phosphorylation

___ 4. The process by which about 10 percent of the energy stored in a sugar molecule is released as it is converted into two small organic-acid molecules is _____. [p.126]
a. photolysis
b. glycolysis
c. fermentation

5. During which of the following phases of respiration is ATP produced directly by substrate-level phosphorylation? [p.127]
   a. Glycolysis
   b. Krebs cycle
   c. Both a and b

6. ATP production by electron transfer phosphorylation involves _____. [p.130]
   a. H$^+$ concentration and electron gradients across a membrane
   b. ATP synthases
   c. both a and b
   d. neither a nor b

7. What is the name of the process by which reduced NADH transfers electrons along a chain of acceptors to oxygen so as to form water and in which the energy released along the way is used to generate ATP? [p.130]
   a. Glycolysis
   b. Acetyl–CoA formation
   c. The Krebs cycle
   d. Electron transfer phosphorylation

8. Pyruvate can be regarded as the end product of _____. [pp.126-127]
   a. glycolysis
   b. acetyl–CoA formation
   c. fermentation
   d. the Krebs cycle

CHAPTER OBJECTIVES/REVIEW QUESTIONS

1. No matter what the source of energy may be, organisms must convert it to _____, a form of chemical energy that can drive metabolic reactions. [p.124]
2. The main energy-releasing pathway is respiration. [p.124]
3. By the end of the second stage of aerobic respiration, which includes the cycle, _____ has been completely degraded to carbon dioxide and water. [pp.125, 129]
4. Which stage of aerobic respiration has the highest yield of ATP? [pp.130-131]
5. Explain, in general terms, the role of oxygen in aerobic respiration. [p.125]
6. Explain the purpose served by the cell investing two ATP molecules into the chemistry of glycolysis. [p.126]
7. Consult Figure 8.6 in the main text. Relate the events that happen during acetyl–coenzyme A formation and explain how the process of acetyl–CoA formation relates glycolysis to the Krebs cycle. [p.129]
8. Explain how the flow of hydrogen ions through the mitochondrion membrane accounts for the production of ATP molecules. [p.130]
9. Briefly describe the process of electron transfer phosphorylation by stating what reactants are needed and what the products are. State how many ATP molecules are produced through operation of the transport system. [p.131]
10. List some environments where very little oxygen is present and where anaerobic organisms might be found. [p.132]
11. List the main anaerobic energy-releasing pathways and the examples of organisms that use them. [pp.132-133]
12. Describe what happens to pyruvate in anaerobic organisms. Then explain the necessity for pyruvate to be converted to a fermentation product. [pp.132-133]
13. List some sources of energy (other than glucose) that can be fed into the respiratory pathways. [pp.134-135]

INTEGRATING AND APPLYING KEY CONCEPTS

1. What problems might humans and other organisms experience if their mitochondria were defective?
2. Human skeletal muscle has both slow-twitch and fast-twitch fibers. Where in the body would you expect each type to predominate?
3. How is the "oxygen debt" experienced by runners and sprinters related to aerobic respiration and fermentation in humans?
4. Predict what your body would do if you switched to a diet of 100 percent protein.

9

HOW CELLS REPRODUCE

INTRODUCTION

We all began life as a fertilized egg. That single cell then reproduced trillions of times; resulting in the complex humans we are today. In this chapter you will examine the principles of cell division, specifically how eukaryotic cells make exact copies of themselves by the process of mitosis. In the next chapter, you will build on this knowledge to examine the principles of meiosis, a more complex process for the production of reproductive cells. In the last section of this chapter you will examine how scientists continue to research the mechanisms of cell division, specifically the control of the cell cycle, since unregulated cell growth can lead to cancer.

FOCAL POINTS

- Table 9.1 [p.142] provides a summary of the major forms of cell division.
- Figure 9.4 [p.144] illustrates how cell division fits into the "life" cycle of the cell.
- Page 147 provides an overview of the stages of mitosis. These stages will be revisited in Chapter 10.

INTERACTIVE EXERCISES

Impacts, Issues: Henrietta's Immortal Cells [p.140]

9.1. AN OVERVIEW OF CELL DIVISION MECHANISMS [pp.142-143]

Selected Words: reproduction [p.142], somatic cells [p.142], asexual reproduction [p.142], gametes [p.142], sexual reproduction [p.142], kinetochore [p.143]

Boldfaced, Page-Referenced Terms

[p.142] mitosis _____

_____

[p.142] meiosis _____

_____

[p.142] sister chromatids _____

_____

[p.143] histones _____

_____

[p.143] nucleosome _____

_____

[p.143] centromere _____

_____

[p.144] cell cycle _____

_____

Matching [pp.142-143]

Choose the most appropriate answer for each term.

1. _____ centromere
2. _____ histones
3. _____ gametes
4. _____ somatic cells
5. _____ sister chromatids
6. _____ mitosis
7. _____ meiosis
8. _____ nucleosome
9. _____ kinetochore

A. Reproductive cells

B. Cell division for sexual reproduction

C. DNA spooled around a histone

D. Small chromosomal region with attachment sites for microtubules

E. The two attached DNA molecules of a duplicated chromosome

F. Body cells that reproduce by mitosis and cytoplasmic division

G. Cell division responsible for growth, tissue repair, and replacement of cells

H. A constriction in the chromosome that can be used as an identifying characteristic

I. Proteins that serve as spools to wind the DNA during condensing of the chromosomes

9.2. INTRODUCING THE CELL CYCLE [pp.144-145]

Selected Words: G_1 interval [p.144], S interval [p.144], G_2 interval [p.144]

Boldfaced, Page-Referenced Terms

[p.144] cell cycle _____

_____

[p.144] interphase _____

_____

[p.144] chromosome number _____

_____

[p.145] diploid _____

_____

[p.145] bipolar spindle _____

_____

Matching [p.144]

Choose the most appropriate answer for each term.

1. _____ Period after duplication of DNA during which the cell prepares for division

2. _____ Synthesis, or duplication of the DNA and proteins

3. _____ Period of cell growth before DNA duplication; a "gap" of interphase

4. _____ Period of cytoplasmic division

5. _____ Period that includes G1, S, G2

6. _____ Period of nuclear division

A. Interphase

B. Mitosis

C. G_1 Interval

D. S Interval

E. G_2 Interval

F. End of Mitosis

Fill-in-the-Blanks [p.145]

During mitosis, a(n) (7) _____ parent cell can produce two diploid (8) _____ cells. This doesn't mean each merely gets forty-six (9) _____. If only the total mattered, then one cell may get two pairs of chromosome 22 and no pairs whatsoever of chromosome 9. But neither cell could function like its parent without (10) _____ of each type of chromosome. A bipolar (11) _____ forms during nuclear division. This dynamic network consists of (12) _____, some of which attached to duplicated (13) _____. Microtubules separate sister (14) _____ and move them to (15) _____ ends of the cell.

Labeling [p.144]

For each number, identify the component of the cell cycle.

16. _____

17. _____

18. _____

19. _____

20. _____

21. _____

22. _____

23. _____

24. _____

25. _____

26. _____

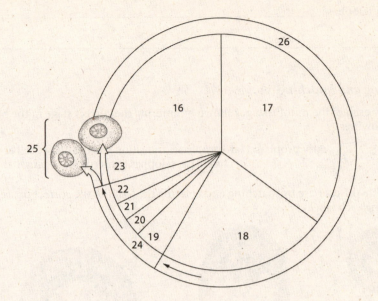

Labeling [p.144]

27. In the preceding diagram, first shade in those stages of the cell cycle that are responsible for the process of mitosis. Second, shade in the stage in which the sister chromatids are formed. [p.144]

9.3. A CLOSER LOOK AT MITOSIS [pp.146-147]

Selected Words: "mitosis"l [p.146], mitos [p.146], meta-l [p.146]

Boldfaced, Page-Referenced Terms

[p.146] prophase_____

_____

[p.146] centrosome_____

_____

[p.146] metaphase _____

_____

[p.146] anaphase_____

_____

[p.146] telophase _____

_____

Labeling and Matching [pp.146-147]

Identify each of the mitotic stages shown by entering the correct stage in the blank beneath the sketch. Select from the following:

early prophase, late prophase, transition to metaphase, cell at interphase, metaphase, anaphase, telophase, interphase—daughter cells

Complete the exercise by matching and entering the letter of the correct phase description in the parentheses following each label.

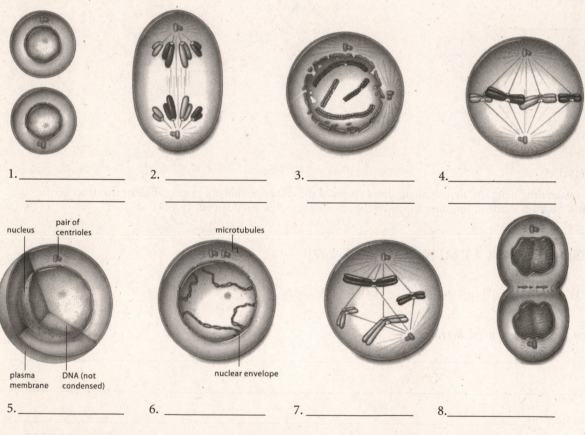

1._____   2._____   3._____   4._____

_____   _____   _____   _____

5._____   6._____   7._____   8._____

_____   _____   _____   _____

A. Attachments between two sister chromatids of each chromosome break; the two are now separate chromosomes that move to opposite spindle poles.

B. Microtubules penetrate the nuclear region and collectively form the spindle apparatus; microtubules become attached to the two sister chromatids of each chromosome.

C. The DNA and its associated proteins have started to condense.

D. All the chromosomes are now fully condensed and lined up at the equator of the spindle.

E. DNA is duplicated and the cell prepares for nuclear division.

F. Two daughter cells have formed, each diploid with two of each type of chromosome, just like the parent cell's nucleus.

G. Chromosomes continue to condense. New microtubules are assembled, and they move one of two centriole pairs toward the opposite end of the cell. The nuclear envelope begins to break up.

H. Patches of new membrane fuse to form a new nuclear envelope around the decondensing chromosomes.

Chronological Order [pp.146-147]

Using the illustration from the previous exercise, place the phases of mitosis in their correct chronological order. Write the correct name of the stage in the answer blank, and then place the number corresponding to the mitosis diagram.

9. _____ ( )

10. _____ ( )

11. _____ ( )

12. _____ ( )

13. _____ ( )

14. _____ ( )

15. _____ ( )

16. _____ ( )

9.4. CYTOPLASMIC DIVISION MECHANISMS [pp.148-149]

Selected Words: cleavage furrow [p.148], cell cortex [p.148], actin and myosin filaments [p.148], primary cell wall [p.149]

Boldfaced, Page-Referenced Terms

[p.148] cytokinesis _____

_____

[p.148] contractile ring _____

_____

[p.149] cell plate _____

_____

Matching [pp.148-149]

Choose the most appropriate answer for each term.

1. _____ Cytoplasmic division
2. _____ Forms from the cell plate
3. _____ Filaments that contract to divide cytoplasm in animal cells
4. _____ Band of actin and myosin that divides cytoplasm in animal cells
5. _____ Indentation in the cell membrane between the cell's poles
6. _____ Formed from vesicles and their wall-building contents in plant cells

A. Contractile ring
B. Cell plate
C. Cleavage furrow
D. Cytokinesis
E. Actin
F. Primary cell wall

Choice [pp.148-149]

For exercises 1-9 on cytoplasmic division, choose from the following forms of division.

   a. plant cell division   b. animal cell division   c.   both plant and animal cell cytoplasmic division

7. _____ A cell plate forms
8. _____ Actin filaments contract
9. _____ Cellulose deposits form a cross-wall between the two daughter cells
10. _____ Involves cells with rigid cell walls
11. _____ A cleavage furrow forms
12. _____ Uses ATP to contract microfilaments
13. _____ The end result is the formation of two daughter cells
14. _____ At the location of a disklike structure, deposits of cellulose accumulate
15. _____ Microtubules orient cellulose fibers prior to cytoplasmic division

9.5. WHEN CONTROL IS LOST [pp.150-151]

Selected Words: free radicals [p.150], mutation [p.150], proto-oncogenes [p.150], benign neoplasms [p.150], malignant neoplasm [p.151], basal cell carcinoma [p.151], squamous cell carcinoma [p.151], malignant melanoma [p.151]

Boldfaced, Page-Referenced Terms

[p.150] growth factors _____

_____

[p.150] tumor _____

_____

[p.150] neoplasms_____

_____

[p.151] cancers _____

_____

Matching [pp.150-151]

Match each of the following definitions with the correct term.

1. _____ Cell masses that may be classified as either benign or malignant

2. _____ Checkpoint genes that stimulate mitosis

3. _____ Abnormally growing and dividing cells of a malignant neoplasm

4. _____ Form of skin cancer

5. _____ Promote tumor formation

6. _____ Checkpoint genes that inhibit mitosis

A. growth factors

B. proto-oncogenes

C. basal cell carcinoma

D. tumor suppressors

E. cancers

F. neoplasms

SELF-TEST

___ 1. The replication of DNA and formation of the sister chromatids occurs _____. [p.144]
a. during a growth phase of interphase
b. immediately before prophase of mitosis
c. during prophase of mitosis
d. during telophase of mitosis

___ 2. Which of the following inhibits the process of mitosis? [p.150]
a. proto-oncogenes
b. actin
c. growth factors
d. tumor suppressors

___ 3. Diploid refers to _____. [p.145]
a. having two chromosomes of each type in somatic cells
b. triple the parental chromosome number
c. half the parental chromosome number
d. having one chromosome of each type in somatic cells

___ 4. Somatic cells differ from gametes in that _____. [p.142]
a. gametes are body cells
b. somatic cells reproduce sexually
c. gametes are important in sexual reproduction
d. somatic cells cannot form from mitosis

___ 5. If a parent cell has sixteen chromosomes and undergoes mitosis, the resulting cells will have _____ chromosomes. [p.145]
a. sixty-four
b. thirty-two
c. sixteen
d. eight
e. four

___ 6. Microtubules bind to the _____, a constriction where sister chromatids attach. [p.143]
a. cell membrane
b. nucleosome
c. centromere
d. mitotic spindle
e. histone

___ 7. The correct order of the stages of mitosis is _____. [pp.146-147]
a. prophase, metaphase, telophase, anaphase
b. telophase, anaphase, metaphase, prophase
c. telophase, prophase, metaphase, anaphase
d. anaphase, prophase, telophase, metaphase
e. prophase, metaphase, anaphase, telophase

___ 8. What is the stage of mitosis that is characterized by the alignment of the chromosomes along a central line in the cell called? [pp.146-147]
a. prophase
b. metaphase
c. prometaphase
d. anaphase
e. telophase

____ 9. During _____, sister chromatids of each chromosome are separated from each other, and those former partners, now chromosomes, move to opposite poles. [pp.146-147]
   a. prophase
   b. metaphase
   c. anaphase
   d. telophase

____ 10. Each histone-DNA spool is a single structural unit called a _____.[p.143]
   a. chromosome
   b. microtubule
   c. centromere
   d. nucleosome

CHAPTER OBJECTIVES/REVIEW QUESTIONS

1. Discuss how HeLa cells can be used to understand cancers today. [p.140]
2. Distinguish between mitosis and meiosis. [p.142]
3. Explain why somatic cells and gametes must use different methods for reproduction. [p.142]
4. Explain how the terms nucleosome, centromere, and sister chromatid relate to chromosome structure. [pp.142-143]
5. List and describe, in order, the various activities occurring in the eukaryotic cell life cycle. [pp.144-145]
6. Understand why a diploid parent cell produces two diploid daughter cells by mitosis. [p.145]
7. Describe the structure and function of the bipolar mitotic spindle. [p.145]
8. Explain the purpose of the centrosome in animal cell division. [p.146]
9. Give a detailed description of the cellular events occurring in the prophase, metaphase, anaphase, and telophase of mitosis. [pp.146-147]
10. Compare and contrast cytokinesis as it occurs in plant and animal cell division; use the following concepts: cleavage furrow, microfilaments at the cell's midsection, and cell plate formation. [pp.148-149]
11. Describe the role of kinases, growth factors, tumor suppressors, and proto-oncogenes in the regulation of the cell cycle. [pp.150-151]
12. Understand the process by which cancers form. [p.151]
13. Describe the three major forms of skin cancer. [p.151]

INTEGRATING AND APPLYING KEY CONCEPTS

1. Runaway cell division is characteristic of cancer. Imagine the various points of the mitotic process that might be sabotaged in cancerous cells in order to halt their multiplication. How might one discriminate between cancerous and normal cells in order to guide those methods of sabotage most effective in combating cancer?
2. As a cancer researcher, how might you use your knowledge of cell cycle regulation to inhibit the division of cancer cells?
3. Why can the process of mitosis also be called "cellular cloning"?
4. The rate at which the cell cycle progresses depends on the kind of cell reproducing. How might a researcher treat damaged cells that have stopped the cell cycle by understanding the behavior of fast cell cycle cells?

10

MEIOSIS AND SEXUAL REPRODUCTION

INTRODUCTION

One of the first things you will notice in this chapter is that meiosis is very similar to mitosis. However, while mitosis is a form of cloning, meiosis represents a mechanism of introducing variation. In order to understand meiosis, you must first recognize why variation is important for a species. Once you have mastered the "why," proceed to the explanation of "how" cells accomplish reduction division. Study the diagrams carefully; the ability to visualize the process is the key to comprehending meiosis.

FOCAL POINTS

- Figure 10.5 [pp.158–159] is the core of this chapter (see also the related animation on the CD).
- Figure 10.11 [pp.164–165] compares mitosis and meiosis, providing an excellent review of Chapters 9 and 10.

INTERACTIVE EXERCISES

Impacts, Issues: Why Sex? [p.154]

10.1. INTRODUCING ALLELES [p.156]

Selected Words: fertilization [p.156], mutations [p.160], new combination of traits [p.160]

Boldfaced, Page-Referenced Terms

[p.156] asexual reproduction _____

_____

[p.156] genes_____

_____

[p.156] clones _____

_____

[p.156] sexual reproduction _____

_____

[p.156] allele _____

_____

Fill-in-the-Blank [p.160]

The first cell of a sexually reproducing eukaryotic organism has (1) _____ genes on (2) _____ chromosomes, where one chromosome is (3) _____ and one is (4) _____. Information in the gene pairs is not identical or sexual reproduction would produce (5) _____. Variations in gene pairs are caused by (6) _____ that are permanently incorporated into the information they carry. Offspring of sexual reproducers inherit new (7) _____ of (8) _____, which is the basis of new combinations of traits.

10.2. WHAT MEIOSIS DOES [pp.156-157]

Selected Words: chromosome number [p.156], *diploid cell* [p.156], *hom-* [p.156], *sperm cell* [p.156], *egg* [p.156]

Boldfaced, Page-Referenced Terms

[p.156] homologous _____

_____

[p.156] meiosis _____

_____

[p.156] germ cells _____

_____

[p.156] gametes _____

_____

[p.157] haploid _____

_____

[p.157] zygote _____

_____

Matching [pp.160-161]

Match each of the following terms with the appropriate statement.

1. _____ chromosome number
2. _____ germ cells
3. _____ diploid
4. _____ haploid
5. _____ gametes
6. _____ meiosis
7. _____ egg
8. _____ sperm cell
9. _____ homologous chromosomes
10. _____ zygote

A. A fertilized egg
B. Mature reproductive structures in animals
C. Male gamete
D. Nuclear division in sexually reproducing eukaryotic species
E. Female gamete
F. A cell that contains one of each type of chromosome
G. A cell that contains two of each type of chromosomes
H. Sum of the chromosomes in a cell
I. Chromosomes with the same length, shape, and collection of genes
J. Immature reproductive cells

True/False [p.157]

Use Figure 10.4 to complete the following statements.

11. _____ The chromosome pairs pictured are homologous pairs.
12. _____ The chromosomes pictured are from a human male.
13. _____ The chromosomes pictured illustrate a haploid cell.
14. _____ One chromosome from each pair comes from the maternal parent.
15. _____ The chromosomes pictured might result from fusion of an egg and sperm cell.
16. _____ Genes on chromosome pair 21 are identical to genes on chromosome pair 22.
17. _____ Genes on one chromosome in pair 21 are identical to genes on the other chromosome in pair 21.

10.3. VISUAL TOUR OF MEIOSIS [pp.158–159]

Selected Words: swap segments [p.158]

Boldfaced, Page-Referenced Term

[p.158] Prophase I _____

_____

[p.158] Metaphase I _____

_____

[p.158] Anaphase I _____

_____

[p.158] Telophase I _____

_____

[p.159] Prophase II _____

_____

[p.159] Metaphase II _____

_____

[p.159] Anaphase II _____

_____

[p.159] Telophase II _____

_____

Choice [pp.158-159]

Recall that in diploid cells the sister chromatids are attached; in haploid cells the sister chromatids have been separated. Use Figure 10.5 to determine whether the individual cells in each phase of meiosis are haploid or diploid.

1. _____   Prophase I
2. _____   Metaphase I
3. _____   Anaphase I
4. _____   Telophase I
5. _____   Prophase II
6. _____   Metaphase II
7. _____   Anaphase II
8. _____   Telophase II

Labeling and Matching [pp.158-159]

Identify each of the meiotic stages in the following diagram by entering the correct stage of either meiosis I or meiosis II in the blank beneath the sketch. Choose from these stages:

prophase I, metaphase I, anaphase I, telophase I, prophase II,

metaphase II, anaphase II, telophase II

Complete the exercise by matching and entering the letter of the correct stage description in the parentheses following each label.

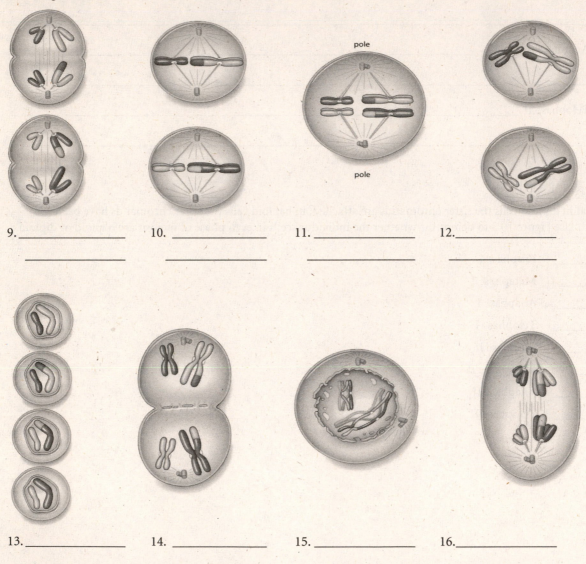

9._____

_____

10._____

_____

11._____

_____

12._____

_____

13._____

_____

14._____

_____

15._____

_____

16._____

_____

A.   The spindle is now fully formed; all chromosomes are positioned midway between the poles of one cell.

B.   In each of two daughter cells, microtubules attach to the kinetochores of chromosomes, and motor proteins drive the movement of chromosomes toward the spindle's equator.

C.   Four daughter nuclei form; when the cytoplasm divides, each new cell has a haploid chromosome number, all in the unduplicated state. The cells may develop into gametes in animals or spores in plants.

D.   In one cell, each duplicated chromosome is pulled away from its homologous partner; the partners are moved to opposite spindle poles.

E. Duplicated chromosomes condense, each chromosome pairs with its homologous partner, and crossing over and genetic recombination (swapping of gene segments) occurs. Each chromosome becomes attached to some microtubules of a newly forming spindle.

F. Motor proteins and spindle microtubule interactions have moved all the duplicated chromosomes so that they are positioned at the spindle equator, midway between the poles.

G. Two haploid cells form, each having one of each type of chromosome that was present in the parent cell; the chromosomes are still in the duplicated state.

H. Attachment between the two chromatids of each chromosome breaks; former "sister chromatids" are now chromosomes in their own right and are moved to opposite poles by motor proteins.

10.4. HOW MEIOSIS INTRODUCES VARIATIONS IN TRAITS [pp.160-161]

Selected Words: "non-sister" chromosomes [p.160], maternal chromosomes [p.160], paternal chromosomes [p.160]

Boldfaced, Page-Referenced Terms

[p.160] crossing over _____

_____

Short Answer

1. What provides for the variation in sexual reproduction? _____

_____

_____

2. In what phase of meiosis is this variation accomplished? _____

_____

_____

3. Why doesn't crossing over occur in mitosis? _____

_____

_____

Matching [p.161]

Choose the most appropriate answer for each term.

4. _____ telophase II

5. _____ 2^{23}

6. _____ paternal chromosomes

7. _____ prophase I

8. _____ nonsister chromatids

9. _____ crossing over

10. _____ metaphase I

11. _____ maternal chromosomes

A. The stage in the cell cycle when the maternal and paternal chromosomes randomly position at the spindle equator

B. The stage in the cell cycle when the chromosomes break at the same places along their length and then exchange corresponding segments

C. The twenty-three chromosomes inherited from your mother

D. The twenty-three chromosomes inherited from your father

E. The number of possible combinations of maternal and paternal chromosomes at metaphase I

F. The molecular exchange of material between nonsister chromatids

G. Chromatids from homologous chromosomes; not linked by a centromere

H. Results in four haploid nuclei

10.5. FROM GAMETES TO OFFSPRING [pp.162-163]

Selected Words: spermatids [p.162], primary oocyte [p.162], secondary oocyte [p.162], polar bodies [p.162], ovum [p.162], zygote [p.162]

Boldfaced, Page-Referenced Terms

[p.162] sporophytes _____

_____

[p.162] gametophyte _____

_____

[p.162] sperm _____

_____

[p.162] egg _____

_____

[p.162] fertilization _____

_____

For questions 1–14, choose from the following:

  a. animal life cycle   b. plant life cycle   c. both animal and plant life cycles

1. _____ Meiosis results in the production of haploid spores

2. _____ A zygote divides by mitosis

3. _____ Meiosis results in the production of haploid gametes

4. _____ Haploid gametes fuse in fertilization to form a diploid zygote

5. _____ A zygote divides by mitosis to form a diploid sporophyte

6. _____ During meiosis, one egg and three polar bodies form

7. _____ A spore divides by mitosis to produce a haploid gametophyte

8. _____ Gametes form by oogenesis and spermatogenesis

9. _____ A haploid gametophyte divides by mitosis to produce haploid gametes

10. _____ A secondary oocyte gets nearly all the cytoplasm

11. _____ A haploid spore divides by mitosis to produce a gametophyte

12. _____ A diploid body forms from mitosis of a zygote

13. _____ A gamete-producing body and a spore-producing body develop during the life cycle

14. _____ One daughter cell of the secondary oocyte develops into a second polar body

Sequence [p.163]

Refer to Figure 10.10 to arrange the following entities in correct order of development, entering a 1 by the stage that appears first and a 4 by the stage that completes the process of spermatogenesis. Then correctly order the stages of development of oogenesis. Complete the exercise by indicating, in the parentheses next to each blank, if each cell is *n* or *2n*.

Spermatogenesis

15. _____ primary spermatocyte

16. _____ sperm

17. _____ spermatid

18. _____ secondary spermatocyte

Oogenesis

19. _____ ovum

20. _____ oogonium

21. _____ secondary oocyte

22. _____ primary oocyte

Short Answer [p.163]

23. Use Figure 10.10 to compare the sperm and egg formed by meiosis. How are the resulting gametes similar?

_____

_____

How are they different? _____

_____

10.6. MITOSIS AND MEIOSIS—AN ANCESTRAL CONNECTION? [pp.164-165]

Selected Words: *mitotic reproduction* [p.164]; *meiotic reproduction* [p.164]

Matching [pp.164-165]

The cell model used in this exercise has two pairs of homologous chromosomes, one long pair and one short pair. Match the descriptions to the letters beneath each picture in the following sketches.

1. _____ One cell at the beginning of meiosis II

2. _____ A daughter cell at the end of meiosis II

3. _____ Metaphase I of meiosis

4. _____ Metaphase of mitosis

5. _____ G1 in a daughter cell after mitosis

6. _____ Prophase of mitosis

A   B   C   D   E   F

Short Answer

The following thought questions refer to the sketches in the preceding exercise; enter answers in the blanks following each question.

7. How many chromosomes are present in cell E? _____

8. How many chromatids are present in cell E? _____

9. How many chromatids are present in cell C? _____

10. How many chromatids are present in cell D? _____

11. How many chromosomes are present in cell F? _____

SELF-TEST

___ 1. Which of the following does not occur in prophase I of meiosis? [pp.158-161]
   a. A cytoplasmic division
   b. A cluster of four chromatids
   c. Homologues pairing tightly
   d. Crossing over

___ 2. Crossing over is one of the most important events in meiosis because _____. [pp.160–161]
   a. it establishes new genetic combinations that were not present in the parents
   b. homologous chromosomes must be separated into different daughter cells
   c. the number of chromosomes allotted to each daughter cell must be halved
   d. it sorts the chromatids into gametes for fertilization

___ 3. Which of the following is the most correct sequence of events in plant life cycles? [p.162]
   a. Fertilization, sporophyte, zygote, meiosis, spores, gametophytes, gametes
   b. Fertilization, zygote, sporophyte, meiosis, spores, gametophytes, gametes
   c. Fertilization, zygote, sporophyte, meiosis, gametes, gametophyte, spores
   d. Fertilization, zygote, gametophyte, meiosis, gametes, sporophyte, spores

___ 4. While observing a given cell under a microscope, you notice that pairs of condensed homologous chromosomes appear to be aligning in the center of the cell. The most likely stage of meiosis for this cell is _____. [pp.158–159]
   a. anaphase II
   b. metaphase II
   c. metaphase I
   d. prophase II
   e. telophase I

___ 5. The end result of meiosis in human females is _____. [p.163]
   a. two diploid clones of the parent cell
   b. four haploid ovum
   c. one haploid ovum and three diploid polar bodies
   d. one haploid ovum and three haploid polar bodies

___ 6. The end result of meiosis in human males is _____. [p.163]
   a. one haploid sperm and three haploid polar bodies
   b. one diploid sperm and three haploid polar bodies
   c. four haploid sperm
   d. four diploid sperm

___ 7. If an organism has a chromosome number of 5, how many different combinations of homologous chromosomes are possible at metaphase I? [p.161]
   a. 2^5
   b. 5^2
   c. 5
   d. 10

___ 8. Which of the following does not increase genetic variation? [pp.160–161]
   a. Crossing over
   b. Random fertilization
   c. Prophase of mitosis
   d. Random homologue alignments at metaphase I

___ 9. Which of the following is the most correct sequence of events in animal life cycles? [p.162]
   a. Meiosis, fertilization, gametes, diploid organism
   b. Diploid organism, meiosis, gametes, fertilization
   c. Fertilization, gametes, diploid organism, meiosis
   d. Diploid organism, fertilization, meiosis, gametes

___ 10. In sexually reproducing organisms, the zygote is _____. [pp.162–163]
   a. an exact genetic copy of the female parent
   b. an exact genetic copy of the male parent
   c. completely unlike either parent genetically
   d. a genetic mixture of male parent and female parents

CHAPTER OBJECTIVES/REVIEW QUESTIONS

1. Give two advantages of sexual reproduction. [p.154]
2. Distinguish between sexual and asexual reproduction. [p.156]
3. Explain the terms gene and allele. [p.156]
4. Explain what is meant by diploid and haploid chromosome number. [pp.156-157]
5. Describe the relationships between homologous chromosomes and sister chromatids. [pp.156-157]
6. Explain the importance of two consecutive cell divisions in meiosis. [p.157]
7. Sketch cell models of the various stages of meiosis I and meiosis II. Using color pencils or crayons to distinguish two pairs of homologous chromosomes (a long pair and a short pair) is helpful. [pp.158-159]
8. Explain the importance and mechanism of crossing over. [pp.160-161]
9. Explain why anaphase I results in haploid cells. [p.161]
10. Discuss the differences between plant and animal life cycles. [p.162]
11. Describe the differences in gamete production in male and female animals. [p.163]
12. Explain how crossing over, the distribution of random mixes of homologous chromosomes into gametes, and fertilization all contribute to variation in the traits of offspring. [pp.160-163]
13. Explain why variation is important to a species. [pp.164-165]

INTEGRATING AND APPLYING KEY CONCEPTS

1. The process of cloning organisms has been in the news for over a decade. It may be possible in the near future to clone humans. Based upon your knowledge of mitosis and meiosis, what are the advantages of cloning to individuals? What are the disadvantages? What about the species as a whole?
2. Imagine a pond where 95% of the fish are green. If a virus that kills green fish infected the pond, what would happen to the population? How does this example illustrate the importance of variation in sexually reproducing organisms?
3. Based upon your knowledge of meiosis, what would provide more variation to a species—a few large chromosomes, or lots of small chromosomes? Explain your answer.

11

OBSERVING PATTERNS IN INHERITED TRAITS

INTRODUCTION

This chapter introduces you to some of the principles of genetics, starting with the work of Gregor Mendel. Advances in the study of genetics are frequently in the news, and most people have an interest in understanding more about inheritance. The key to understanding genetics is the ability to solve problems using probability. In this chapter you will be exposed to many different forms of inheritance, but all use probability to predict the outcome. You need to not only understand the terminology in the chapter but also work the problems if you are going to gain an understanding of genetics.

FOCAL POINTS

- The diagram in Figure 11.4 [p.171], and the terms introduced on this page, are the basic "language" of geneticists. You need to understand these terms before proceeding to the remainder of the chapter.
- Figure 11.7 [p.173] illustrates the Punnett square, a useful tool for predicting outcomes in introductory genetics.
- Figure 11.8 [p.174] shows how the principles of independent assortment are executed during meiosis.
- Figure 11.9 [p.175] outlines how to construct a Punnett square for a dihybrid experiment.

INTERACTIVE EXERCISES

Impacts, Issues: The Color of Skin [p.168]

11.1. MENDEL, PEA PLANTS, AND INHERITANCE PATTERNS [pp.170-171]

Selected Words: heredity [p.170], "blending inheritance" [p.170], discrete units [p.170], Gregor Mendel [p.170], Pisum sativum [p.170], self-fertilizing [p.170], "breed true" [p.171], F1 generation [p.171], F2 generation [p.171], filial [p.171]

Boldfaced, Page-Referenced Terms

[p.171] homologous chromosomes _____

_____

[p.171] gene locus_____

_____

[p.171] alleles _____

_____

[p.171] genes _____

_____

[p.171] locus _____

_____

[p.171] diploid _____

_____

[p.171] mutation _____

_____

[p.171] hybrids _____

_____

[p.171] heterozygous _____

_____

[p.171] homozygous _____

_____

[p.171] dominant _____

_____

[p.171] recessive _____

_____

[p.171] homozygous dominant _____

_____

[p.171] homozygous recessive _____

_____

[p.171] gene expression _____

_____

[p.171] genotype _____

_____

[p.171] phenotype _____

_____

Matching [p.171]

Choose the correct definition for each term.

1. _____ genotype
2. _____ alleles
3. _____ heterozygous
4. _____ dominant
5. _____ phenotype
6. _____ genes
7. _____ homozygous recessive
8. _____ recessive
9. _____ P, F1, F2
10. _____ hybrids
11. _____ diploid organism
12. _____ gene locus
13. _____ homozygous dominant
14. _____ homologous chromosomes

A. Parental, first-generation, and second-generation offspring

B. All the different molecular forms of the same gene

C. Particular location of a gene on a chromosome

D. Describes an individual having a pair of nonidentical alleles

E. An individual with a pair of recessive alleles, such as *aa*

F. Allele whose effect is masked by the effect of the dominant allele

G. Offspring of a genetic cross that inherit a pair of nonidentical alleles for a trait

H. Refers to an individual's observable traits

I. The effect of an allele on a trait masks that of any recessive allele paired with it

J. An individual with a pair of dominant alleles, such as *AA*

K. Units of information about specific traits; passed from parents to offspring

L. A pair of similar chromosomes, one obtained from the father and the other obtained from the mother

M. The particular alleles that an individual carries for a trait

N. Having pairs of genes on homologous chromosomes

Short Answer [pp.170-171]

15. Was Mendel's choice of *Pisum sativum*, the garden pea plant, a good choice to use to study how traits are passed from parent to offspring?

_____

_____

_____

_____

11.2. MENDEL'S LAW OF SEGREGATION [pp.172-173]

Selected Words: predicted ratios [p.173], observed ratios [p.173]

Boldfaced, Page-Referenced Terms

[p.172] testcross _____

_____

[p.172] monohybrid experiments _____

_____

[p.173] probability _____

_____

[p.173] Punnett squares _____

_____

[p.173] segregation _____

_____

Short Answer

1. Why is the predicted ratio likely closer to the observed ratio when a large number of offspring are sampled?

_____

_____

True/False [pp.172-173]

2. _____ A testcross may be used to indicate if an individual is heterozygous or homozygous for a dominant trait.

3. _____ Punnett squares are used to determine the observed ratios of testcrosses.

4. _____ During meiosis, gene pairs are separated into different gametes. This is Mendel's law of segregation.

5. _____ Mendel's experiments did not exactly match his predicted ratios because of errors he made in counting.

6. _____ The larger the sample size, the more closely the predicted ratios will come to the observed ratios.

7. _____ The F_1 generation are offspring of the F_2 generation.

8. _____ Punnett squares calculate the probability of both genotypes and phenotypes.

11.3. MENDEL'S LAW OF INDEPENDENT ASSORTMENT [pp.174-175]

Selected Words: self-fertilized [p.174], exceptions [p.174]

Boldfaced, Page-Referenced Terms

[p.174] dihybrid experiments _____

_____

[p.174] independent assortment _____

_____

Matching [pp.174-175]

Choose the correct definition for each term.

1. _____ independent assortment
2. _____ possible genetic outcomes
3. _____ Punnett-square method
4. _____ testcross
5. _____ dihybrid experiments
6. _____ segregation
7. _____ P generation
8. _____ F_1 generation
9. _____ F_2 generation

A. Used to determine the genotype of an individual when it displays the dominant phenotype

B. The mathematical chance that a given event will occur

C. The separation of traits during a genetic cross

D. Original individuals involved in a testcross

E. The process by which each pair of homologous chromosomes is sorted out into gametes

F. Genetic crosses that examine the inheritance of two traits

G. A graphic means of representing the distribution of gametes and possible zygotes in a genetic cross

H. 3rd generation offspring

I. Offspring of the parent generation

Problems [pp.174-175]

10. In garden pea plants, tall *(T)* is dominant over dwarf *(t)*. In the cross *Tt x tt*, the *Tt* parent would produce a gamete carrying *T* (tall) and a gamete carrying *t* (dwarf) through segregation; the *tt* parent could only produce gametes carrying the *t* (dwarf) gene. Use the Punnett-square method (refer to Figure 11.7 in the text) to determine the following results of a *Tt x tt* cross.

a. The genotypic probabilities of the offspring: _____

_____

b. The phenotypic probabilities of the offspring: _____

_____

11. In fruit flies, the trait vestigial wings *(a)* is recessive to normal wings *(A)*. You wish to determine whether a fly with a dominant phenotype is homozygous dominant or heterozygous. Using a testcross, you mate the fly with a homozygous recessive individual. Following are two possibilities for the F_1 generation. For each, state if the results indicate that the genotype of the parent is *homozygous dominant* or *heterozygous*.

a. 78 normal-winged offspring: _____

_____

b. 37 normal-winged and 41 vestigial-winged offspring: _____

_____

12. Albinos cannot form the pigments that normally produce skin, hair and eye color, so albinos exhibit white hair and pink eyes and skin (because the blood shows through). An albino is homozygous recessive *(aa)* for the pair of genes that code for the key enzyme in pigment production. Suppose a woman of normal pigmentation *(Aa)* with an albino mother marries an albino man. State the possible kinds of pigmentation for this couple's children, and specify the ratio of each kind of child the couple is likely to have.

_____

_____

13. In horses, black coat color is influenced by the dominant allele *(B)* and chestnut coat color is influenced by the recessive allele *(b)*; trotting gait is due to a dominant gene *(T)*, pacing gait to the recessive allele *(t)*. A homozygous black trotter is crossed to a chestnut pacer.

a. What is the predicted appearance and gait of the F_1 and F_2 generations? _____

_____

b. Which phenotype will likely be most common? _____

_____

c. Which genotype will likely be most common?_____

_____

d. Which of the potential offspring will be certain to breed true? _____

_____

11.4. BEYOND SIMPLE DOMINANCE [pp.176-177]

Selected Words: ABO blood typing [p.176], melanin [p.177], fibrillin [p.177], Marfan syndrome [p.177]

Boldfaced, Page-Referenced Terms

[p.176] codominance _____

_____

[p.176] multiple allele systems _____

_____

[p.176] incomplete dominance_____

_____

[p.177] epistasis _____

_____

[p.177] pleiotropy_____

_____

Matching [pp.176-177]

Match the term to its definition.

1. _____ Occurs when one allele of a pair is not fully dominant over the other

2. _____ Occurs when one gene influences multiple traits

3. _____ Effect that occurs when a trait is affected by interactions among different gene products

4. _____ Marfan Syndrome is caused by a defect in these molecules

5. _____ Pigment molecules that denote color variations in different individuals

6. _____ Occurs when three or more alleles of a gene persist among individuals of a population

7. _____ Nonidentical alleles of a gene that are both fully expressed in heterozygotes so neither is dominant or recessive

A. codominance

B. epistasis

C. fibrillin

D. incomplete dominance

E. melanin

F. multiple allele systems

G. pleiotropy

8. List the phenotypes in ABO blood typing. Next to each phenotype list the possible genotypes.

_____

_____

9. Which phenotype in ABO blood typing is homozygous recessive?

_____

_____

Choice [pp.176-177]

Choose the effect that applies to each of the following situations.

a. epistasis   b. pleiotropy   c. multiple allele systems (codominance)   d. incomplete dominance

10. _____   ABO blood typing

11. _____   Pink offspring from a cross between red and white parent plants

12. _____   Comb variation in chickens

13. _____   Defects in fibrillin molecules that cause multiple phenotypic effects

14. _____   Coat color in Labrador retrievers

Problems [pp.176-177]

15. In four o'clock plants, red flower color is determined by gene R and white flower color by R', while the heterozygous condition, RR', is pink. For each of the following crosses, give the phenotypic ratios of the offspring.

a. RR x R'R' _____

_____

b. RR' x RR' _____

_____

16. Using Figure 11.10 as a reference, indicate the possible blood types that may be present in the offspring of the following crosses.

a. AO x AB _____

_____

b. BO x AO _____

_____

c. AA x OO _____

_____

d. OO x OO _____

_____

e. AB x AB _____

_____

17. In sweet peas, there is an epistatic interaction between two genes involved in flower color. Genes *C* and *P* are necessary for colored flowers. In the absence of either (_ _*pp* or *cc*_ _), or both (*ccpp*), the flowers are white. What color and proportions will the offspring of the following offspring exhibit?

a. *CcPp* x *ccpp*_____

_____

b. *CcPP* x *Ccpp*_____

_____

18. In the inheritance of the coat (fur) color of Labrador retrievers, allele *B* specifies that black is dominant over allele *b*, brown (chocolate). Allele *E* permits full deposition of color pigment, but two recessive alleles, *ee*, reduce deposition, and a yellow coat results. Predict the phenotypes of the coat color and their proportions resulting from the cross *BbEe* x *Bbee*.

_____

_____

11.5. LINKAGE GROUPS [p.178]

Selected Words: linked [p.178], probability of a crossover event [p.178]

Boldfaced, Page-Referenced Terms

[p.178] linkage group_____

_____

Fill-in-the-Blanks [p.178]

We now know that there are many (1) _____ on each type of autosome and sex chromosome. All of the genes on one (2) _____ are called a(n) (3) _____ group. If genes on the same chromosome stayed together through (4) _____, then there would be no surprising mixes of parental (5) _____. You could expect parental phenotypes among the F2 offspring of dihybrid experiments to show up in a predictable (6) _____. Genes that have a relatively (7) _____ distance between them are said to be tightly linked.

Short Answer [p. 178]

8. Use the diagram in the right column of page 178 from your text to answer the following: Which pair of genes, AB or CD, would be more likely to be linked in crossing over? Why?

   _____

   _____

   _____

11.6. GENES AND THE ENVIRONMENT [p.179]

Selected Words: tyrosinase [p.179], yarrow plants [p.179], Daphnia pulex [p.179], serotonin [p.179]

Short Answer

1. How does altitude affect the phenotype of yarrow plants?

   _____

   _____

2. How does temperature determine the coat color of Siamese cats?

   _____

   _____

3. What causes a pointed head feature in Daphnia pulex?

   _____

   _____

4. What substance is released when humans are dealing with stress?

   _____

   _____

11.7. COMPLEX VARIATIONS IN TRAITS [pp.180-181]

Selected Words: iris [p.180], camptodactyly [p.181]

Boldfaced, Page-Referenced Terms

[p.180] continuous variation _____

_____

[p.181] bell curves _____

_____

Choice [pp.179-180]

For items 1–5, choose the primary contributing factor from the options below.

   a. environment   b. polygenic inheritance

1. _____   Distribution of height in human populations

2. _____   Continuous variation in a trait

3. _____   Plant height in *Achillea millefolium*

4. _____   The range of eye colors in the human population

5. _____   Heat-sensitive version of one of the enzymes required for melanin production in Himalayan rabbits

---

SELF-TEST

__ 1. Mendel's theory of independent assortment
   is best stated as [pp.174-175]
   a. One allele is always dominant to
      another.
   b. Hereditary units from the male and
      female parents are blended in the
      offspring.
   c. The two hereditary units that
      influence a certain trait separate
      during gamete formation.
   d. Genes on each pair of chromosomes
      are sorted into gametes independent
      of genes on other chromosomes.

__ 2. All the different molecular forms of the
   same gene are called [p.171]
   a. hybrids
   b. alleles
   c. autosomes
   d. loci

__ 3. If two heterozygous individuals are crossed
   in a monohybrid experiment involving
   complete dominance, the expected
   phenotypic ratio is [pp.172–173]
   a. 3:1
   b. 1:1:1:1
   c. 1:2:1
   d. 1:1
   e. 9:3:3:1

__ 4. The phenotypic ratio of a cross between a
   pink-flowered snapdragon and a white-
   flowered snapdragon is [p.176]
   a. 75% red, 25% white
   b. 100% red
   c. 50% pink, 50% white
   d. 100% pink

5. In a testcross, an individual with a dominant phenotype, but unknown genotype, is crossed to an individual known to be homozygous recessive for that trait. If the F1 offspring are all heterozygous, what can you say about the genotype of the dominant parent?
   a. The dominant parent is heterozygous.
   b. The dominant parent is homozygous dominant.
   c. The dominant parent is homozygous recessive.
   d. The dominant parent has no effect on the genotype of the offspring.

6. The tendency for dogs to bark while trailing is determined by a dominant gene, *S*, whereas silent trailing is due to the recessive gene, *s*. In addition, erect ears, *D*, is dominant over drooping ears, *d*. What combination of offspring would be expected from a cross between two erect-eared barkers who are heterozygous for both genes? [pp.174-175]
   a. 1/4 erect barkers, 1/4 drooping barkers, 1/4 erect silent, 1/4 drooping silent
   b. 9/16 erect barkers, 3/16 drooping barkers, 3/16 erect silent, 1/16 drooping silent
   c. 1/2 erect barkers, 1/2 drooping barkers
   d. 9/16 drooping barkers, 3/16 erect barkers, 3/16 drooping silent, 1/16 erect silent

7. If a mother has type O blood and the father has type AB blood, which of the following blood types could not be present in their children? [p.176]
   a. Type A
   b. Type B
   c. Type O
   d. Type AB
   e. More than once answer is possible

8. A single gene that affects several seemingly unrelated aspects of an individual's phenotype is said to be [p.177]
   a. pleiotropic
   b. epistatic
   c. allelic
   d. continuous

9. A bell curve of phenotypes is typical of what form of inheritance? [pp.180-181]
   a. pleiotropic
   b. polygenic
   c. environmental
   d. multiple allele systems

10. What decreases the probability that two genes on the same chromosome will separate during crossing over? [p.178]
    a. The dominance of the alleles
    b. The age of the chromosome
    c. A short distance between the genes
    d. Epistatic interactions
    e. None of the above

CHAPTER OBJECTIVES/REVIEW QUESTIONS

1. Describe the concept of blending inheritance. [p.170]
2. Explain why the pea plant was an ideal model organism for Mendel's studies. [pp.170-171]
3. Distinguish between genes and alleles. [p.171]
4. Explain the terms genotype and phenotype, and relate them to the terms homozygous dominant, homozygous recessive, and heterozygous. [p.171]
5. Explain the significance of Mendel's theory of segregation. [p.172]
6. Determine the genotypic and phenotypic ratios of a monohybrid experiment. [pp.172-173]
7. Explain the purpose and structure of a testcross. [p.172]
8. Determine the genotypic and phenotypic ratios of a dihybrid experiment. [pp.174-175]
9. Explain the significance of Mendel's theory of independent assortment. [pp.174-175]
10. Distinguish among complete dominance, incomplete dominance, and codominance. [pp.176-177]
11. Define *epistatic* and *pleiotropic* inheritance. [p.177]
12. Explain crossing over and the factors that determine the probability of crossing over between two genes on the same chromosome. [p.178]
13. Explain how the environment can influence the expression of a trait. [p.179]
14. Explain what is meant by continuous variation and polygenic inheritance. [pp.180-181]

INTEGRATING AND APPLYING KEY CONCEPTS

1. Solve the following genetics problem: In garden peas, one pair of alleles controls the height of the plant and a second pair of alleles controls flower color. The allele for tall (D) is dominant to the allele for dwarf (d), and the allele for purple (P) is dominant to the allele for white (p). A tall plant with purple flowers crossed with a tall plant with white flowers produces offspring that are 3/8 tall purple, 3/8 tall white, 1/8 dwarf purple, and 1/8 dwarf white. What are the genotypes of the parents?

2. As a breeder of exotic plants, you advertise that you have produced a true-breeding yellow flowered line. Assuming that yellow flower color (Y) is dominant over the less-desirable (w) color, how do you test to ensure that your line is not heterozygous for the trait?

3. Height and eye color in humans are frequently given as examples of continuous variation. What are some other human traits that may display continuous variation? What about in other species?

4. Sickle-cell anemia is a genetic disease in which children that are homozygous for a defective gene ($Hb^{S}Hb^{S}$) produce defective hemoglobin. The genotypes of normal persons are $Hb^{A}Hb^{A}$. If the level of blood oxygen drops below a certain level in a person with the $Hb^{S}Hb^{S}$ genotype, the hemoglobin chains stiffen and cause the red blood cells to form sickle, or crescent, shapes. These cells clog and rupture capillaries, which results in oxygen-deficient tissues where metabolic wastes collect. Several body functions are badly damaged. Severe anemia and other symptoms develop, and death nearly always occurs before adulthood. The sickle-cell gene is considered *pleiotropic*. Persons that are heterozygous ($Hb^{A}Hb^{S}$) are said to possess *sickle-cell trait*. They are able to produce enough normal hemoglobin molecules to appear normal, but their red blood cells will sickle if they encounter oxygen tension (such as at high altitudes). A man whose sister died of sickle-cell anemia married a woman whose blood is found to be normal. What advice would you give this couple about the inheritance of this disease as they plan their family? If a man and a woman, each with sickle-cell trait, planned to marry, what information could you provide for them regarding the genotypes and phenotypes of their future children?

5. Human genetics has reached a point at which it is possible to screen an individual's genome for specific alleles related to disease. What are the positive and negative possibilities of this ability? Are there ethical aspects that should be taken into consideration?

12

CHROMOSOMES AND HUMAN INHERITANCE

INTRODUCTION

This chapter takes the genetics you learned in previous chapters and applies that knowledge to humans. You will also learn about tools that are used to study patterns in human genetics.

FOCAL POINTS

- The preview section [p.184] looks at the relationship of artistic ability to neurobiological disorders.
- Figure 12.2 [p.186] illustrates how sex determination occurs in humans and the subsequent development of the structures characteristic to each sex.
- Figure 12.3 [p.187] looks at karyotypes and how they are used as diagnostic tools.
- Figure 12.8 [p.190] illustrates how a pedigree is used to follow hemophilia inherited by the descendants of Queen Victoria of England. Figures 12.17 and 12.18 [p.197] show pedigrees tracking the inheritance of Huntington's disease and Ellis-van Creveld syndrome, respectively.
- Table 12.1 [p.196] describes many genetic disorders and abnormalities that occur in humans.
- Figure 12.19 [p.198] and 12.20 [p.199] describe prenatal diagnostic tools.

INTERACTIVE EXERCISES

Impacts, Issues: Strange Genes, Tortured Minds [p.184]

12.1. HUMAN CHROMOSOMES [pp.186-187]

Selected Words: environmental factors [p.186], SRY gene [p.187], testosterone [p.187], estrogens [p.187]

Boldfaced, Page-Referenced Terms

[p.186] autosomes _____

_____

[p.186] sex chromosomes _____

_____

[p.187] karyotype _____

_____

Fill-in-the-Blanks [pp.186-187]

Like most animals, humans have (1) _____ chromosomes, meaning that they have two of each

chromosome. Twenty-two of the twenty-three pairs of chromosomes carry genes common to both males and

females. These chromosomes are called (2) _____. One pair is different for males and females. Members of

this pair are called (3) _____ chromosomes. The sex chromosomes of humans are called (4) _____ and

(5) _____. The two sex chromosomes for human females are (6) _____ and the two sex chromosomes

for males are (7) _____. The X and Y chromosomes differ in length, shape, and which genes they carry.

Although human females have identical sex chromosomes, in butterflies, birds, moths, and certain fish it's the (8)

_____ that has identical sex chromosomes. (9) _____ factors determine the sex of some animals such as

certain species of turtles and (10) _____. (11) _____ eggs made by a human female carry an X

chromosome. (12) _____ of the sperm cells carry an X chromosome and (13) _____ of the sperm cells

carry a Y chromosome. The Y chromosome carries the (14) _____ gene – master gene for male sex

determination. This gene controls production of the hormone (15) _____ that triggers formation of the (16)

_____. No SRY gene means less testosterone therefore (17) _____ form instead of testes in human

females.

Short Answer [p.187]

Use Figure 12.3 and the description of karyotyping on to answer the following questions.

18. What substance is used to stop mitosis in metaphase when preparing a sample for karyotyping? How does this substance work?

_____

_____

19. How are the chromosomes arranged into the twenty-three pairs of the final karyotype array?

_____

_____

20. Is the karyotype picture in the above figure that of a male or a female?

_____

_____

12.2. EXAMPLES OF AUTOSOMAL INHERITANCE PATTERNS [pp.188-189]

Selected Words: achondroplasia [p.188], Huntington disease [p.188], expansion mutations [p.188], galactosemia [p.188], neurobiological disorders [p.189]

Matching [pp.188-189]

Match the disorder to its description.

1. _____ This disorder has been linked to mutations on chromosomes 1, 3, 5, 6, 8, 11-15, 18, and 22

2. _____ An autosomal dominant allele causes this disorder of abnormally short arms and legs

3. _____ A neurobiological disorder characterized by periods of depression and elation

4. _____ A metabolic disorder that interferes with the digestion of dairy products

5. _____ A neurobiological disorder characterized by periods of deep despair

6. _____ An autosomal dominant allele causes this disorder that alters a protein necessary for brain development

A. Achondroplasia

B. Huntington's disease

C. Galactosemia

D. Depression

E. Schizophrenia

F. Bipolar disorder

Problems [pp.188-189]

7. The autosomal allele that causes galactosemia (g) is recessive to the allele for normal lactose metabolism (G). A normal woman whose father had galactosemia marries a man with galactosemia who had normal parents. They have three children, two normal and one with galactosemia. List the genotypes for each person involved.

_____

_____

_____

8. Huntington disease is a rare form of autosomal dominant inheritance, *H*; the normal gene is *h*. The disease causes progressive degeneration of the nervous system with onset exhibited near middle age. An apparently normal man in his early twenties learns that his father has recently been diagnosed as having Huntington disorder. What are the chances that the son will develop this disorder?

_____

_____

_____

9. In the accompanying pedigree, people with the trait are indicated by black circles. Is this trait dominant or recessive? Why? What would you expect if the third-generation male with the trait mated with a normal female?

_____

_____

_____

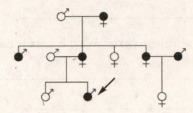

Choice [pp.188-189]

For questions 10-20, choose from the following patterns of inheritance. Some items may require more than one letter.

      a. autosomal recessive   b. autosomal dominant

10. _____ Heterozygotes can remain undetected

11. _____ The trait usually appears in each generation

12. _____ Both parents may be heterozygous normal

13. _____ If one parent is heterozygous and the other homozygous recessive, there is a 50 percent chance that any child of theirs will be affected

14. _____ The allele is usually expressed, even in heterozygotes

15. _____ Heterozygous normal parents can expect that one-fourth of their children will be affected by the disorder

16. _____ The trait is expressed in heterozygotes of either sex

17. _____ Individuals displaying this type of disorder will always be homozygous for the trait

18. _____ The trait is expressed in both the homozygote and the heterozygote

19. _____ Heterozygous unaffected parents have two children affected with the disorder

20. _____ Heterozygous affected parents have a child that is unaffected

12.3. TOO YOUNG TO BE OLD [p.189]

Selected Words: Hutchinson–Gilford progeria syndrome [p.189]

Short Answer

1. Describe the basic characteristics of progeria.

    _____

    _____

    _____

2. What protein is affected in persons with progeria?

    _____

    _____

    _____

3. What is the life expectancy of persons with progeria? What is usually the cause of death?

    _____

    _____

    _____

12.4. EXAMPLES OF X-LINKED INHERITANCE PATTERNS [pp.190-191]

Selected Words: X chromosome [p.190], X-linked gene [p.190], hemophilia A [p.190], color blindness [p.191], red–green color blindness [p.191], Duchenne muscular dystrophy [p.191]

Choice [pp.190-191]

Answer the following questions with one of these choices:

   a. sons   b. daughters   c. sons and daughters   d. fathers   e. mothers   f. fathers and mothers

1. _____   Male humans transmit their Y chromosome to their _____
2. _____   Male humans receive their X chromosome from their _____
3. _____   Human mothers and fathers each provide an X chromosome for their _____
4. _____   Female humans pass their X chromosomes to their _____

12.5. HERITABLE CHANGES IN CHROMOSOME STRUCTURE [pp.192-193]

Selected Words: unequal crossovers [p.192], cri-du-chat [p.192]

Boldfaced, Page-Referenced Terms

[p.192] duplications _____

_____

[p.192] deletion _____

_____

[p.192] inversion _____

_____

[p.192] translocation _____

_____

Labeling and Matching [p.192]

On rare occasions, chromosome structure becomes abnormally rearranged. Such changes may have profound effects on the phenotype of an organism. Label the following diagrams of abnormal chromosome structure as a deletion, a duplication, an inversion, or a translocation. Complete the exercise by matching and entering the letter of the proper description in the parentheses following each label.

1. _____ (   )   2. _____ (   )   3. _____ (   )   4. _____ (   )

A. The loss of a chromosome segment; an example is the cri-du-chat disorder

B. A gene sequence in excess of its normal amount in a chromosome; this alters the position and order of the chromosome's genes; possibly promoted human evolution

C. A chromosome segment that separated from the chromosome and then was inserted at the same place, but in reverse; may cause mispairing in meiosis

D. The transfer of part of one chromosome to a nonhomologous chromosome; an example is when chromosome 14 ends up with a segment of chromosome 8; the Philadelphia chromosome is also an example

12.6. HERITABLE CHANGES IN THE CHROMOSOME NUMBER [pp.194-195]

Selected Words: trisomic [p.194], monosomic [p.194], Down syndrome [p.194], Turner syndrome [p.195], XXX syndrome [p.195], Klinefelter syndrome [p.195], XYY condition [p.195]

Boldfaced, Page-Referenced Terms

[p.194] nondisjunction _____

_____

[p.194] aneuploidy _____

_____

[p.194] polyploidy _____

_____

Short Answer [p.194]

1. If a nondisjunction occurs at anaphase I of the first meiotic division, what will be the proportion of abnormal gametes (for the chromosomes involved in the nondisjunction)?

    _____

    _____

    _____

2. If a nondisjunction occurs at anaphase II of the second meiotic division, what will be the proportion of abnormal gametes (for the chromosomes involved in the nondisjunction)?

    _____

    _____

    _____

3. Define the following terms: *polyploid, trisomic,* and *monosomic.*

    _____

    _____

    _____

Choice [pp.194-195]

For questions 4–13, choose from the following:

    a. Down syndrome   b. Turner syndrome   c. Klinefelter syndrome   d. XYY condition

4. _____ XXY male

5. _____ Females whose ovaries are nonfunctional and who fail to develop secondary sexual traits at puberty

6. _____ Testes smaller than normal, sparse body hair, and some breast enlargement

7. _____ Could only be caused by a nondisjunction in males

8. _____ As a group, they tend to be cheerful and affectionate; many have heart problems

9. _____ XO female; most abort early; distorted female phenotype

10. _____ Males that tend to be taller than average; some mildly retarded but most are phenotypically normal

11. _____ Injections of testosterone reverse feminized traits but not the fertility

12. _____ Trisomy 21; muscles and muscle reflexes weaker than normal

13. _____ At one time these males were thought to be genetically predisposed to become criminals

12.7. HUMAN GENETIC ANALYSIS [pp.196-197]

Selected Words: polydactyly [p.197]

Boldfaced, Page-Referenced Terms

[p.197] pedigrees _____

_____

[p.197] syndrome _____

_____

Matching [p.197]

Using the symbols in Figure 12.18, match each of the following individuals to the description. For each generation row, count across from left to right for each individual.

1. _____ II, 1 & 2   A. Male showing the trait being tracked
2. _____ I, 2   B. Male not showing the trait being tracked
3. _____ IV, 2   C. Female showing the trait being tracked
4. _____ III, 1   D. Female not showing the trait being tracked
5. _____ V, 2   E. Married individuals
6. _____ IV, 1 & 2   F. Siblings
7. _____ III, 5   G. Individual whose sex is not specified

12.8. PROSPECTS IN HUMAN GENETICS [pp.198-199]

Selected Words: genetic disorder [p.198], genetic counseling [p.198], prenatal diagnosis [p.198], embryo [p.198], fetus [p.198], amniocentesis [p.198], chorionic villi sampling [p.198], fetoscopy [p.198], preimplantation diagnosis [p.198], in vitro fertilization [p.198], "test-tube babies" [p.199], abortion [p.199], phenotypic treatments [p.199], phenylketonuria (PKU) [p.199], genetic screening [p.199]

Boldfaced, Page-Referenced Terms

[p.198] pedigree _____

_____

[p.198] syndrome _____

_____

Fill-in-the-Blanks [pp.198-199]

The effects of genetic disorders can often be minimized by (1) _____ treatments such as surgery, prescription drugs, or dietary modification. In one disorder, (2) _____, reduction of the amount of phenylalanine in the diet can slow or even stop deterioration of the brain. Sometimes, (3) _____ diagnosis is used to diagnose disorders before birth. In (4) _____, a syringe is used to draw fluid from the amniotic sac. This fluid can then be analyzed for many genetic disorders. A diagnostic tool that can be used even earlier is (5) _____, in which cells are removed from one of the placental membranes for further analysis. Once a diagnosis is made, parents often face the decision whether or not to have an induced (6) _____.

SELF-TEST

____ 1. A _____ is assembled using chromosomes stopped in metaphase of mitosis. [p.187]
  a. karyotype
  b. bridging cross
  c. wild-type allele
  d. linkage group

____ 2. Chromosomes other than those involved in sex determination are known as _____. [p.186]
  a. nucleosomes
  b. heterosomes
  c. alleles
  d. autosomes

____ 3. Karyotype analysis is _____. [p.187]
  a. a means of detecting and reducing mutagenic agents
  b. a surgical technique separating chromosomes that have failed to segregate properly during meiosis II
  c. used to reveal abnormalities in chromosome structure and number
  d. a process that substitutes defective alleles with normal ones

____ 4. A man with achondroplasia marries a normal woman. What is the chance that their first child will be achondroplastic? [p.188]
  a. 0%
  b. 25%
  c. 50%
  d. 75%

____ 5. Red–green color blindness is a sex-linked recessive trait in humans. A color-blind woman and a man with normal vision have a son. What are the chances that the son is color blind? If the parents ever have a daughter, what is the chance that a daughter will be color blind? (For the latter question, consider only the female offspring.) [p.191]
  a. 100 percent, 0 percent
  b. 50 percent, 0 percent
  c. 100 percent, 100 percent
  d. 50 percent, 100 percent
  e. none of the above

____ 6. Suppose that a hemophiliac male (X-linked recessive allele) and a female carrier for the hemophiliac trait have a non-hemophiliac daughter with Turner syndrome. Nondisjunction could have occurred in. [pp.190, 191, 194]
  a. either parent
  b. neither parent
  c. the father only
  d. the mother only
  e. the non-hemophiliac daughter

____ 7. Nondisjunction involving the X chromosome occurs during oogenesis and produces two kinds of eggs, XX and O (no X chromosome). If normal Y sperm fertilize the two types, which genotypes are possible? [pp.194-195]
  a. XX and XY
  b. XXY and YO
  c. XYY and XO
  d. XYY and YO
  e. YY and XO

____ 8. Of all phenotypically normal males in prisons, the type once thought to be genetically predisposed to becoming criminals was the group with _____. [p.195]
  a. XXY disorder
  b. XYY disorder
  c. Turner syndrome
  d. Down syndrome
  e. Klinefelter syndrome

___ 9. Amniocentesis is _____.[p.198]
  a. a surgical means of repairing deformities
  b. a form of chemotherapy that modifies or inhibits gene expression or the function of gene products
  c. used in prenatal diagnosis; a small sample of amniotic fluid is drawn to detect chromosomal mutations and metabolic disorders in embryos
  d. a form of gene-replacement therapy
  e. a diagnostic procedure; cells for analysis are withdrawn from the chorion

___ 10. John Nash, Pablo Picasso, Virginia Woolf, and Abraham Lincoln were all thought to have _____. [pp. 184 and 200]
  a. phenylketonuria
  b. progeria
  c. hemophilia
  d. achondroplasia
  e. neurobiological disorder

CHAPTER OBJECTIVES/REVIEW QUESTIONS

1. A(n) _____ is a preparation of metaphase chromosomes based on their defining features. [p.187]
2. Describe the Hutchinson–Gilford progeria syndrome and its probable mode of inheritance. [p.189]
3. Define *karyotype*; briefly describe its preparation and value. [p.187]
4. Explain meiotic segregation of sex chromosomes to gametes and the subsequent random fertilization that determines sex in many organisms. [p.186]
5. Name and describe the sex chromosomes in human males and females. [p.186]
6. Human X and Y chromosomes fall in the general category of chromosomes; all other chromosomes in an individual's cells are the same in both sexes and are called _____. [p.186]
7. Carefully characterize patterns of autosomal recessive inheritance, autosomal dominant inheritance, and X-linked recessive inheritance. [pp.188, 190-191]
8. A(n) _____ is a loss of a chromosome segment; a(n) _____ is a gene sequence separated from a chromosome and then inserted at the same place, but in reverse; a(n) _____ is a repeat of several gene sequences on the same chromosome; a(n) _____ is the transfer of part of one chromosome to a non-homologous chromosome. [pp.192-193]
9. _____ is the failure of the chromosomes to separate in either meiosis or mitosis. [p.194]
10. Trisomy 21 is known as _____ syndrome; Turner syndrome has the chromosome constitution _____; XXY chromosome constitution is syndrome; taller-than-average males with sometimes slightly lower-than-average IQs have the condition _____. [pp.194–195]
11. When gametes or cells of an affected individual end up with one extra or one less chromosome than the parental number, the condition is known as _____; relate this concept to monosomy and trisomy.[p.194]
12. A(n) _____ chart or diagram is used to study genetic connections between individuals. [p.197]
13. A genetic is an uncommon but not unhealthy version of a trait, whereas an inherited genetic causes mild to severe medical problems. [pp.196-197]
15. List some benefits to a society of genetic screening and genetic counseling. [pp.198-199]
16. Explain the procedures and purpose of three types of prenatal diagnosis: amniocentesis, chorionic villi analysis, and fetoscopy; compare the risks. [pp.198-199]

INTEGRATING AND APPLYING KEY CONCEPTS

1. The parents of a young boy bring him to their doctor. They explain that the boy does not seem to be going through the same vocal developmental stages as his older brother. The doctor orders a common cytogenetics test to be done, and it reveals that the young boy's cells contain two X chromosomes and one Y chromosome. Describe the test that the doctor ordered and explain how and when such a genetic result, XXY, most logically occurred.
2. Solve the following genetics problem. Show rationale, genotypes, and phenotypes. A husband sues his wife for divorce, arguing that she has been unfaithful. His wife gave birth to a girl with a fissure in the iris of her eye, an X-linked recessive trait. Both parents have normal eye structure. Can the genetic facts be used to argue for the husband's suit? Explain your answer.
3. A test is available to identify heterozygotes for a common, incurable, recessive genetic defect. A man and woman, both of whom have family members with the defect, are trying to decide whether to have the test. What are some of the ethical issues involved?
4. Figure 12.12 in the text shows the relationship of chromosome 2 in humans with two separate chromosomes in apes. How might this relate to evolution? Explain how this change could have occurred.

13

DNA STRUCTURE AND FUNCTION

INTRODUCTION

This chapter looks at the details of DNA structure. It also links the structure of DNA to its replication and function. In later sections, the topic of cloning and how it impacts our lives is discussed.

FOCAL POINTS

- Figure 13.2 [p.204] and Figure 13.3 [p.205] show the classic experiments that led scientists to the understanding that DNA is the hereditary material.
- Figure 13.5 [p.207] details the structure of DNA.
- Figure 13.6 [p.208], Figure 13.7 [p.209], and Figure 13.8 [p.209] show the steps of DNA replication.
- Figure 13.9 [p.210] details the procedure used to transfer cell nuclei in cloning cattle.

INTERACTIVE EXERCISES

Impacts, Issues: Here, Kitty, Kitty, Kitty, Kitty, Kitty [p.202]

13.1. THE HUNT FOR DNA [pp.204–205]

Selected Words: J. Miescher [p.204], F. Griffith [p.204], Streptococcus pneumoniae [p.204], O. Avery [p.204], A. Hershey and M. Chase [p.205], ^{35}S and ^{32}P [p.205]

Boldfaced, Page-Referenced Terms

[p.204] deoxyribonucleic acid, DNA _____

_____

[p.205] bacteriophage _____

_____

Short Answer [pp.204-205]

What are the contributions of each of the following to our knowledge of DNA?

1. Miescher _____

   _____

2. Griffith _____

   _____

3. Avery (et al.) _____

   _____

4. Hershey and Chase _____

   _____

Fill-in-the-Blanks [p.205]

A bacteriophage is a kind of (5) _____ that can infect (6) _____ cells. These infectious particles

carry (7) _____ information about how to make new viruses. Once injected into a host cell, the bacteriophage

injects its (8) _____ material into bacteria and incorporates that new material into the host DNA. In Hershey

and Chase's experiments, the (9) _____ (not the DNA) was labeled with a radioisotope of sulfur, ^{35}S. Hershey

and Chase then labeled the (10) _____ (but not the protein) with the radioisotope of phosphorus known as

(11) _____. When (12) _____ were allowed to infect bacterial cells, the (13) _____ radioisotope

remained outside the bacterial cells; the (14) _____ was part of the material injected *into* the bacterial cells.

Through many experiments, researchers accumulated strong evidence that (15) _____, not (16) _____,

serves as the molecule of inheritance in all living cells.

13.2. THE DISCOVERY OF DNA'S STRUCTURE [pp.206-207]

Selected Words: pyrimidine [p.206], purine [p.206], Erwin Chargaff [p.206], Chargaff's rules [p.206], Rosalind
Franklin [p.206], x-ray crystallography [p.206], James Watson and Francis Crick [p.206], Linus Pauling, Robert
Corey, and Herman Branson [p.206], alpha helix [p.206]

Boldfaced, Page-Referenced Terms

[p.206] nucleotide_____

_____

[p.206] adenine_____

_____

[p.206] guanine _____

_____

[p.206] thymine _____

_____

[p.206] cytosine _____

_____

[p.207] sequence _____

_____

Short Answer [p.206]

1. List the three parts of a DNA nucleotide.

_____

_____

_____

Labeling [p.206]

Four nucleotides are illustrated below. In the blank, label each nitrogen-containing base correctly as guanine, thymine, cytosine, or adenine. In the parentheses following each blank, indicate whether that nucleotide base is a purine (pu) or a pyrimidine (py).

2. _____ ( )   3. _____ ( )   4. _____ ( )   5. _____ ( )

Labeling and Matching [p.207]

Identify each indicated part of the accompanying DNA illustration. Choose from these answers:

phosphate group, purine, pyrimidine, nucleotide, deoxyribose

Complete the exercise by matching and entering the letter of the proper structure description in the parentheses following each label.

The following memory devices may be helpful: Use *pyrCUT* to remember that the pyrimidines are cytosine, uracil (in RNA), and thymine. Use *purAG* to remember that the purines are adenine and guanine; to help recall the number of hydrogen bonds between the DNA bases, remember that AT = 2 and CG = 3.

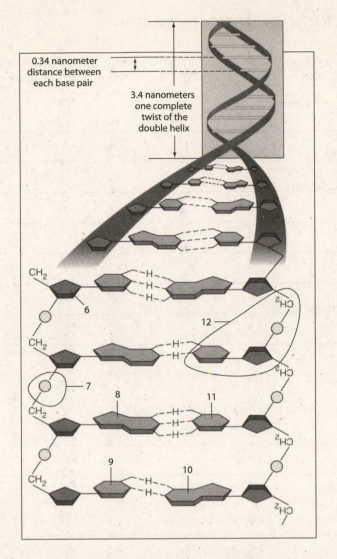

0.34 nanometer distance between each base pair

3.4 nanometers one complete twist of the double helix

6. _____ ( )

7. _____ ( )

8. _____ ( )

9. _____ ( )

10. _____ ( )

11. _____ ( )

12. A complete_____ ( )

A. Thymine

B. Five-carbon sugar

C. Guanine

D. Cytosine

E. Adenine

F. Composed of three smaller molecules: a phosphate group, deoxyribose, and a nitrogenous base

G. A chemical group that joins two sugars in the DNA ladder

13.3. DNA REPLICATION AND REPAIR [pp.208-209]

Selected Words: high-energy phosphate bonds [p.208], DNA proofreading mechanisms [p.209], repair mechanisms [p.209], R. Okazaki [p.209], continuous assembly [p.209], discontinuous assembly [p.209]

Boldfaced, Page-Referenced Terms

[p.208] DNA polymerase _____

_____

[p.208] semiconservative replication _____

_____

[p.208] DNA ligase _____

_____

[p.209] DNA repair mechanisms _____

_____

Short Answer [pp.208-209]

1. Discuss the necessity for Okazaki fragments in DNA synthesis.

    _____

    _____

2. Explain how errors are corrected in DNA.

    _____

    _____

Labeling [p.209]

3. The term *semiconservative replication* refers to the fact that each new DNA molecule resulting from the replication process is "half old, half new." The following sequences represent DNA in the midst of replication. Complete the replication required in the middle of the molecule by adding the required letters for the missing nucleotide bases. Note that ATP energy and the appropriate enzymes are actually required in order to complete this process.

T– _____   _____ –A

G– _____   _____ –C

A– _____   _____ –T

C– _____   _____ –G

C– _____   _____ –G

C– _____   _____ –G

Old New   New Old

True/False [pp.208-209]

If a statement below is true, write T in the blank. If it is false, explain why by changing one or more of the underlined words.

4. _____ The hydrogen bonding of adenine to <u>guanine</u> is an example of complementary base pairing.

5. _____ The replication of DNA is considered a <u>conservative process</u> because each new molecule is really half new and half old.

6. _____ Each <u>parent single strand remains intact</u> during replication, and a new companion strand is assembled on each of those parent strands.

7. _____ DNA <u>ligases</u> govern the assembly of nucleotides on a parent strand.

8. _____ <u>DNA polymerases, DNA ligases, and other enzymes</u> engage in DNA replication.

13.4. USING DNA TO DUPLICATE EXISTING MAMMALS [pp.210-211]

Selected Words: embryo cloning [p.210], embryo splitting [p.210], identical twins [p.210], differentiated cell [p.210], undifferentiated (stem) cell [p.210], adult cloning [p.210], surrogate mother [p.211]

Boldfaced, Page-Referenced Terms

[p.210] reproductive cloning _____

_____

[p.210] somatic cell nuclear transfer (SCNT) _____

_____

[p.211] therapeutic cloning _____

_____

Sequence [p.210]

Correctly order the following events in somatic cell nuclear transfer (SCNT).

1. _____ Egg is treated with an electric current.

2. _____ A micropipette is used to deliver a cell from the donor animal.

3. _____ The receiving egg is held in place by suction through a micropipette.

4. _____ The embryo is transplanted into a surrogate mother.

5. _____ The polar body and chromosomes are removed from the receiving egg.

Matching [pp.210-211]

Match the vocabulary term to its definition.

6. _____ Production of embryos for research   A. surrogate mother

7. _____ Animal that will carry the cloned embryo to term   B. differentiated cell

C. undifferentiated cell

8. _____ Cell that has not been committed to developing into a certain kind of cell   D. identical twins

9. _____ Natural process where an embryo forms identical twins   E. embryo splitting

F. therapeutic cloning

10. _____ Involves removal of the nucleus from an unfertilized egg   G. reproductive cloning

H. somatic cell nuclear transfer (SCNT)

11. _____ Results from a single embryo splitting to form two separate individuals

12. _____ Cell that has been committed to specialize as a certain kind of cell

13. _____ Production of genetically cloned individuals

Short Answer [p.211]

14. How can therapeutic cloning be used to help individuals with spinal cord injuries or who need organ transplants?

_____

_____

_____

13.5. FAME AND GLORY [p.211]

Selected Words: x-ray crystallography [p.211], Maurice Wilkins [p.211], Rudolf Signer [p.211], Nobel Prize [p.211]

Short Answer [p.211]

1. What was Rosalind Franklin's contribution to our understanding of DNA structure?

_____

_____

2. Why didn't Rosalind Franklin share in the 1962 Nobel Prize that went to Watson, Crick, and Wilkins for the discovery of the structure of DNA?

_____

_____

_____

SELF-TEST

___ 1. Each DNA strand has a backbone that consists of alternating _____. [p.206]
   a. purines and pyrimidines
   b. nitrogen-containing bases
   c. hydrogen bonds
   d. sugar and phosphate molecules

___ 2. In DNA, complementary base-pairing occurs between _____. [p.206]
   a. cytosine and uracil
   b. adenine and guanine
   c. adenine and uracil
   d. adenine and thymine

___ 3. Adenine and guanine are _____. [p.206]
   a. double-ringed purines
   b. single-ringed purines
   c. double-ringed pyrimidines
   d. single-ringed pyrimidines

___ 4. Franklin used the technique known as _____ to determine many of the physical characteristics of DNA. [p.211]
   a. transformation
   b. cloning
   c. density-gradient centrifugation
   d. x-ray diffraction

___ 5. The significance of Griffith's experiment in which he used two strains of pneumonia-causing bacteria is that _____. [p.204]
   a. DNA is conservative in its replication
   b. it demonstrated that pathogenic cells had become permanently transformed into harmless cells through a change in the bacteriophage protein coat
   c. it established that pure DNA extracted from disease-causing bacteria and injected into harmless strains transformed them into "pathogenic strains"
   d. it demonstrated that radioactively labeled bacteriophages transfer their DNA but not their protein coats to their host bacteria

6. The significance of the experiments in which ^{32}P and ^{35}S were used is that _____. [p.205]
   a. the semiconservative nature of DNA replication was finally demonstrated
   b. it demonstrated that harmless cells had become permanently transformed through a change in the bacterial hereditary system
   c. it established that pure DNA extracted from disease-causing bacteria transformed harmless strains into "killer strains"
   d. it demonstrated that radioactively labeled bacteriophages transfer their DNA but not their protein coats to their host bacteria

7. Franklin's research contribution was essential in _____. [p.211]
   a. establishing the principle of base pairing
   b. establishing most of the principal structural features of DNA
   c. both a and b
   d. neither a nor b

8. Chargaff's requirement that A binds to T and G binds to C suggested that _____. [p.206]
   a. cytosine molecules pair up with guanine molecules, and thymine molecules pair up with adenine molecules
   b. the two strands in DNA run in opposite directions (are antiparallel)
   c. the number of adenine molecules in DNA relative to the number of guanine molecules differs from one species to the next
   d. the replication process must necessarily be semiconservative

9. A single strand of DNA with the base-pairing sequence C–G–A–T–T–G is compatible only with what sequence? [p.211]
   a. C–G–A–T–T–G
   b. G–C–T–A–A–G
   c. T–A–G–C–C–T
   d. G–C–T–A–A–C

10. Rosalind Franklin's data indicated that the DNA molecule had to be long and thin with a width (diameter) that is 2 nanometers along its length. Double-ringed nucleotides are wider than single-ringed ones, so the uniform width of the DNA molecule results from _____ (as Watson and Crick declared). [p.207]
   a. complementary base-pairing processes that match purine with pyrimidine
   b. semiconservative replication processes
   c. hydrogen bonding of the sugar–phosphate backbones
   d. the antiparallel nature of DNA

CHAPTER OBJECTIVES/REVIEW QUESTIONS

1. The two scientists who assembled the clues to DNA structure and produced the first model were _____ and _____. [pp.206-207]
2. Summarize the research carried out by Miescher, Griffith, Avery, and Hershey and Chase; state the specific advances made by each in the understanding of genetics. [pp.204-205]
3. Draw the basic shape of a deoxyribose molecule, and show how a phosphate group is joined to it when forming a nucleotide. [p.207]
4. Show how a nucleotide base would be joined to the sugar–phosphate combination drawn in objective 3. [p.207]
5. DNA is composed of double-ring nucleotides known as _____ and single-ring nucleotides known as _____; the two purines are _____ and _____, whereas the two pyrimidines are _____ and _____. [p.206]

6. Assume that the two parent strands of DNA have been separated and that the base sequence on one parent strand is A–T–T–C–G–C; the base sequence that will complement that parent strand is _____ .[p.207]
7. List the pieces of information about DNA structure that Rosalind Franklin discovered through her x-ray diffraction research. [p.211]
8. Explain what is meant by the pairing of nitrogen-containing bases (base-pairing), and explain the mechanism that causes bases of one DNA strand to join with bases of the other strand. [pp.206-207]
9. Generally describe how double-stranded DNA replicates from stockpiles of nucleotides. [pp.208-209]
10. Describe the process of making a genetically identical copy of yourself. [pp.210-211]

INTEGRATING AND APPLYING KEY CONCEPTS

1. Review the stages of mitosis and meiosis, as well as the process of fertilization. Relate what was learned in the chapter about DNA replication and the relationship of DNA to a chromosome. As you pass through fertilization and the stages of both types of cell division, use a diploid number of 2n = 4. Show the proper number of DNA threads in each cell at each of the stages of mitosis and meiosis. Include the cells that represent the end products of mitosis and meiosis.
2. How might somatic cell nuclear transfer be used to alleviate health problems in humans? What are the ethical considerations with using SCNT?
3. How do the results of cloning and sexual reproduction differ? How might each interact with evolution?

14

FROM DNA TO PROTEIN

INTRODUCTION

This chapter details two processes: transcription, to make RNA copies of DNA, and translation, to convert the information in the sequence of a nucleic acid into the amino acid sequence of a protein.

FOCAL POINTS

- Figure 14.3 [p.217] compares DNA to RNA.
- Figures 14.4, 14.5, and 14.6 [pp.218–219] outline the steps of transcription.
- Figure 14.7 [p.220] details the modification of mRNA before it leaves the nucleus.
- Figures 14.8 and 14.9 [pp.220-221] explain the mRNA genetic code.
- Figure 14.12 [p.222] outlines the steps of translation.

INTERACTIVE EXERCISES

Impacts, Issues: Ricin and Your Ribosomes [p.214]

14.1. DNA, RNA, AND GENE EXPRESSION [pp.216-217]

Selected Words: genetic information [p.216], uracil [p.216], "genetic words" [p.216], ribosomes [p.216], amino acids [p.216], polypeptide chain [p.217]

Boldfaced, Page-Referenced Terms

[p.216] transcription _____

_____

[p.216] ribosomal RNA (rRNA) _____

_____

[p.216] transfer RNA (tRNA) _____

_____

[p.216] messenger RNA (mRNA) _____

_____

[p.217] translation _____

_____

[p.217] gene expression _____

_____

Fill-in-the-Blanks [pp.216-217]

The (1) _____ of four bases contains the genetic information in DNA. The two steps from genes to proteins are called (2) _____ and (3) _____. In (4) _____, single-stranded molecules of RNA are assembled on DNA templates in the nucleus. In (5) _____, the RNA molecules are shipped from the nucleus into the cytoplasm, where they are used as templates for assembling the sequence of (6) _____of a protein. (7) _____is the multistep process by which genetic information encoded by a gene is converted into a structural or functional part of a cell or body. Proteins – (8) _____– assemble lipids and complex carbohydrates from simple building blocks, replicate DNA, and make RNA.

Complete the Table [p.216]

9. Three types of RNA are transcribed from DNA in the nucleus (two are from genes that code only for RNA). Complete the following table, which summarizes information about these molecules.

RNA Molecule	Abbreviation	Description/Function
a. Ribosomal RNA		
b. Messenger RNA		
c. Transfer RNA		

14.2. TRANSCRIPTION: DNA TO RNA [pp.218-219]

Selected Words: complementary [p.218], 5' to 3' direction [p.219]

Boldfaced, Page-Referenced Terms

[p.218] RNA polymerase _____

_____

[p.219] promoter _____

_____

Short Answer [p.218]

1. List three ways in which a molecule of RNA differs structurally from a molecule of DNA.

_____

_____

2. Cite two similarities between DNA replication and transcription.

_____

_____

3. What are the three key ways in which transcription differs from DNA replication?

_____

_____

14.3. RNA AND THE GENETIC CODE [pp.220-221]

Selected Words: protein-building message [p.221], post-transcriptional modifications [p.220], "cap" [p.220], poly-A tail [p.220], start codon [p.220], stop codons [p.220], peptide bonds [p.221], polypeptide chain [p.221]

Boldfaced, Page-Referenced Terms

[p.220] introns _____

_____

[p.220] exons _____

_____

[p.220] alternative splicing _____

_____

[p.220] codon _____

_____

[p.220] genetic code _____

_____

[p.221] anticodon _____

_____

Short Answer [p.220]

1. In eukaryotes, what modifications to "pre-mRNA" need to be made before it can be used in translation?

_____

_____

_____

Completion [pp.220-221]

2. Given the following DNA sequence, deduce the composition of the mRNA transcript.

    TAC   AAG   ATA   ACA   TTA   TTT   CCT   ACC   GTC   ATC

    ____   ____   ____   ____   ____   ____   ____   ____   ____   ____

                              (mRNA transcript)

3. From the mRNA transcript in question 2, use Figure 14.9 of the text to deduce the composition of the amino acids of the polypeptide sequence.

    ____   ____   ____   ____   ____   ____   ____   ____   ____   ____

                              (amino acids)

Labeling and Matching [Figure 14.7, p.220]

Newly transcribed mRNA contains more genetic information than is necessary to code for a chain of amino acids. Before the mRNA leaves the nucleus for its ribosome destination, an editing process occurs as certain portions of nonessential information are snipped out. Identify each indicated part of the following illustration; use abbreviations for the nucleic acids. Complete the exercise by entering the letter of the description in the parentheses following each label.

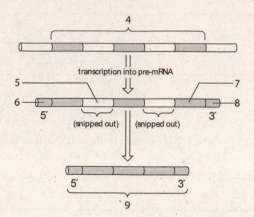

4. _____ ( )

5. _____ ( )

6. _____ ( )

7. _____ ( )

8. _____ ( )

9. _____ ( )

A. An actual coding portion of pre-mRNA

B. A noncoding portion of the newly transcribed mRNA

C. Mature mRNA transcript

D. Poly-A tail

E. The region of the DNA template strand to be copied

F. Cap on the 5' end of mRNA (the first synthesized)

14.4. TRANSLATION: RNA TO PROTEIN [pp.222-223]

Selected Words: small ribosomal subunit [p.222], large ribosomal subunit [p.222], initiation complex [p.222], methionine [p.222], release factors [p.222], polysomes [p.222], biosynthetic process [p.222], GTP [p.222]

Boldfaced, Page-Referenced Terms

[p.222] initiation _____

_____

[p.223] elongation _____

_____

[p.223] termination _____

_____

Fill-in-the-Blanks [p.222]

(1) _____ occurs in the cytoplasm of all cells and consists of three stages: (2) _____, (3) _____, and (4) _____. A small ribosomal subunit binds to (5) _____ RNA and a special initiator (6) _____ RNA base-pairs with the first (7) _____ codon. In (8) _____, the ribosome moves along the mRNA and builds a (9) _____ chain. During the last stage of translation, (10) _____, a STOP codon in the mRNA moves onto the platform, and no tRNA has a corresponding anticodon. Now proteins called (11) _____ factors bind to the ribosome. They trigger (12) _____ activity that detaches the mRNA and the polypeptide chain from the ribosome. When many ribosomes simultaneously translate the same mRNA they are called (13) _____. The energy required for translation is mainly in the form of (14) _____ transfers from the RNA nucleotide (15) _____.

14.5. MUTATED GENES AND THEIR PROTEIN PRODUCTS [pp.224-225]

Selected Words: mutations [p.224], reading frame [p.224], Barbara McClintock [p.225], harmful environmental agents [p.225], ionizing radiation [p.225], free radicals [p.225], nonionizing radiation [p.225], UV light [p.225], thymine dimers [p.225]

Boldfaced, Page-Referenced Terms

[p.224] base-pair substitution_____

_____

[p.224] deletion _____

_____

[p.224] insertion _____

_____

[p.224] transposable elements _____

_____

Short Answer [p.224]

Use Figure 14.13 to answer the following questions.

1. Which is usually the more damaging mutation: base substitution or deletion? Why?

    _____

    _____

2. Would a base substitution mutation in the third position of a codon be more or less likely to cause a serious mutation? Explain. _____

    _____

    _____

3. Would an insertion or deletion mutation be more damaging at the beginning of a mRNA strand or at the end? Why?

    _____

    _____

    _____

Choice [pp.224-225]

Choose the answer that best fits. Some answers will be used more than once.

4. _____ sickle-cell anemia   A. Insertion

5. _____ x-rays   B. Deletion

6. _____ UV radiation   C. Base-pair substitution

7. _____ transposable elements   D. Ionizing radiation

8. _____ thymine dimer   E. Nonionizing radiation

9. _____ free radicals

10. _____ Huntington's disease

Labeling and Matching [p.226]

A summary of the flow of genetic information in protein synthesis is useful as an overview. Identify the numbered parts of the accompanying illustration by filling in the blanks with the names of the appropriate structures or functions. Choose from the following:

DNA, mRNA, rRNA, tRNA, amino acids, anticodon, intron, exon, polypeptide,

ribosomal subunits, transcription, translation

Complete the exercise by matching and entering the letter of the description in the parentheses following each label.

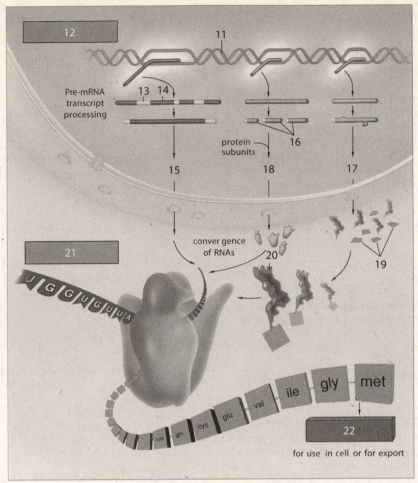

11. _____ ( )

12. _____ ( ) (process)

13. _____ ( )

14. _____ ( )

15. _____ ( )

16. _____ ( )

17. _____ ( )

18. _____ ( )

19. _____ ( )

20. _____ ( )

21. _____ ( ) (process)

22. _____ ( )

A. Coding portion of pre-mRNA that will translate into proteins

B. Carries a form of the genetic code from DNA in the nucleus to the cytoplasm

C. Transports amino acids to the ribosome and mRNA

D. The building blocks of polypeptides

E. Noncoding portion of pre-mRNA

F. Combines with proteins to form the ribosomal subunits

G. Join during the initiation step of protein synthesis

H. Holds the genetic code for protein production

I. Amino acids are joined together

J. RNA synthesized on a DNA template

K. A sequence of three bases that can pair with a specific mRNA codon

L. May serve as a functional protein (enzyme) or structural protein

___ 1. Transcription _____. [p.216]
   a. occurs on the surface of a ribosome
   b. is the final process in the assembly of protein DNA template
   c. occurs during the synthesis of any type of RNA by use of a DNA template
   d. is catalyzed by DNA polymerase

___ 2. _____ carries the actual instructions for a protein's sequence to the ribosome. [p.216]
   a. DNA
   b. mRNA
   c. rRNA
   d. tRNA

___ 3. _____ carries amino acids to ribosomes, where amino acids are linked into the primary structure of a polypeptide. [p.216]
   a. mRNA
   b. tRNA
   c. An intron
   d. rRNA

___ 4. Transfer RNA differs from other types of RNA because it _____. [p.221]
   a. transfers genetic instructions from cell nucleus to cytoplasm
   b. specifies the amino acid sequence of a particular protein
   c. carries an amino acid at one end
   d. contains codons

___ 5. _____ catalyzes the process of transcription. [p.218]
   a. RNA polymerase
   b. DNA polymerase
   c. Phenylketonuria
   d. Transfer RNA

___ 6. _____ is found in RNA but not in DNA. [p.216]
   a. Deoxyribose
   b. Uracil
   c. Phosphate
   d. Thymine

___ 7. Each "word" in the mRNA language consists of _____ letters. [p.220]
   a. three
   b. four
   c. five
   d. more than five

___ 8. The genetic code is used to align _____ amino acids. [p.221]
   a. four
   b. sixteen
   c. twenty
   d. sixty-four

___ 9. The genetic code is composed of _____ codons. [p.220]
   a. three
   b. twenty
   c. sixteen
   d. sixty-four

___ 10. The cause of sickle-cell anemia has been traced to _____. [p.224]
   a. a mosquito-transmitted virus
   b. extra bases inserted into DNA
   c. ionizing radiation
   d. a DNA mutation in a hemoglobin chain

CHAPTER OBJECTIVES/REVIEW QUESTIONS

1. State how RNA differs from DNA in structure and function, and indicate what features RNA has in common with DNA. [p.216]
2. RNA combines with certain proteins to form the ribosome; RNA carries genetic information for protein construction from the nucleus to the cytoplasm; RNA picks up specific amino acids and moves them to the area of mRNA and the ribosome. Identify which type of RNA does each of these jobs. [p.216]
3. Describe the process of transcription, and indicate three ways in which it differs from replication [p.218]
4. Transcription starts at a(n) _____, a specific sequence of bases on one of the two DNA strands that signals the start of a gene. [p.219]
5. What RNA code would be formed from the following DNA code: TAC–CTC–GTT–CCC–GAA? [p.219]
6. Describe how the three types of RNA participate in the process of translation. [p.222]
7. Distinguish introns from exons. [p.220]

8. State the relationship between the DNA genetic code and the order of amino acids in a protein chain. [p.221]
9. Explain how the DNA message TAC–CTC–GTT–CCC–GAA would be used to code for a segment of protein, and state what its amino acid sequence would be. [p.221]
10. What is the advantage of having several different codons encode for most amino acids? [p.220]
11. Cite an example of a change in one DNA base-pair that has profound effects on the human phenotype. [pp.224-225]
12. Briefly describe the spontaneous DNA mutations known as *base-pair substitution, frameshift mutation,* and *transposons.* [pp.224-225]
13. List some of the environmental agents, or mutagens, that can cause mutations. [pp.224–225]

INTEGRATING AND APPLYING KEY CONCEPTS

1. Genes code for specific polypeptide sequences. Not every substance in living cells is a polypeptide. Explain how genes might be involved in the production of a storage starch (such as glycogen) that is constructed from simple sugars.
2. Why is the genetic code almost universal? How did the few alterations occur?
3. In words/diagrams describe the three stages of translation.
4. Why are mutations in germ cells usually more of a problem than mutations in somatic cells?

15

CONTROLS OVER GENES

INTRODUCTION

This chapter looks at the mechanisms of controlling protein synthesis in living organisms, and why such controls are needed.

FOCAL POINTS

- Sections 15.2 [p.230] looks at controls in eukaryotic systems.
- Figure 15.11 [p.237] outlines prokaryotic control.

INTERACTIVE EXERCISES

Impacts, Issues: Between You and Eternity [p.228]

15.1. GENE EXPRESSION IN EUKARYOTIC CELLS [pp.230-231]

Selected Words: polytene chromosomes [p.231], microRNA [p.231], RNA interference [p.231], phosphorylation [p.231]

Boldfaced, Page-Referenced Terms

[p.230] differentiation _____

_____

[p.230] activator _____

_____

[p.230] enhancers _____

_____

[p.230] repressor _____

_____

[p.230] transcription factors _____

_____

Matching [p.230]

Match the situation to the point of control over eukaryotic gene expression.

1. _____ Alternative splicing occurs to modify the mRNA before it leaves the nucleus

2. _____ Enzyme-mediated modifications occur to activate the newly-made protein

3. _____ mRNA stability determines how long it's translated

4. _____ Binding factors must locate special sequences on the DNA

5. _____ Proteins bind to RNA to allow transport through the nuclear pore

A. transcription

B. RNA processing

C. RNA transport

D. translation

E. protein processing

Short Answer [pp.230-231]

6. Although a complex organism such as a human being arises from a single cell, the zygote, differentiation occurs in development. Define *differentiation*. _____

_____

7. How does a cell know which genes to express? _____

_____

_____

8. How do polytene genes affect how the final protein product is made? _____

_____

_____

9. How does a poly-A tail affect how long an mRNA strand is translated? _____

_____

_____

15.2. A FEW OUTCOMES OF EUKARYOTIC GENE CONTROLS [pp.232-233]

Selected Words: Barr bodies [p.232], "mosaic" [p.232], incontinentia pigmenti [p.232], XIST [p.233], Arabidopsis thaliana [p.233]

Boldfaced, Page-Referenced Terms

[p.232] X chromosome inactivation _____

_____

[p.232] dosage compensation _____

_____

[p.233] ABC model _____

_____

[p.232] master genes _____

_____

Fill-in-the-Blanks [p.232]

In mammalian females, the majority of one X chromosome's genes are shut down in a process called (1)

_____. The inactive X is condensed and afterwards called a (2) _____. Since the shutdown is random

in the embryonic cells formed by the time inactivation occurs, all mature mammalian females are a (3) _____

for the expression of X chromosome genes. As an example, human females with the X-linked disorder (4)

_____ have patches of skin with lighter and darker pigmentation. The inactivation of one X in the female is

thought to balance gene expression between males and females, a theory called (5) _____. The shutdown of

the X is accomplished by the (6) _____ gene. This gene's product, a large (7) _____, sticks to the

chromosome and causes it to condense into a Barr body.

Short Answer [pp.232-233]

8.   What causes development of a calico cat?

_____

_____

9.   What controls activation of the ABC master genes in flower formation?

_____

_____

15.3. THERE'S A FLY IN MY RESEARCH [pp.234–235]

Selected Words: drosophila melanogaster [p.234], antenniapedia [p.234], dunce genes [p.234], wingless genes [p.234], wrinkled genes [p.234], minibrain genes [p.234], tinman genes [p.234], groucho genes [p.234], toll genes [p.234], PAX6 [p.234], aniridia [p.234]

Boldfaced, Page-Referenced Terms

[p.234] homeotic genes_____

_____

[p.234] knockout experiments _____

_____

[p.235] pattern formation_____

_____

Matching

Match the gene to what it develops.

1. _____ mutated genes that block development
2. _____ learning and memory
3. _____ heart
4. _____ bristles above the eyes
5. _____ named for the German word for "cool"
6. _____ homologue of the eyeless gene
7. _____ master genes that block development
8. _____ thorax with legs

A. antenniapedia
B. dunce
C. groucho
D. homeotic
E. knockout
F. PAX6
G. tinman
H. toll

Short Answer [pp.234-235]

9. How are knockout experiments used to study gene expression?

_____

_____

10. How do maternal mRNAs distribute protein products in an embryo?

_____

_____

15.4. PROKARYOTIC GENE CONTROL [pp.236-237]

Selected Words: Escherichia coli [p.236], lactose [p.236], lac operon [p.236], allolactose [p.236], cyclic adenosine monophosphate (cAMP) [p.236], lactose intolerance [p.236]

Boldfaced, Page-Referenced Terms

[p.236] operators _____

_____

[p.236] operon _____

_____

True/False [pp.236-237]

Determine if each statement is true or false.

1. _____ Prokaryotes use master genes much like eukaryotes.

2. _____ *E. coli* prefers to use glucose, but can also use other sugars like lactose.

3. _____ cAMP binds to a protein and acts as an activator for transcription of lac operon genes.

4. _____ Allolactose is a conversion product of lactose.

5. _____ Lactose is absorbed directly by the intestine.

6. _____ RNA polymerase transcribes the operon genes when lactose is present.

7. _____ Operators are DNA regions that are binding sites for a promoter.

8. _____ Production of the enzyme lactase increases at about the age of five years.

9. _____ Prokaryotes control their gene expression by adjusting the rate of translation.

10. _____ To some extent, everyone is lactose intolerant.

11. _____ The lac operon is in the *E. coli* chromosome.

12. _____ RNA polymerases cannot bind to twisted promoters and therefore cannot transcribe operon genes.

Labeling and Matching

Escherichia coli, a bacterial cell living in mammalian digestive tracts, uses a negative type of gene control over lactose metabolism. Use the numbered blanks to identify each part of the accompanying diagram of the genes involved. Choose from the following:

lactose, lactose enzyme genes, regulatory gene, repressor–operator complex, promoter, mRNA, lactose operon, repressor protein, RNA polymerase, operator

Complete the exercise by matching and entering the letter of the proper function description in the parentheses following each label. [pp.236–237]

13. _____ ( )

14. _____ ( )

15. _____ ( )

16. _____ ( )

17. _____ ( )

18. _____ ( )

19. _____ ( )

20. _____ ( )

21. _____ ( )

22. _____ ( )

A. Includes promoter, operator, and the genes coding for lactose-metabolizing enzymes

B. Short DNA base sequences on both sides of the promoter

C. A nutrient molecule; a form of it binds to the repressor to allow operation of the lactose operon

D. Major enzyme that catalyzes transcription

E. Capable of preventing RNA polymerases from binding with DNA

F. Genes that code for lactose-metabolizing enzymes

G. Binds to operator and prevents RNA polymerase from binding to DNA and initiating transcription

H. Specific base sequence that signals the beginning of a gene, binding site for RNA polymerase

I. Gene that contains coding for production of repressor protein

J. Carries genetic instructions to ribosomes for production of lactose enzymes

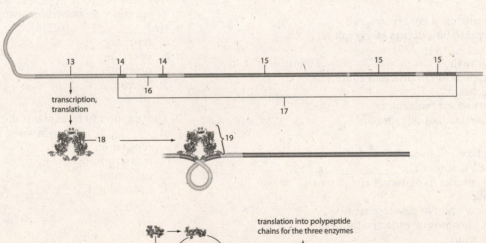

transcription, translation

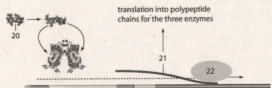

translation into polypeptide chains for the three enzymes

___ 1. Mutated tumor suppressor genes associated with breast and ovarian cancers include _____. [p.228]
   a. BRCA 1 and 2 genes
   b. PAX6 genes
   c. lac operon genes
   d. tinman genes

___ 2. _____ controls govern the rates at which mRNA transcripts that reach the cytoplasm will be translated into polypeptide chains at the ribosomes. [p.231]
   a. Transport
   b. Transcript processing
   c. Translational
   d. Transcriptional

___ 3. _____ refers to the processes by which cells with identical genotypes become structurally and functionally distinct from one another according to the genetically controlled developmental program of the species. [p.230]
   a. Metamorphosis
   b. Metastasis
   c. Cleavage
   d. Differentiation

___ 4. In multicelled eukaryotes, cell differentiation occurs as a result of _____. [p.230]
   a. growth
   b. expression of different subsets of genes
   c. repressor molecules
   d. the death of certain cells

___ 5. One type of gene control discovered in female mammals is _____. [p.232]
   a. a conflict in maternal and paternal alleles
   b. slow embryo development
   c. X chromosome inactivation
   d. operon

___ 6. Due to inactivation of either the paternal or maternal X chromosome, human females with incontinentia pigmenti _____. [p.232]
   a. have mosaic tissues of lighter and darker patches of skin.
   b. develop benign growths
   c. have mosaic patches of skin that lack sweat glands
   d. develop malignant growths

___ 7. Genes that control the information about the basic body plan of eukaryotes are called _____. [p.234]
   a. homeotic genes
   b. knockout genes
   c. repressor genes
   d. toll genes

___ 8. A(n) _____ binds to operator whenever lactose concentrations are low. [p.237]
   a. operon
   b. repressor
   c. promoter
   d. operator

___ 9. Any gene or group of genes together with its promoter and operator sequence is a(n) _____ . [p.236]
   a. repressor
   b. operator
   c. promoter
   d. operon

___ 10. The operon model explains the regulation of _____ in prokaryotes. [p.236]
   a. replication
   b. transcription
   c. induction
   d. Lyonization

CHAPTER OBJECTIVES/REVIEW QUESTIONS

1. List and define the levels of gene control in eukaryotes. [pp.230-231]
2. How do genes relate to cell differentiation in multicelled eukaryotes? [p.230]
3. The condensed X chromosome seen on the edge of each nucleus of female mammals is known as the _____ body. [p.232]
4. Explain how X chromosome inactivation provides evidence for selective gene expression; use the example of incontinentia pigmenti. [p.232]
5. Some control agents bind to _____, which are short regulatory genes on either side of a promoter. [p.230]
6. A(n) _____ speeds up transcription when it binds to a promoter. [p.230]
7. Explain why newly-synthesized polypeptides must be modified before they become functional. [p.231]
8. Genes that control the basic body plans of organisms are called _____. [p.234]

INTEGRATING AND APPLYING KEY CONCEPTS

1. Suppose you have been restricting yourself to a completely vegetarian diet for the past six months. Quite unexpectedly, you find yourself in a social situation that requires you to eat a half-pound sirloin steak. Would you expect to digest the steak as easily as you digest soybean burgers? Explain your yes or no answer in terms of transcriptional controls or feedback inhibition.
2. Describe the sequence of events that occurs on the chromosome of *E. coli* after you drink a glass of milk.
3. What problems would face bacteria that had lost their transcriptional control mechanisms?

16

STUDYING AND MANIPULATING GENOMES

INTRODUCTION

This chapter covers the many modern techniques used in genetics including production of recombinant DNA, using DNA fragments, and producing transgenic organisms.

FOCAL POINTS

- Figure 16.2 [p.242] and Figure 16.4 [p.243] look at the use of restriction enzymes in producing recombinant DNA.
- Figure 16.6 [p.245] illustrates the cycles of polymerase chain reaction (PCR)
- Figure 16.8 [p.246] looks at DNA sequencing
- Figure 16.9 [p.247] look at DNA fingerprinting.
- Figures 16.12–16.15 [pp.251–253] look at genetic engineering and the production of transgenic organisms.

INTERACTIVE EXERCISES

Impacts,Issues: Golden Rice, or Frankenfood? [p.240]

16.1. CLONING DNA [pp.242–243]

Selected Words: "sticky end" [p.242], introns [p.243]

Boldfaced, Page-Referenced Terms

[p.242] restriction enzyme _____

_____

[p.242] recombinant DNA _____

_____

[p.242] DNA cloning _____

_____

[p.242] plasmids _____

_____

[p.242] cloning vectors _____

_____

[p.243] clones _____

_____

[p.243] reverse transcriptase _____

_____

[p.243] cDNA _____

_____

Fill-in-the-Blanks [pp.242-243]

Years after the discovery of the structure of DNA, it was discovered that some bacteria resist infection with (1) _____ due to the activity of (2) _____, enzymes that cut DNA at specific nucleotide sequences. These enzymes are isolated from specific bacteria; for example, (3) _____ is isolated from *E. coli*. In order to produce this (4) _____ DNA, (5) _____ enzymes make staggered cuts in DNA, leaving (6) _____; which can base-pair with other DNA fragments produced by the same enzyme. This is the first step in (7) _____, of a set of laboratory methods using living cells to make many copies of specific DNA fragments. Small circles of bacterial DNA called (8) _____ are used to insert specific DNA into bacteria. These small circles replicate at the same time as the bacterial chromosome and are easily transferred between bacteria, acting as (9) _____ for transfer of the attached genes. The resulting cells are genetically identical (10) _____ that contain a copy of the vector and the foreign DNA it carries. To study eukaryotic genes and their expression, researchers work with (11) _____ because the (12) _____ have already been snipped out. However, restriction enzymes and (13) _____ cut and paste only double-stranded DNA. To make a template for double-stranded DNA, mRNA is placed in a test tube and used with (14) _____, a replication enzyme from certain types of viruses. The resulting strand that base-pairs with the mRNA is called (15) _____. A DNA copy is then made from this molecule using DNA polymerase. Restriction enzymes can then be used for cloning experiments just like an original DNA strand might be used.

Matching [pp.242-243]

Match the steps in the formation of a recombinant DNA with the parts of the accompanying illustration.

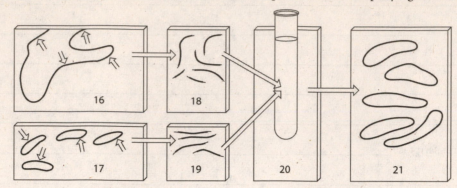

16. _____

17. _____

18. _____

19. _____

20. _____

21. _____

A. Joining of chromosomal and plasmid DNA using DNA ligase

B. Restriction enzyme cuts chromosomal DNA at specific recognition site

C. Cut plasmid DNA

D. Recombinant plasmids containing cloned library

E. Fragments of chromosomal DNA

F. Same restriction enzyme is used to cut plasmids.

16.2. FROM HAYSTACKS TO NEEDLES [pp.244–245]

Selected Words: genomic library [p.244], oligomer [p.244], radioactive phosphate group [p.244], hybridize [p.244], heat-tolerant DNA polymerase [p.244], Taq polymerase [p.245], Thermus aquaticus [p.245]

Boldfaced, Page-Referenced Terms

[p.244] genome_____

_____

[p.244] DNA libraries_____

_____

[p.244] probe_____

_____

[p.244] nucleic acid hybridization_____

_____

[p.244] PCR_____

_____

[p.245] primers_____

_____

Matching [pp.244-245]

Match the most appropriate letter with each numbered partner.

1. _____ DNA polymerase
2. _____ polymerase chain reaction
3. _____ nucleic acid hybridization
4. _____ *Thermus aquaticus*
5. _____ *Taq* polymerase
6. _____ primers

A. Bacterial species from which DNA polymerase used in PCR is extracted.

B. Assembles DNA wherever primers have hybridized with a template in PCR.

C. Oligomers that base-pair with DNA at a certain sequence.

D. Base paring between DNA (or DNA and RNA) from more than one source.

E. Cycled reaction using heat-tolerant DNA polymerase to copy DNA.

F. DNA polymerase used in PCR.

Sequence [p.245]

Order the following events of polymerase chain reaction (PCR) by placing the numbers 1-7 in the appropriate order.

7. _____ The number of target DNA copies reaches the billions
8. _____ DNA is reheated, separating the DNA into single strands; cooled primers hydrogen-bond to DNA
9. _____ Mixture is heated originally causing DNA strands to separate; cooled primers hydrogen- bond to template DNA
10. _____ *Taq* polymerase causes primers to form complimentary strands in the first complete round of PCR
11. _____ *Taq* polymerase causes primers to form complimentary strands in the second complete round of PCR
12. _____ DNA template is mixed with primers, free nucleotides, and heat-tolerant *Taq* DNA polymerase

16.3. DNA SEQUENCING [p.246]

Selected Words: 3' carbon [p.246], dideoxynucleotides [p.246], colored pigment label [p.246], tracers [p.246]

Boldfaced, Page-Referenced Terms

[p.246] DNA sequencing _____

_____

[p.246] electrophoresis _____

_____

Short Answer [p.246]

1. Why do the DNA fragments end up in different lengths?

    _____

    _____

2. How is electrophoresis used to separate DNA fragments of different lengths to determine the DNA sequence?

    _____

    _____

16.4. DNA FINGERPRINTING [p.247]

Selected Words: genetic differences [p.247], population dispersal [p.247]

Boldfaced, Page-Referenced Terms

[p.247] DNA fingerprint_____

_____

[p.247] short tandem repeats _____

_____

Short Answer [p.247]

1. Describe the process of DNA fingerprinting.

    _____

    _____

    _____

2. How is DNA fingerprinting used in crime cases?

    _____

    _____

    _____

16.5. STUDYING GENOMES [pp.248–249]

Selected Words: human genome [p.248], structural genomics [p.249], comparative genomics [p.249], gene-by-gene comparison [p.249], gene APOA5 [p.249], high triglycerides [p.249], microarray [p.249]

Boldfaced, Page-Referenced Terms

[p.249] genomics _____

_____

[p.249] DNA chips _____

_____

Matching [pp.248-249]

Match the term with its description.

1. _____ Sequence of DNA nucleotides coding for a human being

2. _____ Gene used in mice knockout experiments looking at blood lipids

3. _____ Microscopic arrays of different DNA samples separated on small glass plates

4. _____ Used to determine the three-dimensional structure of proteins encoded by a genome

5. _____ Used to compare genomes of different species

6. _____ Consequence of mutations in the APOA5 gene

7. _____ Investigations into the genomes of humans and other species

A. APOA5

B. comparative genomics

C. DNA chips

D. genomics

E. high triglycerides

F. human genome

G. structural genomics

16.6. GENETIC ENGINEERING [p.250]

Selected Words: cross-breeding methods [p.250], gene-swapping [p.250], "factories" [p.250], insulin [p.250], chymotrypsin [p.250]

Boldfaced, Page-Referenced Terms

[p.250] genetic engineering _____

_____

[p.250] transgenic_____

_____

[p.250] genetically modified organisms (GMOs) _____

_____

Short Answer [p.250]

1. List five ways that humans use genetic engineering.

    _____

    _____

    _____

2. What benefit may be achieved from genetically modifying bacteria expressing a fluorescent protein from jellyfish?

    _____

    _____

    _____

16.7. DESIGNER PLANTS [pp.250–251]

Selected Words: Agrobacterium tumefaciens [p.250], Ti plasmid [p.250], genetically-modified crop plants [p.250], Bacillus thuringiensis (Bt) [p.250], USDA Animal and Plant Health Inspection Service (APHIS) [p.251], glyphosphate [p.251], "Frankenfood" [p.251]

Sequence [p.251]

Order the following events in producing a transgenic plant by placing the numbers 1-5 in the correct order.

1. _____   Plant cell divides to form an embryo that expresses the foreign gene

2. _____   A tumefaciens containing a Ti plasmid is produced to use in the transformation

3. _____   The transgenic plant expresses a foreign gene

4. _____   Transgenic plants are grown from genetically-engineered embryos

5. _____   A Ti plasmid-containing bacterium infects a plant cell

True/False

Determine if each statement is true or false. For false statements, correct to make the statement true.

6. _____ "Frankenfood" is a scientific term used to promote genetic engineering.

7. _____ Genetically-modified cotton, canola, corn, alfalfa, soy and sorhum are commonly grown in the United States.

8. _____ Transgenic crops are used to incorporate drought tolerance.

9. _____ *Bt* plasmids are used as vectors to transfer foreign or modified genes into plants.

10. _____ Plants are protected from the devastating effects of insect larvae by insertion of the *Ti* plasmid into the plant cells.

16.8. BIOTECH BARNYARDS [pp.252–253]

Selected Words: cross-breeding [p.252], transgenic animals [p.252], glucose metabolism [p.252], lysozyme [p.252], human interleukin-2 [p.252], organ transplants [pp.252-253], organ rejection [p.253], pandemics [p.253]

Boldfaced, Page-Referenced Term

[p.253] xenotransplantation _____

_____

Short Answer [pp.252-253]

In exercises 1–5, summarize the results/promise of the given experimentation dealing with genetic modifications of animals.

1. rat growth hormone: _____

_____

2. lysozyme: _____

_____

3. glucose metablism: _____

_____

4. xenotransplantation: _____

_____

5. spider silk _____

_____

16.9. SAFETY ISSUES [p.253]

Selected Word: superpathogen [p.253]

1. List the safety guidelines regulated by the USDA for genetic engineering research.

   _____

   _____

   _____

16.10. MODIFIED HUMANS? [p.254]

Selected Words: severe combined immune deficiency (SCID-X1) [p.254], IL2RG gene [p.254], ornithine transcarbamylase [p.254], eugenic engineering [p.254], "superhumans" [p.254]

Boldfaced, Page-Referenced Term

[p.254] gene therapy _____

   _____

Short Answer [p.254]

1. How might age-related diseases like cancer, Parkinson's disease, and diabetes be cured with gene therapy?

   _____

   _____

2. List five diseases currently in testing for gene therapy.

   _____

   _____

3. Summarize the treatment that allowed ten of eleven boys with SCID-X1 to be cured.

   _____

   _____

4. Why wasn't gene therapy successful for Jesse Gelsinger? Why is this a warning for future use of gene therapy with humans?

   _____

   _____

5. What might be the benefits of eugenic engineering? How might eugenic engineering be used inappropriately?

   _____

   _____

SELF-TEST

1. Small, circular molecules of DNA in bacteria are called _____. [p.242]
   a. plasmids
   b. desmids
   c. pili
   d. F particles
   e. transferrins

2. Enzymes used to cut genes in recombinant DNA research are _____. [p.242]
   a. ligases
   b. restriction enzymes
   c. transcriptases
   d. DNA polymerases
   e. replicases

3. The total DNA in a haploid set of chromosomes of a species is its _____. [p.244]
   a. plasmid
   b. enzyme potential
   c. genome
   d. DNA library
   e. none of the above

4. Any DNA molecule that is copied from mRNA is known as _____. [p.243]
   a. cloned DNA
   b. cDNA
   c. DNA ligase
   d. hybrid DNA

5. The study of the genomes of humans and other organisms is called _____. [p.249]
   a. cloning
   b. genetic engineering
   c. gene therapy
   d. genomics

6. A knockout cell is _____.[p.249]
   a. a fertilized egg that can lead to an exceptionally attractive person
   b. a cell in which a particular gene sequence has been excised
   c. any cell that has been removed from an organism
   d. any cell that has been genetically engineered

7. The most commonly used method of DNA amplification is _____. [p.244]
   a. polymerase chain reaction
   b. gene expression
   c. genome mapping
   d. RFLPs

8. Tandem repeats are valuable because _____. [p.247]
   a. they reduce the risks of genetic engineering
   b. they provide an easy way to sequence the human genome
   c. they allow DNA fragmenting without enzymes
   d. they provide DNA fragment sizes unique to each person

9. Which method can rapidly reveal the base sequence of cloned DNA or PCR-amplified DNA fragments? [p.246]
   a. Polymerase chain reaction
   b. Nucleic acid hybridization
   c. Recombinant DNA technology
   d. Automated DNA sequencing

10. A cDNA library is _____. [p.243]
    a. a collection of DNA fragments derived from mRNA and free of introns
    b. cDNA plus the required restriction enzymes
    c. mRNA–cDNA
    d. composed of mature mRNA transcript

CHAPTER OBJECTIVES/REVIEW QUESTIONS

1. How could Golden Rice be beneficial to people in poor countries? [pp.240-255]
2. What is the difference between a genomic library and a cDNA library? [p.243]
3. Some bacteria produce enzymes that cut apart DNA molecules injected into the cell by viruses; such DNA fragments are cut in a staggered way that produces "sticky ends" capable of base-pairing with other DNA molecules cut by the same enzymes. Summarize how these enzymes are used to insert DNA from bacterial plasmids into host bacteria. [p.242]

4. Base-pairing between chromosomal fragments and cut plasmids is made permanent by DNA . [p.243]
5. _____ are small, circular, self-replicating molecules of DNA or RNA within a bacterial cell. [p.242]
6. Define cDNA. [p.243]
7. A special viral enzyme, _____, presides over the process by which mRNA is transcribed into DNA. [p.243]
8. Why do researchers prefer to work with cDNA when working with human genes? [p.243]
9. Polymerase chain reaction is the most commonly used method of DNA _____. [pp.244-245]
10. Explain how gel electrophoresis is used in DNA fingerprinting. [p.247]
11. List some practical genetic uses of DNA fingerprinting. [p.247]
12. How is *Agrobacterium tumefaciens* used in gene transfer in plants? [p.250]
13. How could gene therapy reduce the number of infant deaths and hospital admissions? [p.254]
14. Describe how pig organs can be altered so as not to be rejected by a human recipient. [p.253]

INTEGRATING AND APPLYING KEY CONCEPTS

1. How could scientists guarantee that *Escherichia coli*, the human intestinal bacterium, will not be transformed into a severely pathogenic form and released into the environment if researchers use the bacterium in recombinant DNA experiments?
2. Explain how knowing the genetic makeup of Earth's organisms can help us reconstruct the evolutionary history of life.
3. What problems might be involved in trying to clone extinct animals?
4. Explain how knowing the composition of genes can help scientists derive counterattacks against rapidly mutating organisms.

17

EVIDENCE OF EVOLUTION

INTRODUCTION

This chapter first presents a brief history of evolutionary thought from the time of the ancient Greeks up to Darwin. The chapter then brings Darwin's theory up to date by presenting supporting evidence from a variety of sources- fossils, biogeography, comparative morphology, and comparative biochemistry. As you progress through this chapter, you should note how each class of evidence supports Darwin's theory and helps explain the process of macroevolution.

FOCAL POINTS

- The review of Table 17.1 [p.265] relates Darwin's initial theory to what is currently known from genetics and molecular biology.
- Figure 17.14 [p.270] overlays major biological events with the geologic time scale and introduces the concepts of mass extinctions and adaptive radiations.

INTERACTIVE EXERCISES

Impacts, Issues: Measuring Time [p.258]

17.1. EARLY BELIEFS, CONFOUNDING DISCOVERIES [pp.260-261]

17.2. A FLURRY OF NEW THEORIES [pp.262-263]

17.3. DARWIN, WALLACE, AND NATURAL SELECTION [pp.264-265]

Selected Words: Great Chain of Being [p.260], species [p.260], "fluida" [p.262]

Boldfaced, Page-Referenced Terms

[p.260] naturalist _____

_____

[p.260] biogeography _____

_____

[p.261] comparative morphology _____

_____

[p.261] fossils _____

_____

[p.262] catastrophism _____

_____

[p.262] evolution _____

_____

[p.263] theory of uniformity _____

_____

[p.265] fitness _____

_____

[p.265] adaptive trait _____

_____

[p.265] natural selection _____

_____

Matching [pp.260-261]

Choose the most appropriate answer for each term.

1. _____ comparative morphology
2. _____ biogeography
3. _____ fossils
4. _____ Chain of Being
5. _____ species
6. _____ Aristotle

A. The person that came to view nature as a continuum of organization, from lifeless matter through complex forms of plants and animal life

B. Each kind of being that represents a link in the Chain of Being

C. Extended from the lowest forms of life to humans and spiritual beings

D. Study of similarities and differences in body plans

E. Study of the world distribution of plants and animals

F. Used now as evidence of life in ancient times

Choice [pp.262-263]

For each of the following, choose the individual from the list that is best associated with the statement.

    a. Georges Cuvier   b. Jean Lamarck   c. Charles Lyell

7. _____ Stretching directed "fluida" to the necks of giraffes, which lengthened permanently

8. _____ Supporter of the theory of uniformity

9. _____ Inheritance of acquired characteristics

10. _____ Catastrophism

11. _____ Over great spans of time, geological processes have sculpted the Earth

12. _____ Species improve over generations because of an inherent drive toward perfection

13. _____ First to suggest that the environment is a factor in lines of descent

14. _____ Many species that once existed are now extinct

Choice [p.265]

For each of the following, choose a term from the list below that best completes the sentence. A term may be used more than once.

    a. individual(s)   b. population(s)   c. alleles d. natural selection

15. _____ As a _____ expands the resources become limited

16. _____ Natural _____ have an inherent reproductive capacity to increase in numbers through successive generations

17. _____ _____ will vary in their traits

18. _____ _____ is the outcome of differences in reproduction among individuals of a population that vary in shared traits

19. _____ When resources are limited, members of a _____ will compete

20. _____ An _____ associated with a positive trait tends to increase in frequency

21. _____ As resources become limited _____ will end up competing

17.4. GREAT MINDS THINK ALIKE [p.266]

17.5. ABOUT FOSSILS [pp.266-267]

17.6. DATING PIECES OF THE PUZZLE [p.268]

Selected Words: fossils [p.266], fossilization [p.266], *trace* fossils [p.266], sedimentary rock [p.266], radioisotope [p.268]

Boldfaced, Page-Referenced Terms

[p.267] lineage _____

_____

[p.268] radiometric dating _____

_____

[p.268] half-life _____

_____

Matching [pp.266-268]

Choose the correct term for each of the following statements.

1. _____ fossilization

2. _____ fossil record

3. _____ stratification

4. _____ trace fossils

5. _____ lineage

6. _____ half-life

7. _____ radiometric dating

A. describes the formation of sedimentary rock

B. indirect evidence of past life, such as tracks or burrows

C. a single line of descent

D. a vertical series of fossils that serves as a record of past life

E. E. the transformation by which metal ions replace minerals in bones and hardened tissues

F. F. measuring the radioisotope & daughter element content

G. time it takes for half of a radioisotope's atoms to decay into a product

Fill-in-the-Blanks [pp.266-267]

The fossil record consists of approximately (8) _____ known specimens. The odds of finding fossil evidence is extremely (9) _____. The remains of an organism usually (10) _____ or are (11) _____ by scavengers. Fossils are more likely to form in the absence of (12) _____. Typically, (13) _____ organisms do not fossilize even though they were most likely present in ancient communities.

Short Answer [p.268]

14. Carbon-14 has a half-life of 5,370 years. Assume that a sample of organic material contains 4.0 grams of carbon-14. How many grams would be left after three half-lives? How many years would have elapsed?

_____

_____

_____

17.7. A WHALE OF A STORY [p.269]

17.8. PUTTING TIME INTO PERSPECTIVE [pp.270-271]

17.9. DRIFTING CONTINENTS, CHANGING SEAS [pp.272-273]

Selected Words: artiodactyls [p.269], missing links [p.269], *Rhodocetus* [p.269], *Dorudon* [p.269], Pangea [p.272], continental drift [p.272]

Boldfaced, Page-Referenced Terms

[p.270] geologic time scale _____

_____

[p.273] plate tectonics _____

_____

[p.273] Gondwana _____

_____

Matching [p.270]

Refer to Figure 17.14 in the text. For each of the following events, choose the most appropriate time period from the list provided. Some answers may be used more than once.

1. _____ Origin of the vascular plants
2. _____ Gymnosperms dominate the land plants; origin of the mammals
3. _____ Origin of the angiosperms
4. _____ Origin of the reptiles
5. _____ Origin of amphibians
6. _____ Origin of eukaryotic cells
7. _____ Modern humans evolve
8. _____ Origins of life

A. Cretaceous
B. Quaternary
C. Triassic
D. Devonian
E. Carboniferous
F. Proterozoic
G. Archean
H. Silurian
I. Jurassic

Choice [pp.270-273]

For the following, identify the category of evidence to which the statement best applies.

a. plate tectonics theory b. geological time scale

9. _____ *Lystrosaurus* (therapsid) fossils in Africa and South America

10. _____ Transitions between layers mark the boundaries of intervals of time

11. _____ Studies of Gondwana and Pangea

12. _____ Geological processes like wind, water, and fire have altered the surface of the Earth

13. _____ The movements of continents over long periods of time

14. _____ Comparison of iron compasses in North America and Europe

---

SELF-TEST

___ 1. The patterns, trends, and changes in lines of descent is known as _____. [p.262]
   a. taxonomy
   b. classification
   c. biogeography
   d. evolution

___ 2. Which is not part of Natural Selection? [p.265]
   a. Populations will increase in size over time.
   b. Resources will be limited.
   c. Individuals in a population will vary in their traits.
   d. The individual can use "fluida" to change their physical features.

___ 3. Who co-discovered the process of evolution by natural selection? [p.266]
   a. Lyell
   b. Cuvier
   c. Lamarck
   d. Wallace

___ 4. Study of the distribution of similar species around the world is called _____. [p.260]
   a. comparative morphology
   b. biogeography
   c. stratification
   d. catastrophism

___ 5. The idea that gradual, repetitive changes shaped the surface of Earth is called _____. [p.263]
   a. the theory of uniformity
   b. natural selection
   c. catastrophism
   d. the Chain of Being

___ 6. Which scientist developed catastrophism? [p.262]
   a. Lyell
   b. Lamarck
   c. Darwin
   d. Cuvier

___ 7. The method of using half-lives of radioactive compounds to date fossils is called _____. [p.268]
   a. nucleic acids hybridization
   b. radiometric dating
   c. catastrophism
   d. plate tectonics

___ 8. An increase in adaptation to an environment is called _____. [p.265]
   a. morphological divergence
   b. mutation
   c. fitness
   d. speciation

9. How many mass extinctions have occurred on Earth? [p.270]
   a. 5
   b. 4
   c. 2
   d. 1

10. Which of the following is used as evidence of plate tectonics? [pp.272-273]
    a. Spreading of the ocean floor at mid-ocean ridges
    b. Iron compasses
    c. Fossils of ancient organisms such as Lystrosaurus
    d. The supercontinents of Pangea and Gondwana
    e. All of the above

CHAPTER OBJECTIVES / REVIEW QUESTIONS

1. Explain how biogeography contributes to the study of evolution. [p.260]
2. Give the basic premises of Cuvier's and Lamarck's theories to explain the change in organisms over time. [p.262]
3. Explain the theory of uniformity and why it played an important role in the study of evolution. [p.263]
4. Describe the evidence Darwin used to develop the theory of natural selection. [pp.264-265]
5. Explain how fitness relates to natural selection. [p.265]
6. Explain how stratification and fossilization are related. [pp.266-267]
7. Define *lineage*. [p.267]
8. Describe the importance of radiometric dating to the study of fossils. [p.268]
9. Calculate the amount of radioactive material that would remain after a given number of half-lives. [p.268]
10. Be able to place major biological events on a geologic time scale. [p.270]
11. List the evidence that supports the plate tectonics theory. [p.272-273]

INTEGRATING AND APPLYING KEY CONCEPTS

1. Species have previously been defined as groups of interbreeding organisms that are reproductively isolated from other groups. What problem does this definition hold for scientists who study macroevolutionary processes?
2. As a researcher, you have discovered two groups of plants on two isolated islands in the Pacific. These plants have very similar characteristics, but also minor differences. You are not sure whether your observations can be explained by divergence or convergence. How would you determine if these two plants are members of the same lineage?
3. Draw an evolutionary timeline and indicate when the appearance of the major groups of animals occurred (fish, reptiles, amphibians, mammals, birds, etc.). Also indicate when the Earth experienced the mass extinctions.

18

PROCESSES OF EVOLUTION

INTRODUCTION

Having examined the "big" processes that influenced evolution in the previous chapter, it is time now to focus on how populations change over time. This is called microevolution. This chapter first presents some important terminology regarding variation, and then establishes how to measure change in a population using the Hardy-Weinberg equilibrium equation. Following the introduction to microevolution, the chapter proceeds into a discussion of the many forms of selection, from sexual to disruptive. Perhaps one of the more important aspects of the chapter is the discussion of genetic drift, interbreeding, and gene flow and how these factors work on the allele frequencies in a population. This is a detailed chapter, but one that provides a good working knowledge of how scientists study evolutionary processes.

FOCAL POINTS

- For the Hardy-Weinberg equilibrium [pp.280-281], note especially the five conditions of a population at equilibrium, as this provides the foundation for studying evolution in populations.
- Figure 18.4 [pp.281] shows three modes of natural selection. The text devotes several sections towards understanding the material presented in this figure.

INTERACTIVE EXERCISES

Impacts, Issues: Rise of the Super Rats [pp.276]

18.1. INDIVIDUALS DON'T EVOLVE, POPULATIONS DO [pp.278-279]

18.2. A CLOSER LOOK AT GENETIC EQUILIBRIUM [pp.280-281]

Selected Words: morphological traits [p.278], morpho- [p.278], physiological traits [p.278], polymorphism [p.278], behavioral traits [p.278], qualitative differences [p.278], quantitative differences [p.278], phenotype [p.278], natural selection [p.279], genetic drift [p.279], gene flow [p.279], Hardy-Weinberg equilibrium equation[p.280], hemochromatosis [p.281]

Boldfaced, Page-Referenced Terms

[p.278] population _____

_____

[p.278] gene pool _____

_____

[p.278] alleles _____

_____

[p.279] lethal mutation _____

_____

[p.279] neutral mutation _____

_____

[p.279] allele frequencies _____

_____

[p.279] genetic equilibrium _____

_____

[p.279] microevolution _____

_____

Matching [pp.278-281]

1. _____ microevolution
2. _____ gene pool
3. _____ population
4. _____ dimorphism
5. _____ alleles
6. _____ polymorphism
7. _____ quantitative differences
8. _____ qualitative differences
9. _____ lethal mutations
10. _____ morphological traits
11. _____ neutral mutation
12. _____ allele frequencies
13. _____ genetic equilibrium
14. _____ physiological traits
15. _____ behavioral traits

A. Genes that have slightly different molecular forms

B. A trait with three or more forms in a population

C. A range of incrementally small variations in a specified trait

D. Traits that define the form of the body

E. A change in the DNA that does not affect survival or reproduction

F. Traits that determine how an individual will respond to stimuli

G. Small scale change in a population's allele frequencies

H. A group of individuals of the same species in a specified area

I. A trait that disrupts a phenotype and causes death

J. The genes of all individuals and their offspring in a population

K. Traits that help the body function in its environment

L. A trait with two forms in a population

M. The relative abundances of an allele among all individuals of a population

N. A trait that has distinct forms

O. A population that is not evolving with regard to a certain allele

Short Answer [pp.280-281]

16. List the five conditions that must be met if the gene pool is to remain stable and the population is not evolving.

_____

_____

_____

17. For the following situation, assume that the conditions listed in question 16 do exist; therefore, there should be no change in gene frequency generation after generation. Consider a population of hamsters in which the dominant gene B produces a black coat color and the recessive gene b produces a gray coat color (two alleles will produce the coat color). The dominant gene has a frequency of 80% or (.80). The recessive gene will then have a frequency of 20% or (.20). From this, the assumption is made that 80% of all sperm and eggs have gene B and 20% of the sperm and eggs have gene b.

a. Calculate the probabilities of all possible matings in the Punnett square. (see left-hand diagram)

_____

_____

b. Summarize the genotype and phenotype frequencies on the F_1 generation. (see center diagram)

_____

_____

c. Further assume that the individuals of the F_1 generation produce another generation and the assumptions of the Hardy-Weinberg rule still hold. What are the frequencies of the sperm produced? (see right-hand diagram)_____

_____

_____

	Sperm		Genotypes	Phenotypes
	0.80 B	0.20 b	____ BB	
Eggs 0.80 B	BB	Bb	____ Bb	____% black
0.20 b	Bb	bb	____ bb	____% gray

Parents	B sperm	b sperm
____ BB	____	____
____ Bb	____	____
____ bb	____	____
Totals= ____	____	____

The egg frequencies may be similarly calculated. Note that the gamete frequencies of the F_2 generation are the same as the gamete frequencies of the previous generation. Phenotype percentage also remains the same. Thus, the gene frequencies did not change between the F_1 and F_2 generation. Again, given the assumptions of the Hardy-Weinberg equilibrium, gene frequencies do not change generation after generation.

18. In a population, 81% of the organisms are homozygous dominant, and 1% are homozygous recessive. Find the following:

a. The percentage of heterozygotes

_____

b. The frequency of the dominant allele

_____

c. The frequency of the recessive allele

_____

19. In a population of 200 individuals, determine the following for a particular locus if $p = 0.80$.

a. The number of homozygous dominant individuals

_____

b. The number of heterozygous individuals

_____

c. The number of heterozygous individuals

_____

20. If the percentage of gene D is 70% in a gene pool, find the percentage of gene d.

_____

21. If the frequency of gene R in a population is 0.60, what percentage of the individuals will be heterozygous Rr?

_____

_____

18.3. NATURAL SELECTION REVISITED [p.281]

18.4. DIRECTIONAL SELECTION [pp.282-283]

18.5. SELECTION AGAINST OR IN FAVOR OF EXTREME PHENOTYPES [pp. 284-285]

Selected Words: directional selection [p.281], stabilizing selection [p.281], disruptive selection [p.281], mark-release-recapture method [p.282]

Boldfaced, Page-Referenced Terms

[p.281] natural selection _____

_____

[p.282] directional selection _____

_____

[p.284] stabilizing selection _____

_____

[p.285] disruptive selection _____

_____

Labeling [p.281]

1. For each of the following three curves, first identify the curve as an example of stabilizing selection, directional selection, or disruptive selection; then give the general characteristics of each form of selection.

  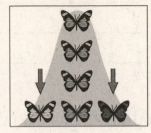

a. _____

_____

_____

b. _____

_____

_____

c. _____

_____

_____

Choice [pp.282-285]

For each of the following, choose from one of the forms of selection in the lettered list.

   a. stabilizing selection   b. directional selection c. disruptive selection

2. _____ Intermediate forms of a trait in a population are favored

3. _____ The range of variation in a phenotype tends to shift in a consistent direction

4. _____ The most frequent wing color of peppered moths shifted from a light form to a dark form as tree trunks became soot-darkened because coal was being used for fuel during the English industrial revolution

5. _____ Intermediate forms of a phenotype are selected against, while forms at the end of the range of variation are favored

6. _____ Antibiotic resistance favors resistant bacterial populations

7. _____ An example is the body weight of sociable weavers

8. _____ An example is the selection of bill size in finches of West Africa

9. _____ Coat color of pocket mice favors their environmental surroundings

10. Explain the problems that are arising from our misuse of antibiotics.

_____

_____

_____

18.6. MAINTAINING VARIATION [pp.286-287]

18.7. GENETIC DRIFT – THE CHANCE CHANGES [pp.288-289]

18.8. GENE FLOW [p.289]

18.9. REPRODUCTIVE ISOLATION [pp.290-291]

Selected Words: sickle-cell anemia [p.287], malaria [p.287], probability [p.288], Ellis-van Creveld syndrome [p.288], emigration [p.288], immigration [p.289], Prezygotic Isolating mechanism [p.290], Postzygotic Isolating mechanism [p.291]

Boldfaced, Page-Referenced Terms

[p.286] sexual dimorphism_____

_____

[p.286] sexual selection _____

_____

[p.287] balanced polymorphism _____

_____

[p.288] genetic drift _____

_____

[p.288] fixation _____

_____

[p.288] bottleneck _____

_____

[p.289] founder effect _____

_____

[p.289] inbreeding _____

_____

[p.289] gene flow _____

_____

[p.290] reproductive isolation _____

_____

[p.290] speciation _____

_____

Choice [pp.286-289]

For each of the following, choose the most appropriate category form the list below.

a. balanced polymorphism   b. genetic drift   c. sexual selection   d. gene flow   e. inbreeding

1. _____ The random change of allele frequencies over time
2. _____ Immigration and emigration
3. _____ Adaptive traits increase reproductive success
4. _____ The physical flow of alleles between populations
5. _____ This is leading to the spread of genes from transgenic organisms to wild species
6. _____ May result in the fixation of alleles over time
7. _____ The relationship in hemoglobin structure between sickle-cell anemia and malaria
8. _____ An example is the courtship rituals of many species
9. _____ Examples are bottleneck and founder effect
10. _____ The increased frequency of Ellis-van Creveld syndrome is caused by this in Amish populations
11. _____ Is primarily driven by female choice
12. _____ Increases the frequency of homozygous individuals and lowers genetic diversity

Fill-in-the-Blanks [pp.288-289]

Random change in allele (13) _____ leads to the (14) _____ condition and a(n) (15)

_____ of genetic diversity over time. This is genetic drift's outcome in all (16) _____; it simply happens

faster in (17) _____ ones. Once alleles from the parent population have become (18) _____, their (19)

_____ will not change again unless mutation or (20) _____ flow introduces new alleles.

For the following statements, choose the appropriate type of isolating mechanism

    a. prezygotic isolating mechanism   b. postzygotic isolating mechanism

21. _____ temporal isolation

22. _____ hybrid sterility

23. _____ mechanical isolation

24. _____ behavioral isolation

25. _____ hybrid inviability

26. _____ ecological isolation

18.10. ALLOPATRIC SPECIATION [pp.292-293]

18.11. OTHER SPECIATION MODELS [pp.294-295]

18.12. MACROEVOLUTION [pp.296-297]

Selected Words: archipelago [p.292], honeycreepers [p.293], autopolyploid [p.294], microevolution [p.296], macroevolution [p.296]

Boldfaced, Page-Referenced Terms

[p.292] allopatric speciation _____

_____

[p.294] sympatric speciation _____

_____

[p.294] polyploidy _____

_____

[p.295] parapatric speciation _____

_____

[p.296] coevolution _____

_____

[p.296] stasis _____

_____

[p.296] exaptation _____

_____

[p.296] adaptive radiation _____

_____

[p.297] key innovation _____

_____

[p.297] extinct _____

_____

[p.297] mass extinction _____

_____

Matching [pp.292-297]

1. _____ allopatric speciation
2. _____ sympatric speciation
3. _____ polyploidy
4. _____ parapatric speciation
5. _____ coevolution
6. _____ stasis
7. _____ exaptation
8. _____ key innovation

A. species doesn't change over time
B. 3 or more sets of chromosomes
C. physical barrier separates a population
D. new structure evolves from an existing one
E. two species evolve in response to each other
F. a population extends across a variety of habitats
G. new species form within the home range of the original species
H. structural modification that increases an organism's chance for survival

Choice [pp.292-295]

For the following statements, choose the appropriate type of speciation.

a. allopatric speciation   b. sympatric speciation   c. parapatric speciation

9. _____ Genetic differences in trees and shrubs due to the Great Wall of China
10. _____ Giant velvet walking worm & blind velvet walking worm
11. _____ Greenish warblers of Siberia
12. _____ Hawaiian honeycreepers
13. _____ Thatch palms on Lord Howe Island
14. _____ Polyploidy resulting in *T. aestivum*

SELF-TEST

___ 1. Selection for the intermediate form of a trait is called _____. [p.284]
    a. stabilizing selection
    b. directional selection
    c. disruptive selection
    d. wayward selection

___ 2. Differences in the molecular structure of a gene are called _____. [p.278]
    a. a gene pool
    b. a bottleneck
    c. gene flow
    d. alleles

___ 3. The sum of all genes in the entire population is the _____. [p.278]
    a. gene pool
    b. genetic variation
    c. gene flow
    d. allele frequency

___ 4. The relative abundance of each type of allele in a population is the _____. [p.279]
    a. gene pool
    b. genetic variation
    c. gene flow
    d. allele frequency

___ 5. According to the Hardy-Weinberg rule, the allele frequencies of a population will not change over successive generations if which of the following is true? [p.280]
    a. The population is infinitely large and all individuals survive and produce equal number of offspring.
    b. There is random mating.
    c. There is no mutation.
    d. The population is isolated.
    e. All of the above.

___ 6. A trait that exists in only two forms in a population is said to be _____. [p.278]
    a. lethal
    b. dimorphic
    c. polymorphic
    d. fixed

___ 7. A _____ is a group of individuals of the same species in a specified area. [p.278]
    a. gene pool
    b. balanced polymorphism
    c. population
    d. species

___ 8. An insect population that becomes increasingly resistant to a class of insecticides is an example of _____ selection. [p.282]
    a. directional
    b. sexual
    c. disruptive
    d. stabilizing

___ 9. If one sex of a species favors a trait in the opposite sex of the species, this is called _____. [p.286]
    a. sexual selection
    b. directional selection
    c. a bottleneck
    d. genetic drift

___ 10. Selection for traits at both ends of a range of variations is called _____ selection. [p.285]
    a. polymorphic
    b. directional
    c. sexual
    d. disruptive
    e. balancing

___ 11. Which isolating mechanism is considered a postzygotic mechanism? [p.290]
    a. hybrid sterility
    b. ecological
    c. behavioral
    d. mechanical

CHAPTER OBJECTIVES / REVIEW QUESTIONS

1.   What is a population? [p.278]
2.   Distinguish among morphological, physiological, and behavioral traits. [p.278]
3.   Distinguish between dimorphism and polymorphism. [p.278]
4.   Explain the difference between gene pool and alleles. [p.278]
5.   Explain the difference between a lethal and a neutral mutation. [p.279]
6.   Define microevolution. [p.279]
7.   List the five conditions that must be met for the Hardy-Weinberg rule to apply. [p.280]
8.   Calculate allele and other genotype frequencies when provided with the homozygous recessive genotype frequency. [pp.280-281]
9.   Define and provide an example of directional selection, stabilizing selection, and disruptive selection. [pp.282-285]
10. Define and give an example of sexual selection. [p.286]
11. Define balanced polymorphism and explain why the relationship between malaria and sickle-cell anemia is used as an example. [p.287]
12. Explain how genetic drift can lead to the fixation of alleles in a population. [p. 288]
13. Distinguish the founder effect from a bottleneck. [pp.288-289]
14. Explain the consequences of inbreeding. [p.289]
15. Explain the influence that gene flow has on the gene pool of a population. [p.289]
16. List the different types of isolating mechanisms. [p.290]
17. Explain the difference between allopatric, sympatric, and parapatric speciation. [p.292-295]
18. Explain the difference between microevolution and macroevolution. [p.296]

INTEGRATING AND APPLYING KEY CONCEPTS

1.   What type of selection do diseases such as HIV and SARS exhibit on the human population? What effects do vaccines have on the gene pool?
2.   Endangered animals are frequently confined to zoos for protection. How does this contribute to a bottleneck effect? How could a gene pool be increased and what would be the benefit of doing so?
3.   What are the similarities between bacterial resistance and insect resistance to pesticides? What should society do in order to avoid the production of "super" bacteria and insects?
4.   Give an example of a neutral trait in humans. For this trait predict what conditions may have existed in the past that would have made this trait advantageous.
5.   Indicate the positives and negatives of creating pure bred breeds of dogs, horses, cattle, etc.
6.   Explain how mutations and gene flow can counter the effects of genetic drift.

19

ORGANIZING INFORMATION ABOUT SPECIES

INTRODUCTION

This chapter explains how recent evidence has caused many species to be reclassified according to shared ancestry instead of the traditional ranking systems. The various lines of evidence include comparing body form, patterns of development, and biochemistry to help form evolutionary trees.

FOCAL POINTS

- Figure 19.9 [p.308] shows the genetic similarity, based upon amino acid sequences, between various species.
- Figure 19.14 [p.312] shows a proposed evolutionary tree for the 6 kingdoms that all life is currently classified within.

INTERACTIVE EXERCISES

Impacts, Issues: Bye Bye Birdie [p.300]

19.1. TAXONOMY AND CLADISTICS [pp.302-303]

19.2. COMPARING BODY FORM AND FUNCTION [pp.304-305]

Selected Words: Carolus Linnaeus [p.302], node [p.303], cladogram [p.303], *Klados* [p.303]

Boldfaced, Page-Referenced Terms

[p.302] taxonomy _____

_____

[p.302] taxon _____

_____

[p.303] phylogeny _____

_____

[p.303] cladistics _____

_____

[p.303] characters _____

_____

[p.303] cladogram _____

_____

[p.303] monophyletic group _____

_____

[p.303] evolutionary tree diagram _____

_____

[p.303] sister groups _____

_____

[p.304] homologous structures _____

_____

[p.304] morphological divergence _____

_____

[p.305] morphological convergence _____

_____

[p.305] analogous structures _____

_____

Matching [pp.302-305]

1. _____ phylogeny
2. _____ cladistics
3. _____ cladogram
4. _____ clade
5. _____ homologous structures
6. _____ analogous structures
7. _____ morphological divergence
8. _____ morphological convergence

A. Evolution of similar body parts in different lineages

B. Diagram that shows a network of evolutionary relationships

C. Similar body parts that reflect shared ancestry

D. Evolutionary relationships between groups

E. Group of species that share a set of characters

F. Body parts are similar in appearance but did not evolve from a common ancestor

G. Classification method that groups species based upon shared characteristics

H. Change in body form from that of a common ancestor

Fill-in-the-Blanks [p.303]

In order to understand evolutionary relationships better biologists collect data about species. They are looking for (9) _____ in specific characters. Collecting and analyzing this type of data is referred to as (10) _____. The basic principle is that the (11) _____ number of steps is the most likely pathway to have occurred. Finding the simplest pathway is called (12) _____. A (13) _____ or grid of information is established to compare specific characters. The proposed pattern of evolution creates an evolutionary diagram called a (14) _____.

Matching [pp.302-303]

15. _____ taxonomy

16. _____ taxon

17. _____ characters

18. _____ monophyletic group

A. Quantifiable features of an organism

B. Science of naming and classifying species

C. Evolutionary group that includes an ancestor and all of its descendants

D. Group of organisms

19.3. COMPARING PATTERNS OF DEVELOPMENT [pp.306-307]

19.4. COMPARING DNA & PROTEINS [pp.308-309]

19.5. MAKING DATA INTO TREES [pp.310-311]

19.6. PREVIEW OF LIFE'S EVOLUTIONARY HISTORY [pp.312-313]

Selected Words: homeotic genes [p.306], *Dlx* gene [p.306], *Hox* gene [p.306], protein primary structure [p.308], cytochrome b [p.308], character matrix [p.310], tree of life [p.312]

Boldfaced, Page-Referenced Terms

[p.308] molecular clock _____

_____

[p.310] parsimony analysis _____

_____

[p.312] six kingdom classification system _____

_____

[p.312] three domain system _____

_____

Fill-in-the-Blanks [pp.306-307]

The majority of vertebrate embryos will develop in similar ways. All will have a stage in which they have four (1) _____ and a (2) _____. The buds form when the (3) _____ gene is expressed. This gene encodes for a signal that tells the cells to (4) _____ and give rise to an (5) _____. The (6) _____ gene helps sculpt the details of the body's form. When (7) _____ is expressed the (8) _____ gene is suppressed. A (9) _____ is a random change in the gene sequence of an individual. (10) _____ mutations are ones that have little or no effect upon an individual's (11) _____. The accumulation of mutations in the DNA can be likened to the ticks of a (12) _____.

Short Answer [pp.308-309]

Answer the following questions based upon the DNA sequences.

Species	DNA Sequence
A	TTATCGTACCGTAT
B	TTATTGTAGCGTAT
C	TTATTGTACCGTAT
D	TTTTTGTACCGAAT

13. Which species are most closely related? _____
14. Which species are the least related? _____
15. Assuming that mutations occur individually and at constant intervals, indicate the evolutionary timeline for the species. _____

Choice [p.306]

Indicate the correct letter next to each choice.

  a. homeotic gene   b. *Dlx* gene   c. *Hox* gene

16. _____   Forms legs, arms and wings of various species
17. _____   Suppressor gene that blocks the formation of appendages
18. _____   A master gene that guides the formation of body parts
19. _____   Apetala 1 gene
20. _____   Encodes for the formation of appendages
21. _____   Reason why pythons do not possess legs
22. _____   Most likely the gene that evolved later

___ 1. Who established a simple naming system for species in the 18th century? [p.302]
   a. Darwin
   b. Lyell
   c. Linnaeus
   d. Wallace

___ 2. Evolutionary relationships are referred to as _____. [p.303]
   a. phylogeny
   b. cladistics
   c. taxons
   d. sister groups

___ 3. A _____ is a diagram that shows evolutionary relationships. [p.303]
   a. cladogram
   b. clade
   c. monophyletic group
   d. taxon

___ 4. Structures that are similar due to common ancestry are referred to as _____ structures. [p.304]
   a. analogous
   b. homologous
   c. vestigial
   d. morphological

___ 5. The wing surfaces of birds, bats, and insects are _____ structures. [p.305]
   a. homologous
   b. vestigial
   c. morphological
   d. analogous

___ 6. Which gene signals embryonic cells to give rise to an appendage? [p.306]
   a. Hox
   b. Apetala 1
   c. Dlx
   d. Control

___ 7. _____ is the science of naming and classifying species. [p.302]
   a. cladistics
   b. evolution
   c. taxonomy
   d. phylogeny

___ 8. Which type of DNA is least likely to be used in a nucleotide comparison? [p.309]
   a. Mitochondrial
   b. Nucleic
   c. Chloroplast
   d. Ribosomal

CHAPTER OBJECTIVES / REVIEW QUESTIONS

1. Explain the advantages and disadvantages of the Linnaean naming system. [p.302]
2. Explain the advantages of using cladistics to develop an evolutionary diagram. [p.303]
3. Distinguish between morphological divergence and morphological convergence. Explain how each process results in homologous or analogous structures. [pp.304-305]
4. Compare the functions of the *Hox* gene to the *Dlx* gene. [pp.306-307]
5. Analyze the DNA sequences in Figures 19.9 & 19.10 for similarities and differences between the species. [pp.308-309]
6. Explain what information the tree of life can provide for us. [p.312]

INTEGRATING AND APPLYING KEY CONCEPTS

1. Design a matrix with 10 characters that you feel are important to determine the relatedness of 4 different mammal species. Fill in the matrix and evaluate the relatedness of the species. Based upon your results, construct an evolutionary tree for the species.
2. Consider the various groups that are currently classified under Kingdom Protista. Develop new Protista into several new kingdoms and explain the features necessary for an organism to belong in the new kindoms.

20

LIFE'S ORIGIN AND EARLY EVOLUTION

INTRODUCTION

As scientists study our planet they continue to find new forms of life in a wide array of habitats. One of the most challenging questions for them to answer is "How did life 1ˢᵗ arise on Earth"? Some scientists believe that life initially arose from nonliving matter while others believe the earliest forms of life were carried to Earth by a meteor. In this chapter you will take a quick tour through the history of life while examining the key evolutionary events that have occurred throughout the Earth's history. What you should recognize is how each evolutionary event was in response to an environmental change and how that event changed life on the planet.

FOCAL POINT

• Figure 20.11 [pp.326-327] provides a summary of the major concepts of the chapter. You should understand the key events portrayed on the diagram, as well as the general time line of evolution.

INTERACTIVE EXERCISES

Impacts, Issues: Looking for Life in All the Odd Places [p.316]

20.1. IN THE BEGINNING… [pp.318-319]

20.2. HOW DID CELLS EMERGE? [pp.320-321]

Selected Words: Thermus aquaticus [p.316], Bacillus infernos [p.316], hydrothermal vents [p.320]

Boldfaced, Page-Referenced Terms

[p.318] big bang model _____

_____

[p.320] proto-cells _____

_____

[p.321] RNA world _____

_____

Choice [pp.316-321]

For each of the following statements choose the category about the study of the early evolution of life that fits the best.

   a. conditions on the early Earth   b. origin of building blocks of life
   c. origin of protein and metabolism   d. origin of genetic material
   e. origin of the plasma membranes   f. life in extreme environments

1. _____ Proto-cells were membrane enclosed sacs that captured energy, concentrated materials, and replicated themselves

2. _____ Organisms such as Thermus aquaticus and Cyanidium caldarium are examples showing that life can exist under these conditions

3. _____ There was little or no free Oxygen in the atmosphere

4. _____ Stanley Miller's experiment attempted to replicate this

5. _____ One hypothesis predicts that simple organic compounds may have formed in outer space

6. _____ Early life may have existed as an RNA world

7. _____ The first proteins may have originated on clay rich tidal flats

8. _____ The helically coiled, double-stranded DNA structure is less susceptible to breakage than RNA

9. _____ The first biomolecules may have originated near thermal vents

10. _____ Water vapor, carbon dioxide, and nitrogen gas were present with Hydrogen gas

11. _____ Simple metabolic pathways evolved at hydrothermal vents

12. _____ A bilayer is the basis of all cell membranes

13. _____ Life can adapt to nearly any environment that has carbon and energy

20.3. LIFE'S EARLY EVOLUTION [pp.322-323]

20.4. WHERE DID ORGANELLES COME FROM? [pp.324-325]

20.5. TIME LINE FOR LIFE'S ORIGIN AND EVOLUTION [pp.326-327]

20.6. ABOUT ASTROBIOLOGY [p.328]

Selected Words: Bangiomorpha pubescens [p.323], endo- [p.324], symbiosis [p.324], Amoeba discoides [p.325]

Boldfaced, Page-Referenced Terms

[p.323] stromatolites _____

_____

[p.324] endosymbiosis _____

_____

Choice [pp.322-323]

For each of the following statements, choose from one of the two categories.

    a. rise of eukaryotes   b. age of prokaryotes

1. _____ An atmosphere containing oxygen starts to form
2. _____ May have originated near deep-sea thermal vents
3. _____ The red alga Bangiomorpha pubescens is an early example
4. _____ The first fossils of these are evident in the Proterozoic
5. _____ Stromatolites are fossilized remains

Fill-in-the-Blanks [p.323]

      An atmosphere enriched with free (6) _____ had three irreversible effects. First, it stopped the further (7) _____ origin of living cells. Second, (8) _____ respiration evolved and in time became the dominant energy-(9) _____ pathway. Third, as oxygen enriched the atmosphere, an (10) _____ formed. This layer blocks much of the sun's (11) _____ from reaching the Earth's surface. This layer enabled life to move onto (12) _____.

Choice [pp.324-325]

For the following statements, choose the most appropriate category from the list below.

    a. Evolution of mitochondria and chloroplasts
    b. Origin of the nucleus and ER
    c. Evidence of endosymbiosis

13. _____ Amoeba discoides is a living example of this process
14. _____ Occurred after the noncyclic pathway of photosynthesis changed the atmosphere
15. _____ Photosynthetic cells may have been engulfed by aerobic bacteria
16. _____ May have evolved to protect metabolic machinery from uninvited guests
17. _____ May have originated as infoldings of the plasma membrane
18. _____ This evolved to help protect the cell's hereditary material from foreign DNA

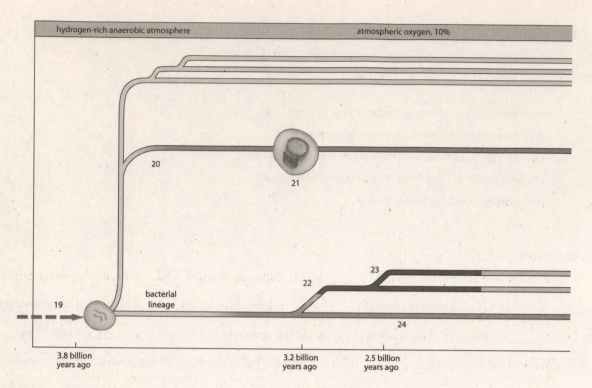

hydrogen-rich anaerobic atmosphere   atmospheric oxygen, 10%

20

21

19

bacterial
lineage

22

23

24

3.8 billion
years ago

3.2 billion
years ago

2.5 billion
years ago

Labeling [pp.326-327]

For each of the following, match each label from the diagram with the correct event in the history of life.

19. _____

20. _____

21. _____

22. _____

23. _____

24. _____

25. _____

26. _____

27. _____

28. _____

A. Start of aerobic respiration

B. Cyclic pathway of photosynthesis

C. Noncyclic pathway of photosynthesis

D. Origin of prokaryotes

E. Endomembrane systems and nucleus

F. Ancestors of eukaryotes

G. Endosymbiotic origin of the chloroplasts

H. Origin of mitosis and meiosis

I. Endosymbiotic origin of the mitochondria

J. Origin of the first protists

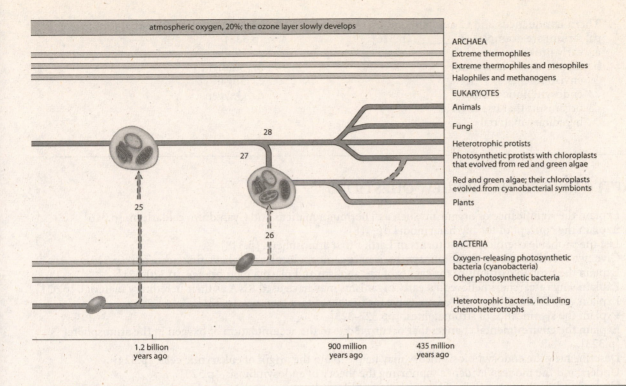

atmospheric oxygen, 20%; the ozone layer slowly develops

ARCHAEA
Extreme thermophiles
Extreme thermophiles and mesophiles
Halophiles and methanogens

EUKARYOTES
Animals
Fungi
Heterotrophic protists
Photosynthetic protists with chloroplasts that evolved from red and green algae
Red and green algae; their chloroplasts evolved from cyanobacterial symbionts
Plants

28

27

25

26

BACTERIA
Oxygen-releasing photosynthetic bacteria (cyanobacteria)
Other photosynthetic bacteria

Heterotrophic bacteria, including chemoheterotrophs

1.2 billion years ago

900 million years ago

435 million years ago

SELF-TEST

___ 1. The concept of endosymbiosis is used to explain which of the following? [p.324]
a. The origin of Archaebacteria
b. The evolution of organelles such as mitochondria and chloroplasts
c. The major trend in the evolution of the animal kingdom
d. The formation of the first replicating proteins
e. None of the above

___ 2. When did the first Prokaryotic cells arise? [p.326]
a. 2.8 – 2.0 billion years ago
b. 4.3 – 3.2 billion years ago
c. 2.0 – 1.6 billion years ago
d. 1.4 – 1.1 billion years ago

___ 3. Which of the following gases was not present in the atmosphere of the early Earth? [p.318]
a. Hydrogen
b. Carbon dioxide
c. Oxygen
d. Methane
e. All of the above were present

___ 4. The hereditary material of early cells was most likely _____. [p.321]
a. DNA
b. Proteins
c. Lipids
d. RNA
e. Carbohydrates

___ 5. When did the fist Eukaryotic cell arise? [p.323]
a. 2.8 – 2.0 billion years ago
b. 4.3 – 3.2 billion years ago
c. 2.0 – 1.6 billion years ago
d. 1.4 – 1.1 billion years ago

___ 6. Stromatolites are_____. [pp.322-323]
a. the first eukaryotes
b. remnants of mitochondria in living cells
c. an example of the first plasma membrane
d. fossilized remains of early prokaryotes
e. none of the above

_ 7. Thermus aquaticus and Cyanidium caldarium are examples of _____. [p.316]
   a. early prokaryotes
   b. life that exists in extreme environments
   c. endosymbiotic events
   d. organisms that use RNA as hereditary material

_ 8. What substance is essential for life to exist? [p.328]
   a. Water
   b. Sunlight
   c. Sugar
   d. Oxygen

CHAPTER OBJECTIVES / REVIEW QUESTIONS

1. Explain the significance of organisms such as Thermus aquaticus and Cyanidium caldarium. [p.316]
2. Explain the concept of the big bang model. [p.318]
3. List the probable chemical constituents of Earth's first atmosphere. [p.318]
4. Give two locations where the abiotic synthesis of organic molecules may have first occurred. [p.319]
5. Explain the probable origins of the agents of metabolism and plasma membranes. [p.320-321]
6. Explain why researchers believe that early organisms may have used RNA as their hereditary material. [p.321]
7. Explain why DNA is the preferred molecule for the hereditary material. [p.321]
8. Explain the significance of stromatolites. [pp.322-323]
9. Explain the environmental changes that occurred due to the accumulation of oxygen in the atmosphere. [p.323]
10. Describe how the endosymbiosis theory may help explain the origin of eukaryotic cells. [p.324]
11. Understand the modern evidence supporting the theory of endosymbiosis. [p.325]
12. Understand the basic timeline of the evolution of life and key events along the time line. [pp.326-327]
13. Explain how studying conditions on other planets can provide clues to how life may have arisen on Earth. [p.328]

INTEGRATING AND APPLYING KEY CONCEPTS

1. As Earth's atmosphere and oceans become increasingly loaded with carbon dioxide and various industrial waste products, how do you think life on Earth will evolve?
2. The evolution of life on this planet has been shaped by the dynamic nature of the Earth's surface. Scientists are currently studying the landscape of other planets in our solar system. What conditions do you think they should be looking for if they want to find a planet that has or had life on it?
3. Endosymbiosis was a major event in the evolution of life on the planet. What do you think life would look like today if endosymbiosis had never occurred?

21

VIRUSES AND PROKARYOTES

INTRODUCTION

This chapter discusses viruses, bacteria, and the Archaea. The features of each group are discussed as well as the evolution of disease causing agents.

FOCAL POINTS

- The introduction [p.332] discusses the impact that HIV/AIDS is having upon the human population.
- Table 21.1 and Figure 21.2 [p.334] show the characteristics of viruses.
- Table 21.2 [p.336] and Figure 21.4 & 21.5 [p.227] illustrate how viruses replicate.
- Figure 21.10 [p.341] and Figure 21.11 [p.337] illustrate DNA and cell replication in bacteria, and the DNA movement between bacteria that occurs in conjugation.
- Table 21.5 [p.346] lists major viral pathogens and the diseases they cause.
- Section 21.8 [pp.346-347] provides an overview of the Evolution of Infectious Diseases.

INTERACTIVE EXERCISES

Impacts, Issues: The Effects of AIDS [p.332]

21.1. VIRAL CHARACTERISTICS AND DIVERSITY [pp.334-335]

21.2. VIRAL REPLICATION [pp.336-337]

21.3. VIROIDS & PRIONS [p.338]

Selected Words: pathogen [p.334], virus [p.334], bacteriophage [p.335], adenovirus [p.335], vaccine [p.335], viral replication cycle [p.336], prions [p.338]

Boldfaced, Page-Referenced Terms

[p.336] lytic cycle _____

_____

[p.336] lysogenic cycle _____

_____

[p.336] reverse transcriptase _____

_____

[p.336] lysis _____

_____

Matching [pp.335-338]

1. _____ Adenovirus
2. _____ Bacteriophage
3. _____ HIV
4. _____ RNA virus
5. _____ Viroid
6. _____ Herpes virus

A. Enveloped DNA virus
B. Causes yellow fever, dengue fever, rabies, mumps, and measles
C. Naked virus that infects animals
D. Infects bacteria or Archaeans
E. Retrovirus
F. Circle of RNA without a protein coat

Short Answer [pp.335-336]

7. List the 3 hypotheses about viral origins. _____
_____
_____

8. List the steps of viral replication. _____
_____
_____

9. Explain the difference between the lytic and lysogenic cycle of viral reproduction. _____
_____
_____

10. Describe the principal features of a virus. _____
_____
_____

21.4. PROKARYOTES – ENDURING, ABUNDANT, & DIVERSE [pp.339]

21.5. PROKARYOTIC STRUCTURE AND FUNCTION [pp.340-341]

Selected Words: Prokaryotes [p.339], strain [p.339], *Escherichia coli* [p.339], nucleiod_[p.340]

Matching [p.339]

Match the correct mode of nutrition to the energy and carbon source.

1. _____ Photoautotrophic
2. _____ Chemoautotrophic
3. _____ Photoheterotrophic
4. _____ Chemoheterotrophic

A. Light & organic compounds
B. Organic compounds & organic compounds
C. Inorganic substances & carbon dioxide
D. Light & carbon dioxide

Short Answer [pp.340-341]

5. List the four main features found in Prokaryotic cells. _____

_____

_____

_____

6. Explain the difference between transduction and transformation. _____

_____

_____

_____

Labeling [p.340]

For each of the following, match each label from the diagram with the correct prokaryotic cell structure.

7. _____   A. Outer capsule

8. _____   B. Cytoplasm

9. _____   C. Nucleiod region

10. _____   D. Flagellum

11. _____   E. Plasma membrane

12. _____   F. Cell wall

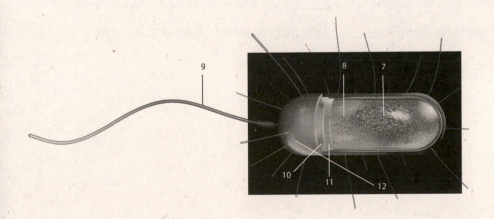

Fill-in-the-Blanks [pp.340-342]

Prokaryotes have an amazing reproductive capacity. One cell can divide into two, these two can then divide into 4, the four can divide into eight, and so on. In some Prokaryotes the process may only take (13) _____ minutes. Each descendant cell will contain one (14) _____. The most common method of cell reproduction is called (15) _____. Some Prokaryotes can inherit DNA directly from another type; this process is called (16) _____. When a plasmid is exchanged between two Prokaryotic cells it is called (17) _____. Having such a variety of reproductive methods has enabled the survival of Prokaryotes for billions of years.

21.6. THE BACTERIA [pp.342-343]

21.7. THE ARCHAEANS [pp.344-345]

21.8. EVOLUTION AND INFECTIOUS DISEASE [pp.346-347]

Selected Words: *Aquifex* [p.342], cyanobacteria [p.342], *Anabaena* [p.342], *Helicobacter pylori* [p.342], disease [p.346], emerging disease [p.347]

Fill-in-the-Blanks [pp.342-343]

Some cyanobacteria use a process called (1) _____ to incorporate atmospheric nitrogen into ammonia. A (2) _____ is the cell that is capable of running this process. (3)_____ are the closest living relatives of ancient prokaryotic cells that evolved into mitochondria. A variety of stains are used in order to identify bacteria. Bacteria that have a thick cell wall and will retain a purple dye are called (4) _____ while those that do not retain the dye because of their thin walls are called (5) _____. When conditions turn unfavorable, several types of Gram-positive bacteria can form (6) _____ in order to avoid to survive. Many of the pathogenic bacteria require a (7) _____ to transport them from host to host.

Choice [pp.344-345]

For the following statements, choose the most appropriate category from the list below.

    a. Archaean   b. Methanogens   c. Extreme halophiles   d. Extreme thermophiles

8. _____ Live in the guts of termites, cattle and other animals

9. _____ Lives in extremely salty environments

10. _____ Can act as chemoautotrophs that metabolize sulfur

11. _____ Strictly anaerobic, free oxygen will kill them

12. _____ Most recent domain that is recognized by the scientific community

13. _____ Contains a purple pigment called bacteriorhodopsim

14. _____ Release carbon-containing gas that is impacting the global carbon cycle

15. _____ Live beside hydrothermal vents

16. _____ Can use aerobic reactions & photosynthesis to produce ATP

17. _____ Among the smallest known cells

18. _____ Group of cells that are the most closely related to eukaryotes

Matching [pp.346-347]

Match the following terms with the health concerns.

19. _____ AIDS

20. _____ tuberculosis

21. _____ malaria

22. _____ epidemic

23. _____ endemic

24. _____ pandemic

25. _____ emerging disease

26. _____ Ebola

27. _____ SARS

A. Disease that spreads worldwide

B. Disease that abruptly spreads through a large section of the population

C. Caused by a virus, has lead to 2.7 million deaths per year

D. This brief pandemic began in China in 2002–2003 and infected 8000 people

E. Caused by a bacteria, has lead to 1.6 million deaths per year

F. Disease that occurs continuously but is localized

G. Pathogen that has not coevolved with humans

H. Produces high fever and flu like conditions, kills 50–90 % of the people infected

I. Caused by a Protists, has lead to 1.3 million deaths per year

SELF-TEST

___ 1. Which of the following is a noncellular particle that consists of protein & nucleic acids? [p.334]
   a. virus
   b. bacteria
   c. archaean
   d. Protista

___ 2. Which ones are single-celled organisms that do not contain a nucleus? [p.333]
   a. Eukaryotes
   b. Prokaryotes
   c. Viruses
   d. Protistas

___ 3. In which pathway is the host cell destroyed in order to release the new copies of the virus? [p.336]
   a. lysogenic
   b. dormant
   c. lytic
   d. reverse transcriptase

___ 4. Which virus will specifically attack bacteria? [p.335]
   a. adenovirus
   b. HIV
   c. tobacco mosaic
   d. bacteriophage
   e. all of the above

___ 5. Which Prokaryotic cell will use organic compounds for energy? [p.339]
   a. photoautrophic
   b. chemoautotrophic
   c. chemoheterotrophic
   d. none of the above
   e. all of the above

___ 6. Organisms that breakdown wastes or remains are called _____. [p.339]
   a. saprobes
   b. photosynthesizers
   c. heterotrophs
   d. autotrophs
   e. none of the above

___ 7. What is the correct term for a bacterial cell that has a round shape? [p.340]
   a. bacillus
   b. spirillium
   c. helical
   d. coccus
   e. none of the above

___ 8. Which process involves the transfer of a plasmid between two Prokaryotic cells? [p.341]
   a. conjugation
   b. transduction
   c. transformation
   d. translocation

___ 9. Which is an aquatic cyanobacteria that carries out nitrogen fixation? [p.342]
   a. *Helicobacter pylori*
   b. *Thermus aquaticus*
   c. *Anabaena*
   d. *Chondromyces crocatus*

___ 10. What term is used to describe when a pathogen breaches the body's surface? [p.346]
   a. infection
   b. disease
   c. pandemic
   d. epidemic

CHAPTER OBJECTIVES / REVIEW QUESTIONS

1. Distinguish between chemoautotrophs and photoautotrophs. [p.339]
2. Describe the principal body forms of bacteria (inside and outside). [p.340]
3. Explain how, with no nucleus and very few membrane bound organelles, bacteria are capable of reproduction as well as running metabolism. [pp.340-341]
4. State the ways in which Archaea differ from bacteria. [p.344]
5. Draw examples of the various types of viruses and explain how they are adapted to infecting a specific host. [pp.334-335]
6. Distinguish between the lytic and lysogenic cycle of viral reproduction. Give examples of diseases that follow each of the cycles. [pp.336-337]

INTEGRATING AND APPLYING KEY CONCEPTS

1. Various types of Archaea produce methane as a waste product of their metabolism. Methane is a combustible gas that could be harvested and used as a potential source of energy. Design a methane production system that could be used as a form of alternate energy. Be sure to identify the limiting factors and come up with realistic solutions to overcome them.
2. What is the most realistic explanation for the origin of various viral pathogens? What concerns should the human species have in regards to the emerging diseases?

22

"PROTISTS"—THE SIMPLEST EUKARYOTES

INTRODUCTION

This chapter introduces you to the most varied group of eukaryotes, the Protista. Their metabolism, structure, and chromosomes are often more closely related to members of other kingdoms than members of their own.

FOCAL POINTS

- Figure 22.2 [p.352] shows a family tree of the Protista and gives examples of organisms in each.
- Figure 22.3 [p.353] shows the generalized Protist life cycle.
- Figures 22.4 and 22.5 [p.354] illustrate the structure of flagellated Protists.
- Figure 22.13 [p.359] depicts the life cycle of the plasmodia that cause malaria.
- Figure 22.20 [p.362] diagrams the life cycle of the green algae *Chlamydomonas*, which is similar to the life cycle of green plants.
- Figure 22.25 [p.365] look at the slime mold life cycle.

INTERACTIVE EXERCISES

Impacts, Issues: The Malaria Menace [p.350]

22.1. THE MANY PROTIST LINEAGES [pp.352-353]

Selected Words: Clade [p.352], monophyletic [p.352], lineage [p.352]

Boldfaced, Page-Referenced Terms

[p.352] Protists _____

_____

[p.352] primary endosymbiosis _____

_____

[p.352] secondary endosymbiosis _____

_____

[p.352] mixotrophs _____

_____

Choice [pp.352-353]

For each of the following terms choose P if it is a Prokaryotic (bacterial) characteristic, E if it is a Eukaryotic characteristic, and B if it is a characteristic of both groups.

1. _____ Multicellular

2. _____ Circular DNA (recall Ch. 21)

3. _____ Can be photoautotrophs (recall Ch. 21)

4. _____ Can develop into a cyst

5. _____ Contain a nucleoid region (recall Ch. 21)

6. _____ Can reproduce sexually and asexually

Fill-in-the-Blanks [pp.352-353]

The vast majority of Protists are (7) _____ celled organisms. Most of them are (8) _____ that live in (9) _____ or (10) _____. Although tiny, they have had a huge impact upon the Earth. (11) _____ Protists contain chloroplasts that enable them to run (12) _____. Some Protists are adaptable and can shift between being auto or (13) _____. The (14) _____ will determine which mode of energy production that will be used. Protists will reproduce (15) _____ when environmental conditions are favorable and (16) _____ when environmental conditions become less favorable. (17) _____ cells will dominate the life cycle of most Protists while only the (18) _____ is diploid.

22.2. FLAGELLATED PROTOZOANS [pp.354-355]

22.3. FORAMINIFERANS & RADIOLARIANS [p.356]

22.4. CILIATES [p.357]

22.5. DINOFLAGELLATES [p.358]

22.6. CELL-DWELLING APICOMPLEXANS [p.359]

22.7. THE STRAMENOPILES [pp.360-361]

Selected Words: flagellated protozoans [p.354], hydrogenosomes [p.354], *Giardia lamblia* [p.354], *Trichomonas vaginalis* [p.354], plankton [p.356], macronucleus [p.357], micronucleus [p.357], *Paramecium* [p.357], alga bloom [p.358], *Karenia brevis* [p.358], *Plasmodium* [p.359], *Anopheles* [p.359], Fucoxanthin [p.360]

Boldfaced, Page-Referenced Terms

[p.355] kinetoplastids _____

_____

[p.354] pellicle _____

_____

[p.355] trypanosomes _____

_____

[p.355] euglenoids _____

_____

[p.355] contractile vacuole _____

_____

[p.356] foraminiferans _____

_____

[p.356] radiolarians _____

_____

[p.357] alveolates _____

_____

[p.357] ciliates _____

_____

[p.358] dinoflagellates _____

_____

[p.359] Apicomplexans _____

_____

[p.360] stramenopiles _____

_____

[p.361] water molds _____

_____

[p.360] diatoms _____

_____

[p.360] brown algae _____

_____

Fill-in-the-Blanks [pp.354-359]

(1) _____ & (2) _____ have multiple flagella and are capable of living in oxygen poor water and will typically lack (3) _____. (4) _____ are long, tapered cells with an undulating membrane. The tsetse fly will act as a (5) _____, helping spread African sleeping sickness from one person to the next. Photosynthetic euglenoids can detect light levels with an (6) _____ that is located near the base of the flagellum. Forams have a (7) _____ shell that makes up their body plan. Many forams form a significant component of the marine (8) _____. Ciliates, dinoflagellates, & apicomplexans are all members of a group known as the (9) _____. (10) _____ are freshwater ciliates that are covered with cilia. Some of the photosynthetic (11) _____ will have an endosymbiotic relationship with corals. Various (12) _____ will infect (parasitize) a variety of animals. The female (13) _____ mosquito acts as a vector and transmits the (14) _____ to a human host.

Choice [pp.355-359]

Put as many letters in each blank as are applicable.

15. _____ Euglenoids
16. _____ Foraminiferans
17. _____ Paramecium
18. _____ Plasmodium
19. _____ Giardia lamblia
20. _____ Trichomonas vagunalis
21. _____ Trypanosoma brucei
22. _____ Dinoflaellates

A. red tide
B. photosynthetic
C. flagellated
D. African sleeping sickness
E. malaria
F. transmitted by contaminated water
G. apicomplexan
H. primary component of many ocean sediments
I. ciliated
J. transmitted by biting insects
K. sexually transmitted

22.8. THE PLANT DESTROYERS [p.361]

22.9. GREEN ALGAE [pp.362-363]

22.10. RED ALGAE DO IT DEEPER [p.364]

22.11. AMOEBOID CELLS AT THE CROSSROADS [p.365]

Selected Words: *Phytophthora infestans* [p.361], chlorophytes [p.362], *Chlamydomonas* [p.362], *Volvox* [p.362], green algae [p.362], *Spirogyra* [p.363], carrageenan [p.364], phycobilins [p.364], cellular slime mold [p.365], *Dictyostelium discoideum* [p.365], *Amoeba proteus* [p.365]

Boldfaced, Page-Referenced Terms

[pp.362-363] charophyte algae _____

_____

[p.364] red algae _____

_____

[p.364] agar _____

_____

[p.365] plasmodium _____

_____

[p.362] Melvin Calvin _____

_____

Labeling and Matching [p.362]

Identify each indicated part of the following illustration by entering its name in the appropriate numbered blank. Choose from the following terms:

asexual reproduction, resistant zygote, gametes meet, meiosis and germination,

zygote, gamete production

1. _____   A. A device to survive unfavorable environmental conditions

2. _____   B. More spore copies are produced

3. _____   C. Haploid cells form smaller haploid gametes when nitrogen levels are low

4. _____   D. Formed after fertilization

5. _____   E. Two haploid gametes coming together

6. _____   F. Reduction of the chromosome number

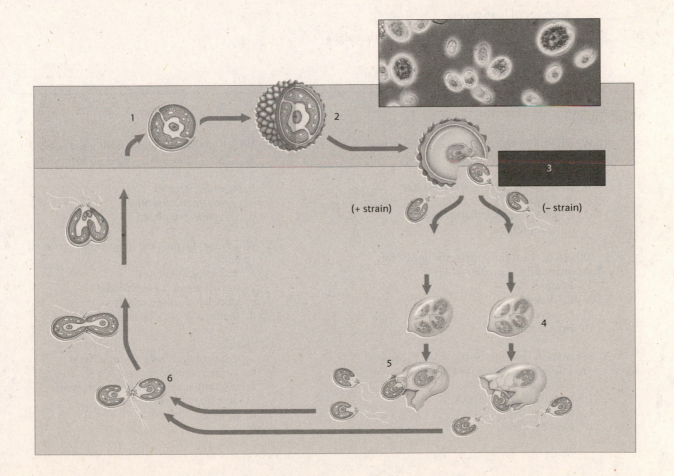

(+ strain)   (– strain)

Choice [pp.362-365]

For the following statements, choose the most appropriate category from the list below.

    a. green algae   b. red algae   c. amoeboid cells

7. _____ Chlamydomonas

8. _____ Carrageenan extraction

9. _____ Use pseudopodia for movement

10. _____ Mutualistic relationship with fungi to form a lichen

11. _____ Chorella

12. _____ Dictyostelium discoideum

13. _____ Agar is extracted from the cell walls

14. _____ Porphyra

15. _____ Cladophora

16. _____ Contains phycobilins

17. _____ Uses cyclic AMP for cell signaling

18. _____ Volvox

19. _____ Ancestor of land plants

20. _____ Survive at deeper depths that most other algae

SELF-TEST

___ 1. Which is not regarded as one of the major protest lineages? [p.352]
    a. Green algae
    b. Blue-green algae (cyanobacteria)
    c. Red algae
    d. Ciliates, dinoflagellates, & apicomplexans

___ 2. Which of the following is not one of the stramenopiles? [pp.360-361]
    a. Diatom
    b. Euglenoid
    c. Oomycote
    d. Brown algae

___ 3. Population "blooms" of _____ cause "red tides" and extensive fish kills? [p.358]
    a. *Euglena*
    b. specific dinoflagellates
    c. diatoms
    d. *Plasmodium*

___ 4. Which of the following is not a parasitic flagellated protozoan? [pp.354-355]
    a. *Plasmodium sp.*
    b. *Giardia lamblia*
    c. *Trichomonas vaginalis*
    d. *Trypanosoma cruzi*

___ 5. Which two groups are most closely related? [p.352]
    a. red algae & ciliates
    b. brown algae & euglenoids
    c. amoebas & slime molds
    d. diatoms & radiolarians

___ 6. Red algae _____. [p.364]
    a. are primarily marine organisms
    b. contain chlorophylls a & b
    c. contain xanthophylls as their main accessory pigment
    d. all of the above

7. Because of their pigmentation, cellulose walls, and starch storage similarities, the _____ algae are thought to be ancestral to more complex plants. [pp.362-363]
   a. Red
   b. Brown
   c. Blue-green
   d. Green

8. Which group contains calcium carbonate shells? [p.356]
   a. Foraminiferans
   b. Diatoms
   c. Dinoflagellates
   d. Ciliates

9. Which is not a chlorophyte? [pp.362-363]
   a. Volvox
   b. Chlamydomonas
   c. *Ulva*
   d. Spirogyra

CHAPTER OBJECTIVES / REVIEW QUESTIONS

1. State the principal characteristics of the radiolarians and foraminiferans. Indicate how they generally move from one place to another and how they obtain food. [p.356]
2. Two flagellated protozoans that cause human disease and misery are _____ & _____. [p. 354]
3. Characterize the apicomplexan group, identify the group's most prominent representative, and describe the life cycle of that organism. [p.359]
4. List the features common to most ciliated protozoans. [p.357]
5. Explain what causes red tides and how they can impact human health. [p.358]
6. State the outstanding characteristics of the red, brown, and green algae. [pp.360-364]

INTEGRATING AND APPLYING KEY CONCEPTS

1. Explain why classifying the Protists is such a difficult process. Which phyla are currently recognized by the scientific community? Indicate the possible evolutionary connection between these phyla and phyla from other Kingdoms.
2. What problems could arise if various algae were introduced into communities that they are not native to?

23

PLANT EVOLUTION

INTRODUCTION

This chapter introduces you to the evolutionary relationships that exist between members of the plant kingdom. In order to understand the major divisions of plants we must first examine the evolutionary trends that exist in plants, presented in Section 23.2. An understanding of the general life cycle of a plant [Figure 23.5] is also important. As you progress through the chapter, you should appreciate the diversity and relationship that plants have to the other kingdoms.

FOCAL POINTS

- Figure 23.5 [p.372] shows the general life cycle of plants. A general understanding of this diagram will help with the more detailed diagrams later in the chapter.
- Figure 23.23 [p.384] outlines the life cycle of a flowering plant. Of particular importance is the introduction of double fertilization and the formation of endosperm.

INTERACTIVE EXERCISES

Impacts, Issues: Beginnings, And Endings [p.368]

23.1. EVOLUTION ON A CHANGING WORLD STAGE [pp.370-371]

23.2. EVOLUTIONARY TRENDS AMONG PLANTS [pp.372-373]

Selected Words: Cooksonia [p.370], ozone [p.370], gymnosperm [p.371], angiosperm [p. 371], roots [p.372], alternation of generations [p.372]

Boldfaced, Page-Referenced Terms

[p.370] embryophytes_____

_____

[p.372] xylem _____

_____

[p.372] phloem _____

_____

[p.372] gametophyte _____

_____

[p.372] sporophyte _____

_____

[p.372] sporangium _____

_____

[p.372] cuticle_____

_____

[p.372] vascular tissue_____

_____

[p.372] lignin _____

_____

[p.373] seed _____

_____

[p.373] pollen grain _____

_____

Choice [p.371]

Choose the most appropriate description for each of the numbered time frames in the following diagrams.

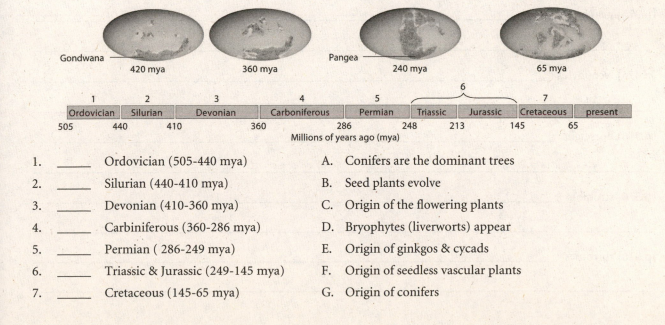

1. _____ Ordovician (505-440 mya)
2. _____ Silurian (440-410 mya)
3. _____ Devonian (410-360 mya)
4. _____ Carbiniferous (360-286 mya)
5. _____ Permian (286-249 mya)
6. _____ Triassic & Jurassic (249-145 mya)
7. _____ Cretaceous (145-65 mya)

A. Conifers are the dominant trees
B. Seed plants evolve
C. Origin of the flowering plants
D. Bryophytes (liverworts) appear
E. Origin of ginkgos & cycads
F. Origin of seedless vascular plants
G. Origin of conifers

Matching [pp.372-373]

Match each of the following statements to the appropriate term.

8. _____ Haploid multicelled body in which the haploid gamete forms   A. sporangium

9. _____ Spore forming structure   B. lignin

10. _____ Immature gametophyte that gives rise to sperm   C. gametophyte

11. _____ Moves sugars and other photosynthetic products   D. xylem

12. _____ Embryo and nutritive tissue in a tough, waterproof coat   E. phloem

13. _____ Moves water and mineral ions from roots to all parts of the plant   F. pollen grain

14. _____ Organic compound adds structural support   G. seed

23.3. THE BRYOPHYTES [pp.374-375]

23.4. SEEDLESS VASCULAR PLANTS [pp.376-377]

23.5. ANCIENT CARBON TREASURES [p.378]

Selected Words: thallus [p.374], *Marchantia* [p.374], *Sphagnum* [p.374], *Lycopodium* [p.376], *Equisetum* [p.376], fiddlehead [p.377], *epiphyte* [p.377], "fossil fuels" [p.378]

Boldfaced, Page-Referenced Terms

[p.374] bryophytes _____

_____

[p.374] rhizoids _____

_____

[p.375] peat bogs _____

_____

[p.376] lycophytes _____

_____

[p.376] horsetails _____

_____

[p.376] strobilus _____

_____

[p.376] rhizomes _____

_____

[p.377] ferns _____

_____

[p.377] sori _____

_____

[p.378] coal _____

_____

Fill-in-the-Blanks [pp.374-375]

Modern bryophytes contain 24,000 species of (1) _____, (2) _____, and hornworts. None of these (3) _____ plants is taller that twenty centimeters due to the lack of (4) _____. They have leaflike, stemlike, and usually rootlike parts called (5) _____. These are elongated cells or threadlike structures that absorb (6) _____ and nutrients, and anchor the (7) _____ to substrates. All three groups will produce motile (8) _____ that are flagellated and swim through water to the egg. The (9) _____are attached and generally dependant on the (10) _____. (11) _____ will take place when the sperm and egg unite. This generally leads to the formation of a (12) _____.

Choice [p.375]

Indicate if the step / structure occurs during the (H) haploid or (D) diploid stage of the life cycle.

13. _____ Sporophyte
14. _____ Gametophyte
15. _____ Rhizoids
16. _____ Spores germinate
17. _____ Zygote develops into a sporophyte
18. _____ Rain will disperse the spores

Choice [pp.376-378]

For questions 23-33, choose from the following:

a. whisk ferns   b. lycophytes   c. horsetails   d. ferns   e. all of the above

19. _____ The members of the genus Equisetum are the surviving species
20. _____ The club mosses belong to this group
21. _____ Seedless vascular plants
22. _____ The genus Psilotum is an example
23. _____ Clusters of sporangia called sori are found on the lower surface of the fronds
24. _____ Stems were used by pioneers of the American West to scrub cooking pots

25. _____ The sporophyte contains xylem and phloem

26. _____ The sporophyte is the predominant phase of the life cycle

27. _____ The genus Lycopodium is an example

28. _____ The young leaves are coiled into the shape of a fiddlehead

29. _____ "Amphibians" of the plant kingdom, life cycle requires water

Labeling [p.377]

Each of the following numbers corresponds to the indicated structure in the diagram. For each item, name the structure and within the parentheses indicate whether it is diploid (2*n*) or haploid (*n*).

30. _____ ( )   34. _____ ( )

31. _____ ( )   35. _____ ( )

32. _____ ( )   36. _____ ( )

33. _____ ( )

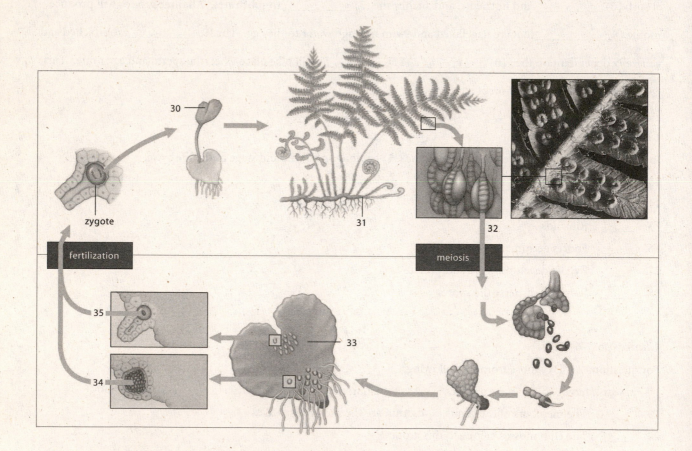

23.6. SEED-BEARING PLANTS [p.379]

23.7. GYMNOSPERMS—PLANTS WITH NAKED SEEDS [pp.380-381]

Selected Words: pollen [p.379], seed [p.379], evergreen [p.380], deciduous [p.380]

Boldfaced, Page-Referenced Terms

[p.379] microspores _____

_____

[p.379] megaspores _____

_____

[p.379] ovule _____

_____

[p.380] gymnosperms _____

_____

[p.380] conifers _____

_____

[p.380] cycads _____

_____

[p.380] ginkgos _____

_____

[p.380] gnetophytes _____

_____

Matching [p.379]

Choose the most appropriate statement for each term.

1. _____ microspore
2. _____ pollen
3. _____ megaspore
4. _____ ovule
5. _____ gymnosperms

A. Develops into the female gametophyte
B. Sporangium enclosed within the ovule
C. Develops into the male gametophyte
D. Consists of a few cells, one of which produces sperm
E. Ancestor of the angiosperms (flowering plants)

Choice [pp.380-381]

For the following questions choose the most appropriate category from the following:

    a. cycads   b. ginkgos   c. gnetophytes   d. conifers

6. _____ Fleshy-coated seeds of female trees produce an awful smell

7. _____ Includes redwoods and bristlecone pines

8. _____ Look like palms or ferns but are not that closely related

9. _____ Only a single species, Ginkgo biloba, survives

10. _____ Includes the genera Welwitschia and Ephedra

11. _____ The favored male trees are now planted; they have attractive, fan-shaped leaves and are resistant to insects, disease, and air pollutants

12. _____ The oldest of the trees belong to this group

13. _____ Most species are "evergreen" with needlelike or scale-like leaves

23.8. ANGIOSPERMS—THE FLOWERING PLANTS [pp.382-383]

23.9. FOCUS ON A FLOWERING PLANT LIFE CYCLE [p.384]

23.10. THE WORLD'S MOST NUTRITIOUS PLANT [p.385]

Selected Words: herbaceous [p.383], cotyledons [p.383], ovary tissue [p.384], double fertilization [p.384]

Boldfaced, Page-Referenced Terms

[p.382] angiosperms _____

_____

[p.382] ovaries _____

_____

[p.382] fruit _____

_____

[p.382] flower _____

_____

[p.382] coevolution _____

_____

[p.382] pollinators _____

_____

[p.383] magnoliids _____

_____

[p.383] eudicots _____

_____

[p.383] monocots _____

_____

[p. 384] endosperm _____

_____

Matching [pp.382-384]

Choose the most appropriate definition for each term.

1. _____ coevolution

2. _____ flowers

3. _____ pollinators

4. _____ examples of magnoliid plants

5. _____ examples of eudicot plants

6. _____ examples of monocots

A. Orchids, palms, lilies, and grasses, including sugarcane, corn, rice, and wheat

B. Specialized reproductive shoots

C. Agents that deliver pollen of one species to female parts of the same species

D. Most herbaceous plants, such as cabbages and dandelions, most flowering shrubs and trees, such as oaks, maples, and fruit trees

E. Avocados and magnolias

F. The evolution of two or more species due to ecological interactions

Labeling [p.384]

Label each of the following structures indicated in the diagram of a flowering plant life cycle.

7. _____

8. _____

9. _____

10. _____

11. _____

12. _____

13. _____

14. _____

15. _____

16. _____

17. _____

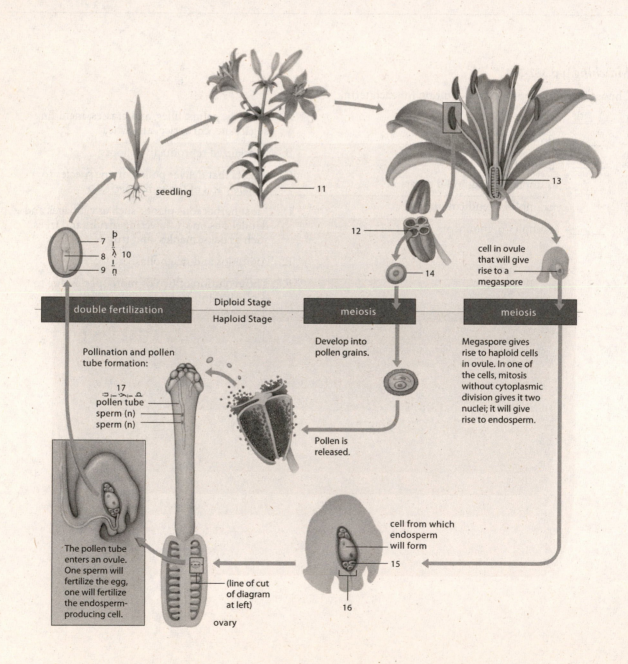

seedling

11

13

12

14

cell in ovule that will give rise to a megaspore

7
8   } 10
9

| double fertilization | Diploid Stage | meiosis | meiosis |
| | Haploid Stage | | |

Develop into pollen grains.

Megaspore gives rise to haploid cells in ovule. In one of the cells, mitosis without cytoplasmic division gives it two nuclei; it will give rise to endosperm.

Pollination and pollen tube formation:

17
pollen tube
sperm (n)
sperm (n)

Pollen is released.

cell from which endosperm will form

15

The pollen tube enters an ovule. One sperm will fertilize the egg, one will fertilize the endosperm-producing cell.

(line of cut of diagram at left)

ovary

16

Quinoa is a leafy (18) _____ that originated in the Andes. It produces highly nutritious (19)

_____. Quinoa seeds contain (20) _____ protein compared to 12% found in wheat and 8% found in

rice. It also contains all of the necessary (21) _____ that humans require. Quinoa is also highly resistant to

(22) _____, (23) _____, and (24) _____ which makes it an easy plant to cultivate.

SELF-TEST

___ 1. The plants around you today are descendants of ancient species of _____. [p.370]
   a. brown algae
   b. green algae
   c. bryophytes
   d. red algae

___ 2. The _____ move(s) water and mineral ions from the roots to all plant parts. [p.372]
   a. spores
   b. seeds
   c. flowers
   d. phloem
   e. xylem

___ 3. Which of the following contain vascular tissue? [pp.374-376]
   a. Horsetails
   b. Mosses
   c. Liverworts
   d. Hornworts

___ 4. The _____ distributes sugars that are made in the photosynthetic tissue. [p.372]
   a. spores
   b. seeds
   c. flowers
   d. phloem
   e. xylem

___ 5. Bryophytes _____. [p.374]
   a. have vascular systems that enable them to live on land
   b. include lycophytes, horsetails, and ferns
   c. have true roots but not stems
   d. include mosses, liverworts, and hornworts

___ 6. Which is a seedless vascular plant? [pp.374, 376-377, 380-383]
   a. Mosses
   b. Gymnosperms
   c. Angiosperms
   d. Hornworts
   e. Ferns

___ 7. _____ are underground stems found in ferns. [p.376]
   a. Rhizomes
   b. Rhizoids
   c. Strobilus
   d. Sori

___ 8. Which of the following is an example of a gymnosperm? [p.380]
   a. Cycad
   b. Conifer
   c. Ginkgo
   d. All of the above

___ 9. Coal deposits are formed primarily from the remains of _____. [p. 378]
   a. angiosperms
   b. bryophytes
   c. green algae
   d. gymnosperms
   e. seedless vascular plants such as club mosses

___ 10. Flowers are the defining characteristics of the _____. [p.382]
   a. gymnosperms
   b. bryophytes
   c. ferns
   d. angiosperms
   e. all of the above

___ 11. Which of the following is not an angiosperm? [p.382]
   a. Magnoliid
   b. Cycad
   c. Eudicot
   d. Monocot

CHAPTER OBJECTIVES / REVIEW QUESTIONS

1.   Understand the evolutionary time line of plants. [pp.370-371]
2.   Distinguish between xylem and phloem. [p.372]
3.   Define *sporophyte* and *gametophyte*. [p.372]
4.   List the common features shared by all of the bryophytes. [pp.374-375]
5.   State the functions of the rhizoids. [p.374]
6.   Understand the general life cycle of a bryophyte. [p.375]
7.   List the four groups of seedless vascular plants. [pp.376-377]
8.   Explain the function of a rhizome. [p.376]
9.   Understand the major aspects of a fern life cycle. [p.377]
10.   Explain the meaning of the general term *epiphyte*. [p.377]
11.   Describe the process by which coal was formed. [p.378]
12.   Distinguish between microspores and megaspores. [p.379]
13.   Explain the differences between pollination and fertilization. [p.379]
14.   Describe the key features of conifers, cycads, ginkgos, and gnetophytes. [pp.380-381]
15.   Describe the general life cycle of a typical gymnosperm. [p.381]
16.   Explain how coevolution relates to pollination. [p.382]
17.   Name and cite examples of the three major classes of flowering plants. [p.383]
18.   Understand the key aspects of an angiosperm life cycle. [p.384]
19.   Describe the sequence of events in a typical angiosperm life cycle. [p.384]
20.   Explain the benefits Quinoa can provide to human society. [p.385]

INTEGRATING AND APPLYING KEY CONCEPTS

1.   Why is coal considered a nonrenewable resource? Describe the environmental conditions necessary to renew the Earth's supply of coal.
2.   Explain the ecological consequences if one species of a coevolved pair were to go extinct?   What type of ripple effect can this produce throughout an entire ecosystem?

24
FUNGI

INTRODUCTION

Fungi represent one of the four kingdoms of eukaryotic organisms, yet they are frequently overlooked in the study of biology. In this chapter you will be introduced to the anatomy, physiology, and classification of the fungi. Typically the greatest challenge for students is understanding the basic life cycle of fungus with its haploid and diploid states. As you progress through the chapter you should notice the role fungi play in various ecosystems as well as the relationship humans and various fungi have.

FOCAL POINTS

- Figure 24.4 [p.392] shows the life cycle of a zygomycete.
- Figure 24.8 [p.394] outlines the life cycle of a sac fungus. Note the similarities to and differences from that of the zygomycete.
- Figure 24.12 [p.396] depicts the general life cycle of a club fungus. Note the similarities to and differences from that of the other two classes of fungi.

INTERACTIVE EXERCISES

Impacts, Issues: High-Flying Fungi [p.388]

24.1. FUNGAL TRAITS AND CLASSIFICATION [pp.390-391]

Selected Words: spores [p.390], chitin [p.390]

Boldfaced, Page-Referenced Terms

[p.390] fungi _____

_____

[p.390] dikaryotic _____

_____

[p.390] mycelium _____

_____

[p.390] hypha _____

_____

[p.391] septae _____

_____

Matching [pp.390-391]

1. _____ lichen
2. _____ chytrids
3. _____ glomeromycetes
4. _____ fungi
5. _____ mycelium
6. _____ hypha
7. _____ spores

A. Reproductive cells or multicelled structures of fungi

B. Meshes of branched filaments with a high surface area for food absorption

C. Live inside plant roots without harming the plant

D. The filaments in a mycelium; each consists of cells of interconnecting cytoplasm reinforced walls

E. Spore producing heterotroph with chitin cell walls

F. Fungal association with photosynthetic cells

G. Produce flagellated spores

Fill-in-the-Blanks [pp.390-391]

Fungi are spore-producing (8) _____ that contain (9) _____ within their cell wall. Some are single celled like (10) _____. Those that are multicelled grow as a mesh of filaments called the (11) _____. Each filament is called a (12) _____. Fungi will grow over organic matter, secrete (13) _____, and absorb the nutrients. This method of obtaining nutrients classifies them as a (14) _____. Due to this role, fungi play a major part in keeping nutrients (15) _____ in any ecosystem.

24.2. THE FLAGELLATED FUNGI [p.391]

24.3. ZYGOSPORE FUNGI AND RELATIVES [pp.392-393]

Selected Words: *Butrachochytrium dendobatidis* [p.391], *Rhizopus stolonifer* [p.392], *Pilobolus* [p.392]

Boldfaced, Page-Referenced Terms

[p.391] chytrids_____

_____

[p.392] zygote fungi _____

_____

[p.392] gametangium _____

_____

[p.392] zygospore _____

_____

[p.392] mycosis _____

_____

[p.393] microsporidians _____

_____

[p.393] glomeromycetes _____

_____

Labeling [p.392]

The numbered items in the following illustration (Rhizopus life cycle) represent missing information. Indicate the name of each numbered structure or process on the corresponding line. For each structure, indicate in the parentheses whether it is haploid (n) or diploid (2n).

1. _____ ( )

2. _____ ( )

3. _____ ( )

4. _____ ( )

5. _____ ( )

6. _____ ( )

7. _____ ( )

8. _____ ( )

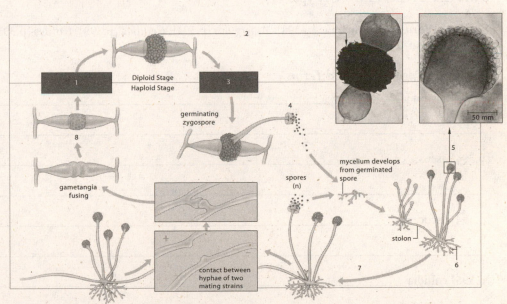

Choice [pp.391-393]

For the following statements chose the correct group;

    a. chytrid   b. zygote   c. microsporidian   d. glomeromycetes

9. _____ Ancient fungal lineage

10. _____ Enterocytozoon bieneusi

11. _____ Pilobus

12. _____ Intracellular parasites

13. _____ Flagellated spores

14. _____ Form zygospore during sexual reproduction

15. _____ Rely on host for ATP

16. _____ Parasite of amphibians

17. _____ Rhizopus stolonifer

24.4. SAC FUNGI- ASCOMYCETES [pp.394-395]

24.5. CLUB FUNGI- BASIDIOMYCETES [pp.396-397]

24.6. THE FUNGAL SYMBIONTS [pp.398-399]

24.7. AN UNLOVED FEW [p.399]

Selected Words: septae hyphae [p.394], *Aspergillus* [p.394], *Candida* [p.395], truffles [p.395], *Armilaria ostoyae* [p.396], ergotism [p.399]

Boldfaced, Page-Referenced Terms

[p.394] sac fungi _____

_____

[p.394] ascus _____

_____

[p.394] ascocarp _____

_____

[p.395] conidia _____

_____

[p.396] club fungi _____

_____

[p.396] basidiocarp _____

_____

[p.398] lichens _____

_____

[p.398] mutualism _____

_____

[p.398] endophytic fungi _____

_____

[p.399] mycorrhiza _____

_____

Matching [pp.394-399]

Match the correct term with the definition or example.

1. _____ sac fungi
2. _____ club fungi
3. _____ lichen
4. _____ endophytic fungi
5. _____ mycorrhiza

A. Sac fungi that resides within leaves and stems of most plants

B. Fungi that lives in association with tree roots

C. Ascomycota

D. Mutualistic relationship between fungus and green algae

E. Basidiomycetes

Labeling [p.395]

The numbered items in the following illustration of the club fungi life cycle represent missing information. Indicate the name of each numbered structure or process on the corresponding line. For each structure, indicate in the parentheses whether it is haploid (n) or diploid (2n).

6. _____

7. _____

8. _____ ( )

9. _____

10. _____ ( )

11. _____ ( )

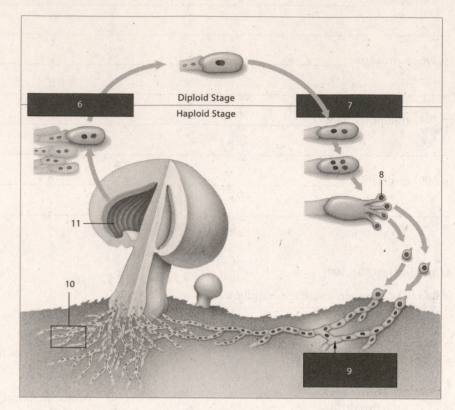

Matching [pp.394-395]

Choose the most appropriate statement for each term.

12. _____ Saccharomyces cerevisiae

13. _____ truffles

14. _____ Candida albicans

15. _____ Penicillium

16. _____ Aspergillus

17. _____ Neurospora

A. "Flavor" Camembert and Roquefort cheeses; produce antibiotics

B. Has uses in genetic research

C. Causes infections in humans

D. Highly prized edible fungi

E. Makes citric acid for candles and soft drinks; ferments soybeans for soy sauce

F. Baking yeast

Matching [p.399]

Choose the most appropriate statement for each of the following terms.

18. _____ ergotism
19. _____ histoplasmosis
20. _____ "blind staggers"
21. _____ Cryphonectria parasitica
22. _____ Aspergillus and penicillium
23. _____ Claviceps purpurea

A. Household molds that can cause sinus congestion, sneezing, and a raw throat

B. A lung disease caused by the fungus Histoplasma capsularum

C. The parasitic fungus that is responsible for widespread destruction of the chestnut trees

D. A disease that causes vomiting, diarrhea, and hallucinations

E. A form of ergotism that affected the armies of Peter the Great

F. The species responsible for the disease ergotism

SELF-TEST

___ 1. Most true fungi send out cellular filaments called _____. [p.390]
   a. mycelia
   b. hyphae
   c. mycorrhizae
   d. asci

___ 2. The cell wall of fungi is composed of _____. [p.390]
   a. cellulose
   b. starch
   c. chitin
   d. glycogen

___ 3. Fungi are most closely related to which of the following groups? [pp.390]
   a. Animals
   b. Plants
   c. Protista
   d. Bacteria

For questions 4 – 11, choose from the following;
   a. club fungi [pp.396-397]
   b. chytrids [p.391]
   c. sac fungi [pp.394-395]
   d. zygomycetes [pp.392-393]

___ 4. The group that includes Rhizopus stolonifer, the notorious black bread mold.

___ 5. The group that includes edible morels and truffles as well as bakers' and brewer's yeast.

___ 6. The group that includes the commercial mushroom *Agaricus bisporus*.

___ 7. The group that includes *Penicillium*, which has a variety of species that produce penicillin and substances that flavor Camembert and Roquefort cheeses.

___ 8. The group whose spore-producing structure form inside a saclike structure called an ascus.

___ 9. The group that forms a thin, clear covering around the zygote called the zygospore.

___ 10. Plays an important role as a decomposer of forest plants.

___ 11. Parasite of amphibians, may be contributing to the decline in amphibians worldwide.

CHAPTER OBJECTIVES / REVIEW QUESTIONS

1. Define mutualism. [p.398]
2. Explain the process by which fungi absorb their nutrients. [p.390]
3. Distinguish among the meanings of the following terms: *hypha, spore,* and *mycelium.* [p.390]
4. Explain the significance of the chytrids, microsporidians, and glomeromycetes. [pp.391, 393]

5. List the common names for the major groups of fungi. [p.391]
6. Understand the general life cycle of a zygomycete. [p.392]
7. Understand the general life cycle of a club fungus. [p.396]
8. Understand the economic importance of the sac fungi. [p.395]
9. Recognize selected disease-causing fungi by name. [p.399]
10. Describe the mycorrhizae relationship. [p.399]

INTEGRATING AND APPLYING KEY CONCEPTS

1. In the past fungi have been classified with the plants, but are now considered to be more animal-like. List several characteristics of the fungi that would lead scientists to consider them more closely related to the animals than the plants.
2. Explain what could happen to a pond ecosystem if there was an explosion in the chytrid population.
3. Since fungal spores can be dispersed by wind, what are some potential ways to prevent widespread fungal invasions from Africa?

25

ANIMAL EVOLUTION–THE INVERTEBRATES

INTRODUCTION

This chapter is a quick trip through the various invertebrate phyla.

FOCAL POINTS

- Figures 25.3 and 25.4 [p.405] describe the body symmetry and the different types of body cavities of various animals.
- Figure 25.7 [p.407] shows two proposed evolutionary trees for animal classification.
- Each section after that looks at the incredible diversity of invertebrate animals.

INTERACTIVE EXERCISES

Impacts, Issues: Old Genes, New Drugs [p.402]

25.1. ANIMAL TRAITS & BODY PLANS [pp.404-405]

25.2. ANIMAL ORIGINS & ADAPTIVE RADIATION [pp.406-407]

Selected Words: Conus [p.402], heterotroph [p.404], digestive system [p.405], circulatory system [p.405], segmentation [p.405], colonial theory [p.406], adaptive radiation [p.406], morphology [p.407]

Boldfaced, Page-Referenced Terms

[p.404] animals _____

_____

[p.404] gut _____

_____

[p.404] radial symmetry _____

_____

[p.404] bilateral symmetry _____

_____

[p.404] cephalization _____

_____

[p.405] protostomes _____

_____

[p.405] deuterostomes _____

_____

[p.405] coelom _____

_____

[p.405] pseudocoel _____

_____

[p.406] choanoflagellates _____

_____

[p.407] molt _____

_____

Fill-in-the-Blanks [p.402]

Cone snails produce a cocktail of paralytic secretions called (1) _____. These peptides shut down (2) _____ in the cells of their victims. One peptide is one thousand times more potent than (3) _____, yet not addictive. All cone snails produce (4) _____, which catalyzes an early step in conotoxin production. Humans also produce this enzyme, but in us it functions in (5) _____. This enzyme is also found in fruit flies. It is estimated the gene has been around for at least (6) _____ years, when these three groups shared a common ancestor.

Matching [pp.404-407]

Choose the most appropriate description for each term.

7. _____ animals

8. _____ gut

9. _____ ectoderm, endoderm, mesoderm

10. _____ anterior end

11. _____ radial symmetry

12. _____ bilateral symmetry

13. _____ pseudocoel

14. _____ coelom

15. _____ posterior end

16. _____ segmentation

17. _____ cephalization

18. _____ protostomes

19. _____ deuterostomes

20. _____ choanoflagellates

A. An evolutionary process whereby sensory structures and nerve cells became concentrated in a head

B. Digestive sac or tube that opens at the body surface

C. Animal body cavity lined with tissue derived from mesoderm

D. Head end

E. Animals having body parts arranged regularly around a central axis

F. Primary tissue layers that give rise to all adult animal tissues and organs

G. Cavity that is not fully lined by mesoderm

H. Animal lineage in which the 1st opening becomes the mouth

I. Animals having appendages that are paired

J. Series of animal body units that may or may not be similar to one another

K. Tail end

L. Animal lineage in which the 1st opening becomes the anus

M. Multicellular organisms, diploid body cells, hetertorphic, most are motile at some point in their lives

N. Modern protests most closely related to animals

25.3. THE SIMPLEST LIVING ANIMAL [p.408]

25.4. THE SPONGES [pp.408-409]

25.5. CNIDARIANS—TRUE TISSUES [pp.410-411]

Selected Words: asymmetrical [p.408], "filter feeding" [p.408], spicules [p.408], budding [p.409], fragmentation [p.409], Trichoplax adhaerens [p.408], medusa [p.410], polyp [p.410], gastrodermis [p. 411], Hydra [p.410], Obelia [p.410], Physalia [p.411]

Boldfaced, Page-Referenced Terms

[p.408] sponges _____

_____

[p.408] placozoans _____

_____

[p.409] larva (plural, larvae) _____

_____

[p.409] hermaphrodite _____

_____

[p.410] cnidarians _____

_____

[p.410] nematocysts _____

_____

[p.410] nerve net _____

_____

[p.410] hydrostatic skeleton _____

_____

Matching [pp.408-411]

Choose the most appropriate answer for each term.

1. _____ medusa
2. _____ larva
3. _____ amoeboid cells
4. _____ sponge skeletal elements
5. _____ Trichoplax
6. _____ scyphozoans
7. _____ collar cells

A. Flagellated cells that absorb and move water through a sponge as well as engulf food

B. Distribute food to cells throughout the sponge's body

C. Closest animal relative to choanoflagellates

D. Jellyfish shaped like a bell

E. Group known as true jellyfish

F. Protein fibers and glasslike spicules of silica

G. Free living sexually immature stage

Fill-in-the-Blanks [pp.410-411]

All (8) _____ posses radial symmetry; they include the jellyfishes, sea anemones, corals, and animals such as *Hydra*. Most of these animals live in a marine environment. (9) _____ are one of the distinguishing features of the Cnidarians. They are capsules capable of discharging threads that will entangle or pierce prey. Cnidarians have two common body plans, the (10) _____ looks like a bell or an umbrella while the (11) _____ has a tube-like body with a tentacle fringed mouth at one end. The saclike cnidarian gut processes food within the (12) _____. Cells of the (13) _____ will secrete digestive enzymes. The nervous system is a(n) (14) _____, a simple nervous system to control movement and changes in shape. The (15) _____ is a layer of gelatinous secreted material that lies between the epidermis and gastrodermis. Some cnidarians will have a symbiotic relationship with photosynthetic (16) _____. If the coral loses its protist symbiont, this will result in an event called (17) _____.

Labeling [p.411]

Identify each indicated part of the following illustration.

18. _____

19. _____

20. _____

21. _____

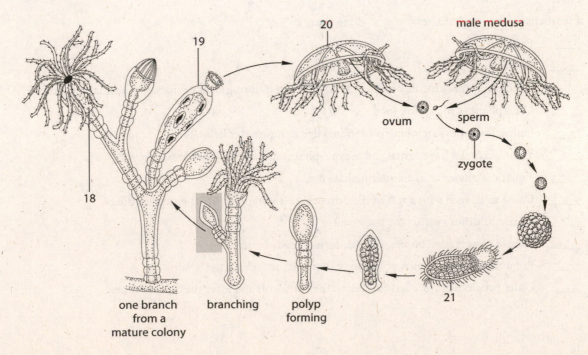

one branch from a mature colony   branching   polyp forming   ovum   male medusa   sperm   zygote

25.6. FLATWORMS—SIMPLE ORGAN SYSTEMS [pp.412-413]

Selected Words: organ [p.412], organ system [p.412], hermaphrodites [p.412], turbellarians [p.412], flukes [p.412], tapeworms [p.412], *definitive* host [p.412], *intermediate* host [p.412], *schistosoma* [p. 412], scolex [p.413]

Boldfaced, Page-Referenced Terms

[p.412] flatworms _____

_____

[p.412] pharynx _____

_____

[p.412] nerve cord _____

_____

[p.412] ganglia _____

_____

[p.413] proglottids _____

_____

Choice [pp.412-413]

For questions 1 – 12, choose from the following. Letters can be used more than once and blanks can have more than one letter.

     a. turbellarians   b. flukes   c. tapeworm

1. _____ Parasitic
2. _____ Posses a scolex
3. _____ Ancestral forms probably had a gut but later lost it during their evolution in animal intestines
4. _____ Use a pharynx to capture food
5. _____ Only a few types (including planarians) live in freshwater habitats
6. _____ Their life cycles have sexual and asexual phases and at least two kinds of hosts
7. _____ Aquatic snails serve as an intermediate host
8. _____ Flame cells, each with a tuft of cilia, drive excess water out into the surroundings
9. _____ Nutrients diffuse across the body wall
10. _____ Cluster of nerve cell bodies, ganglia, form a crude brain
11. _____ Proglottids are new units of the body that bud just behind the head
12. _____ Older proglottids store fertilized eggs; they break off and leave the body in feces

Labeling [p.412]

Identify the parts of the animal shown dissected in the accompanying drawings.

13. _____

14. _____

15. _____

16. _____

17. _____

18. _____

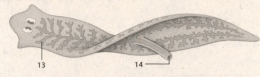

protonephridia

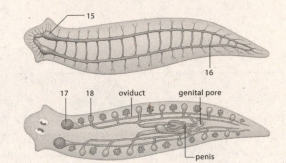

15

16

17   18   oviduct   genital pore

penis

Short Answer [p.412]

Answer exercises 19-22 with reference to the drawings accompanying the exercise set above.

19. What is the common name of the animal dissected? _____

20. Is the animal parasitic? _____

21. Is the animal hermaphroditic? _____

22. Name the coelom type exhibited by this animal. _____

23. What type of symmetry does this animal have? _____

Labeling [p.413]

The numbered items in the following illustration represent missing information; fill in the corresponding answer blanks to complete the narrative of the life cycle of the beef tapeworm.

24. _____   29. _____

25. _____   30. _____

26. _____   31. _____

27. _____   32. _____

28. _____   33. _____

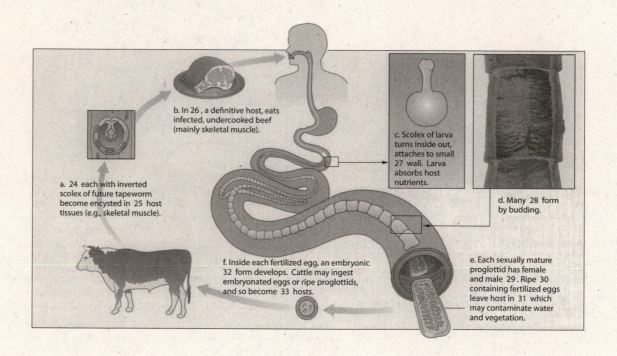

a. 24 each with inverted scolex of future tapeworm become encysted in 25 host tissues (e.g., skeletal muscle).

b. In 26 , a definitive host, eats infected, undercooked beef (mainly skeletal muscle).

c. Scolex of larva turns inside out, attaches to small 27 wall. Larva absorbs host nutrients.

d. Many 28 form by budding.

e. Each sexually mature proglottid has female and male 29 . Ripe 30 containing fertilized eggs leave host in 31 which may contaminate water and vegetation.

f. Inside each fertilized egg, an embryonic 32 form develops. Cattle may ingest embryonated eggs or ripe proglottids, and so become 33 hosts.

25.7. ANNELIDS—SEGMENTED WORMS [pp.414-415]

Selected Words: polychaetes [p.414], leeches [p.414], oligochaetes [p.414], chaetae [p.414], parapodia [p. 414], ganglia [p. 415]

Boldfaced, Page-Referenced Terms

[p.414] annelids _____

_____

[p.415] nephridia (singular, nephridium) _____

_____

Matching [pp.414- p415]

Choose the appropriate answer for each term.

1. _____ cuticle
2. _____ annelids
3. _____ earthworms
4. _____ marine polychaetes
5. _____ brain
6. _____ nephridia
7. _____ nerve cords
8. _____ chaetae
9. _____ hydrostatic skeleton
10. _____ leeches
11. _____ clitellum

A. Fluid filled coelomic chambers
B. Oligochaete
C. Extensions of nerve cell bodies leading away from the brain
D. Contain parapodia
E. Secretory region that produces mucus
F. Outer layer of secreted protein
G. Fused pair of ganglia that coordinates activities
H. Lack bristles and has a sucker at either end
I. Chitin-reinforced bristles on each side of the body on nearly all segments
J. Segmented worms
K. Regulate volume and composition of body fluid

Short Answer [p.415]

12. Describe how an earthworm burrows into the soil.

_____

_____

13. What purpose does *Hirudo medicinalis* have in medicine?

_____

_____

Labeling [p.415]

Identify each part of the following illustrations.

14. _____

15. _____

16. _____

17. _____

18. _____

19. _____

20. _____

21. _____

22. _____

23. _____

24. _____

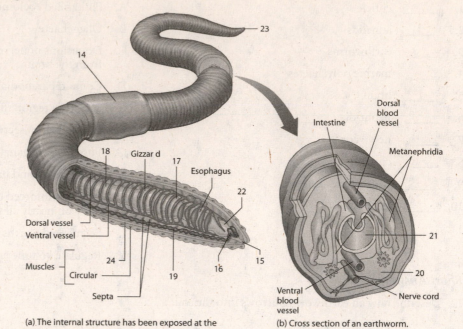

(a) The internal structure has been exposed at the anterior end of an earthworm.

(b) Cross section of an earthworm.

Short Answer [p.415]

Answer items 25-30 with reference to the drawings labeled in the preceding section.

25. Name this animal. _____

26. Name this animal's phylum. _____

27. Name two distinguishing characteristics of this group.

_____

28. Is this animal segmented? (yes / no)

29. What type of symmetry does the adult have? _____

30. Does this animal have a true coelom? (yes / no)

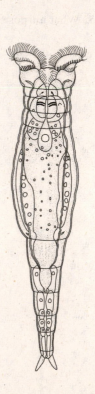

25.8. MOLLUSKS—ANIMALS WITH A MANTLE [pp.416-417]

25.9. CEPHALOPODS FAST AND BRAINY [p.418]

Selected Words: siphon [p.417], mantle cavity [p.416], jet propulsion [p.418]

Boldfaced, Page-Referenced Terms

[p.416] mollusks _____

_____

[p.416] mantle _____

_____

[p.416] gills _____

_____

[p.416] radula _____

_____

[p.416] gastropods _____

_____

[p.416] chitons _____

_____

[p.416] torsion _____

_____

[p.417] bivalves _____

_____

[p.417] cephalopods _____

_____

Fill-in-the-Blanks [pp.416-417]

A(n) (1) _____ is a bilateral animal with a reduced coelom. Most have a(n) (2) _____ of calcium carbonate and protein, which were secreted from cells of a tissue that drapes over the body mass. This tissue, the (3) _____, is unique to mollusks. Special respiratory organs, the (4) _____, contain cilia that help cause water to flow through the cavity. Most mollusks have a fleshy (5) _____. Many have a(n) (6) _____, a tongue-like organ hardened with chitin. The mollusks with a well-developed head often posses (7) _____ and (8) _____. To protect themselves from predators, (9) _____ sometimes incorporate (10) _____ into their tissues, while cephalopods use (11) _____ to quickly escape predators.

Matching [pp.416-418]

Identify the animals in the following drawings by matching each with the appropriate description.

12. _____ Animal A   I. Bivalve

13. _____ Animal B   II. Cephalopod

14. _____ Animal C   III. Gastropod

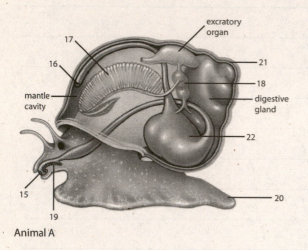

Animal A

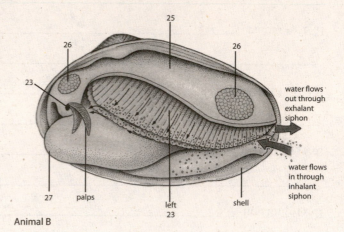

Animal B

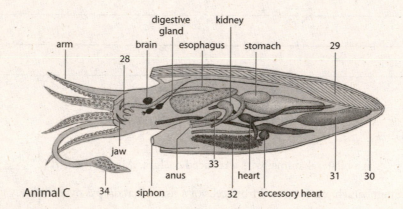

Animal C

Labeling

Identify each numbered part in the preceding drawings by writing its name in the appropriate blank.

15. _____

16. _____

17. _____

18. _____

19. _____

20. _____

21. _____

22. _____

23. _____

24. _____

25. _____

26. _____

27. _____

28. _____

29. _____

30. _____

31. _____

32. _____

33. _____

34. _____

Choice [pp.412-413]

For questions 35-46, choose from the following classes of the mollusks.

    a. chitons   b. gastropods   c. bivalves   d. cephalopods

35. _____ Class with the swiftest invertebrates, the squids

36. _____ The "belly foots"

37. _____ Possess a dorsal shell divided into eight plates

38. _____ Class with the smartest invertebrates, the octopuses

39. _____ Class with the largest invertebrates, the giant squids

40. _____ Torsion causes most internal organs to twist during development

41. _____ Includes clams, scallops, oysters, and mussels

42. _____ Most diverse of the mollusks

43. _____ Move rapidly with a system of jet propulsion

44. _____ Most can discharge a dark fluid from an ink sac, used to distract predators

45. _____ The anus dumps wastes near the mouth

46. _____ When disturbed they will suction themselves to a hard substrate

25.10. ROTIFERS & TARDIGRADES—TINY & TOUGH [p.419]

25.11. ROUNDWORMS—UNSEGMENTED WORMS THAT MOLT [p.420]

25.12. ARTHROPODS—ANIMALS WITH JOINTED LEGS [p.421]

Selected Words: molt [p.419], nematodes [p.420], Caenorhabditis elegans [p.420], Ascaris lumbricoides [p.420], Trichinella spiralis [p.420], *trichinosis* [p.420], Wucheria bancrofti [p.420], elephantiasis [p.420], pinworms [p.420], Enterobius vermicularis [p.420], hookworms [p.420], *jointed appendages* [p.421]

Boldfaced, Page-Referenced Terms

[p.419] rotifers _____

_____

[p.419] tardigrades _____

_____

[p.420] roundworms _____

_____

[p.421] arthropods _____

_____

[p.421] exoskeleton_____

_____

[p.421] antennae_____

_____

[p.421] metamorphosis _____

_____

Matching [pp.419-421]

Match the group with the appropriate term.

1. _____ rotifer
2. _____ tardigrades
3. _____ roundworms
4. _____ *Trichinella spiralis*
5. _____ *Ascaris lumbricoides*
6. _____ Pinworms
7. _____ *Wuchereria bancrofti*

A. Causes lymphatic filariasis
B. Posses cilia on their head
C. Forms cysts within the muscle of pork
D. Common infection in children
E. Commonly called the water bears
F. Bilateral, unsegmented worm with a cuticle
G. A large infection can clog the digestive tract of the host

Short Answer [pp.419-421]

Answer exercises 8-10 for the following drawing of a dissected animal.

8. What is the common name of the dissected animal? _____

9. Is the animal hermaphroditic? _____

10. Name the coelom type exhibited by this animal. _____

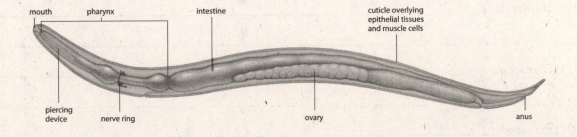

Choice [p.421]

For questions 11-21, choose from the following six adaptations that contributed to the success of arthropods:

    a. hardened exoskeleton   b. highly modified segments   c. jointed appendages

    d. specialized developmental stages   e. specialized sensory structures

11. _____ Tissues get remodeled as the juveniles mature into adults

12. _____ A cuticle of chitin, proteins, and waxes

13. _____ Each stage of the life cycle is specialized for a different task

14. _____ Might have evolved as a defense against predation

15. _____ The immature forms are considered eating machines

16. _____ In the ancestor of insects, different segments became fused into a head, thorax, and an abdomen

17. _____ This is shed in order to provide room for new growth

18. _____ The appendages became modified for specialized tasks

19. _____ The exoskeleton of the legs is modified for walking

20. _____ These structures can detect touch & waterborne chemicals

21. _____ The eyes are compound with many lenses

25.13. CHELICERATES AND THEIR RELATIVES [p.422]

25.14. THE MOSTLY MARINE CRUSTACEANS [p.423]

25.15. MYRIAPODS---LOTS OF LEGS [p.424]

Selected Words: cephalothorax [p.422], book lungs [p.422], copepods [p.423]

Boldfaced, Page-Referenced Terms

[p.422] Malphigian tubules _____

_____

[p.422] chelicerates _____

_____

[p.423] crustaceans _____

_____

[p.424] myriapod _____

_____

Dichotomous Choice [pp.422-424]

Circle one of two possible answers given between parentheses in each statement.

1. Nearly all arthropods possess (strong claws / an exoskeleton).

2. (Spiders / Crabs) exchange gases through book lungs.

3. Chelicerae are found in (ticks / barnacles).

4. The largest arthropods are (lobsters and crabs / barnacles and pillbugs).

5. (Barnacles / Copepods) are part of the marine zooplankton.

6. Of all the arthropods, only (copepods / barnacles) have a calcified "shell."

7. Adult (barnacles / copepods) cement themselves to wharf pilings, rocks, and similar surfaces.

8. As is true of other arthropods, crustaceans undergo a series of (rapid feedings / molts) and so shed the exoskeleton during their lifecycle.

9. (Millipedes / Centipedes) have fused segments and each one contains two pairs of legs.

10. Adult (millipedes / centipedes) have a flattened body with a pair of walking legs per segment.

11. (Millipedes / Centipedes) are fast moving, aggressive predators, outfitted with fangs and venom glands.

12. Spiders posses (Malphigian tubules / gills) that filter waste from the body tissues.

13. (Horseshoe crabs / Lobsters) posses a spine-like telson for steering.

Labeling [p.422]

Identify each numbered body part in this illustration, then answer question 19.

14. _____

15. _____

16. _____

17. _____

18. _____

19. Name the subgroup of arthropods to which the animal belongs. _____

Identify each numbered part of the animal pictured at the right.

20. _____

21. _____

22. _____

23. _____

24. _____

25. _____

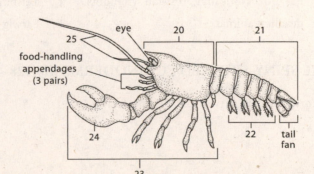

25.16. A LOOK AT INSECT DIVERSITY [pp.424-425]

25.17. INSECT DIVERSITY & IMPORTANCE [pp.426-427]

Selected Words: nymphs [p.425], larva [p.425], *incomplete* metamorphosis [p.425], *complete* metamorphosis [p.425], malphigian tubules [p.425]

Boldfaced, Page-Referenced Terms

[p.425] metamorphosis _____

_____

Matching [pp.424-427]

Choose the most appropriate answer for each term.

1. ___ complete metamorphosis
2. ___ the most successful species of insect
3. ___ incomplete metamorphosis
4. ___ malphigian tubules
5. ___ insect life-cycle stages
6. ___ shared insect adaptations

A. Head, thorax, and abdomen; paired sensory antennae and mouthparts; three pairs of legs and two pairs of wings

B. Involves gradual, partial change from the first immature form until the last molt

C. Winged insects, also the only winged invertebrates

D. Larva-nymph-pupa-adult

E. Structures that help eliminate toxic metabolic wastes

F. Larva grows into a pupa and then reorganizes into the adult form

Matching [pp.424-427]

7. _____ scavenger with a flattened body
8. _____ wingless blood suckers
9. _____ winged and sucks plant juices
10. _____ has prokaryotic and protistan symbionts
11. _____ deadliest animal

A. Mosquito
B. Aphid
C. Lice
D. Termites
E. Earwig

25.18. THE SPINY-SKINNED ECHINODERMS [pp.428-429]

Selected Words: bilateral larvae [p.428], tube feet [p.428]

Boldfaced, Page-Referenced Terms

[p.428] echinoderm _____

_____

[p.428] water-vascular system _____

_____

Fill-in-the-Blanks [pp.428-429]

The second lineage of coelomate animals is referred to as the (1) _____. The body wall of all echinoderms will contain protective spines, spicules, or plates made rigid with (2) _____. Oddly, adult echinoderms have (3) _____ symmetry with some bilateral features, but many produce larvae with (4) _____ symmetry. Adult echinoderms have no (5) _____, but a decentralized (6) _____ system allows them to respond to information about food, predators, and so forth. Sea stars feed on mollusks by sliding their (7) _____ into the bivalves shell. They then secrete (8) _____ that will kill the mollusk.

The (9) _____ feet of sea stars are used for walking, burrowing, clinging to rocks, or gripping a meal of clam or snail. These "feet" are part of a(n) (10) _____ system that is unique to echinoderms. Tube feet change shape constantly as (11) _____ action redistributes fluid through the water-vascular system. It is the (12) _____ that will redistribute the fluid among the tube feet. Coarse, indigestible remnants are regurgitated back through the mouth. Their small (13) _____ is of no help in getting rid of empty clam or snail shells. Another echinoderm, the (14) _____, has long pointed spines. Another, the (15) _____, sometimes evades predation by shooting its (16) _____ out of its anus.

Labeling [p.428]

Identify each indicated part of the two following illustrations.

17. _____

18. _____

19. _____

20. _____

21. _____

22. _____

23. _____

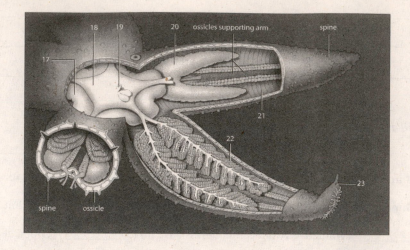

Short Answer [p.428]

Answer questions 25 and 26 for the animal just pictured.

24. Name the animal shown. _____

25. Is it a protostome or deuterostome? _____

SELF-TEST

1. Which of the following is true for sponges? They have _____. [pp.408-409]
   a. distinct cell types
   b. symmetry
   c. muscles
   d. complete digestive systems

2. Bilateral symmetry is characteristic of _____. [p.412]
   a. cnidarians
   b. sponges
   c. jellyfish
   d. flatworms

3. Flukes and tapeworms are parasitic _____. [pp.412-413]
   a. leeches
   b. flatworms
   c. jellyfish
   d. roundworms

4. In which group does the mouth appear before the anus during development? [p.405]
   a. Protostomes
   b. Deuterostomes
   c. Neither
   d. Both groups

5. Which two groups are the closest related? [p.407]
   a. Chordates & echinoderms
   b. Chordates & cnidarians
   c. Arthropods & echinoderms
   d. Mollusks & roundworms

6. Which group of mollusks are the most intelligent? [p.418]
   a. Gastropods
   b. Bivalves
   c. Cephalopods
   d. Chitons

7. Torsion is a process characteristic of _____. [p.416]
   a. chitons
   b. bivalves
   c. gastropods
   d. cephalopods
   e. echinoderms

8. The _____ have bilateral symmetry, cylindrical bodies tapered on both ends, a tough protective cuticle, and a false coelom, and they present the simplest example of a complete digestive system. [p.420]
   a. roundworms
   b. cnidarians
   c. flatworms
   d. echinoderms

9. A complete digestive tract with a mouth and an anus is not seen in _____. [p.412]
   a. annelids
   b. flatworms
   c. mollusks
   d. roundworms

10. What are the hair-like structures found on polychaetes called? [p.414]
    a. Setae
    b. Parapodia
    c. Cilia
    d. Fimbrae

Matching [pp.408-429]

Match each of the following phyla with the corresponding characteristics (letters a-i) and representatives (A-M). A phylum may match with more than one letter from the group of representatives.

11. _____ , _____ Annelids

12. _____ , _____ Arthropods

13. _____ , _____ Cnidarians

14. _____ , _____ Echinoderms

15. _____ , _____ Mollusks

16. _____ , _____ Nematodes

17. _____ , _____ Flatworms

18. _____ , _____ Sponges

19. _____ , _____ Rotifers

a. Choanovytes (collar cells) + spicules

b. Jointed legs + an exoskeleton

c. Pseudocoelomate + wheel organ + soft body

d. Soft body + mantle; may or may not have radula or shell

e. Bilateral symmetry + blind-sac gut

f. Radial symmetry + blind sac-gut; stinging cells

g. Body compartmentalized into repetitive segments; coelom containing nephridia (primitive kidneys)

h. Tube feet + calcium carbonate structures in skin

i. Complete gut + bilateral symmetry = cuticle; includes many parasitic species, some of which are harmful to humans

A. Small animals with a crown of cilia and two "exuding toes"

B. Corals, sea anemones, and *Hydra*

C. Tapeworms and planaria

D. Insects

E. Jellyfish and the Portuguese man-of-war

F. Sand dollars and starfishes

G. Earthworms and leeches

H. Lobsters, shrimp, and crayfish

I. Organisms with spicules and collar cells

J. Scorpions and millipedes

K. Octopuses and oysters

L. Flukes

M. Hookworms, pinworms

20. Explain why Placozoans are classified as animals. [p.408]

_____

_____

_____

_____

CHAPTER OBJECTIVES / REVIEW QUESTIONS

1. List the six general characteristics that define an "animal." [pp.404-405]
2. Distinguish between radial symmetry and bilateral symmetry, describe the various animal gut types that are found in both groups. [pp.404-405]
3. List the characteristics that distinguish sponges from other animal groups. [pp.408-409]
4. State what nematocysts are used for and explain how they function. [p.410]
5. Two cnidarian body types are the _____ and the _____. [p.410]
6. Describe the structure typical of a cnidarian, using terms such as *epithelium, nerve cells, nerve net, mesoglea,* and *hydrostatic skeleton.* [p.410]
7. Define *organ-system* level of construction and relate this to flatworms. [p.412]
8. List the three main types of flatworms and briefly describe each; name groups that are parasitic. [pp. 412-413]
9. Describe the body plan of roundworms, comparing its various systems with those of the flatworm body plan. [p.420]
10. Describe, by their characteristics, *protostome* and *deuterostome* lineages, and cite examples of animal groups belonging to each lineage. [pp.405, 428]
11. List and generally describe the major groups of mollusks and their members. [pp.416-418]
12. Explain why cephalopods came to have such a well-developed sensory and motor system and are able to learn. [p.418]
13. List four different lineages of arthropods; briefly describe each. [pp.421-425]
14. Name some common types of crustaceans. [p.423]
15. Name the features shared by all insects. [pp.424-425]
16. List five examples of animals known as echinoderms. [pp.428-429]
17. Describe how locomotion and predation occur in sea stars. [pp.428-429]

INTEGRATING AND APPLYING KEY CONCEPTS

1. Most highly evolved invertebrates have bilateral symmetry, a complete gut, a true coelom, and segmented bodies. Why do you suppose that having a true coelom and a segmented body is considered more highly evolved than the condition of lacking a coelom or possessing a false coelom and having an unsegmented body? Cite evidence from the chapter that would support or reject this idea.
2. What features make arthropods so successful?
3. What is the advantage for an organism to go through metamorphosis?

26
ANIMAL EVOLUTION- THE CHORDATES

INTRODUCTION

This chapter looks at trends in the development of vertebrates with special emphasis on human origins.

FOCAL POINTS

- Figure 26.5 [p.436] looks at the family tree of chordates with some of the innovations at branch points.
- Figure 26.16 [p.442] shows the family tree of amniotes.
- Sections 26.13-26.15 [pp.452-457] describe the evolution of humans from early primates.
- Figure 26.41 [p.458] looks at the appearance and extinction of various hominid species.

INTERACTIVE EXERCISES

Impacts, Issues: Transitions Written in Stone [p.432]

26.1. THE CHORDATE HERITAGE [pp.434-435]

26.2. VERTEBRATE TRAITS AND TRENDS [pp.436-437]

26.3. JAWLESS LAMPREYS [p.438]

26.4. THE JAWED FISH [pp.438-439]

Selected Words: "*missing links*" [p.432], Archaeopteryx [p.432], nerve cord [p.434], cranium [p.435], gill slits [p.435], hagfish [p.435], jawless fish [p.435], placoderms [p.436], lampreys [p.438], *teleosts* [p.439], coelacanths [p.439]

Boldfaced, Page-Referenced Terms

[p.434] chordates _____

_____

[p.434] notochord _____

_____

[p. 434] vertebrates _____

_____

[p.434] lancelets _____

_____

[p.435] tunicates _____

_____

[p.435] craniates _____

_____

[p.436] endoskeleton _____

_____

[p.436] vertebral column _____

_____

[p.436] jaws _____

_____

[p.436] vertebrae _____

_____

[p.437] fins _____

_____

[p.437] gills _____

_____

[p.437] lungs _____

_____

[p.437] kidneys _____

_____

[p.438] scales _____

_____

[p.439] swim bladder _____

_____

[p.439] cartilaginous fishes _____

_____

[p.439] bony fish _____

_____

Fill-in-the-Blanks [pp.432-437]

One obstacle in Darwin's time to the acceptance of evolution was the lack of fossils of (1) "_____" (2) _____ is one such fossil that bridges the gap between reptiles and birds. (3) _____ dating has dated fossils of this organism to 150 million years ago. It and many similar fossils provide support for (4) _____.

Four major features distinguish the embryos of chordates from those of all other animals: a tubular dorsal (5) _____, a pharynx with (6) _____ in its wall, a(n) (7) _____, and a tail that extends past the anus during some point of the organism's lifetime. The invertebrate chordates are represented today by tunicates and (8) _____, which obtain their food by (9) _____. Lancelets draw in plankton-laden water through the mouth, pass it over mucus coated cells that will trap the particulate matter and direct it towards the (10) _____. The water is then forced out through the gill slits in the pharynx. (11) _____ are among the most primitive of all living chordates. This group derives its name from the carbohydrate rich (12) _____ that covers the adult body. Most will remain attached to rocks or hard substrates in marine habitats after their larvae undergo (13) _____.

In ancestors of the vertebrate line, the notochord was replaced by a bony (14) _____, which was the foundation for fast moving (15) _____. The evolution of (16) _____ started an evolutionary arms race between predators and prey. Fish evolved (17) _____ to help them swim faster as well as paired (18) _____ evolved in early aquatic vertebrates to help enhance the exchange of gases.

As the ancestors of land vertebrates began spending less time immersed in water and more time on land exposed to air, their dependency on gill declined and (19) _____ evolved; more elaborate and efficient (20) _____ systems evolved along with more complex and efficient lungs.

Labeling [p.434]

21. Name the organism in the accompanying diagram. _____

22. Is this an invertebrate of vertebrate chordate? _____

Name the numbered structures in the diagram.

23. _____
24. _____
25. _____
26. _____

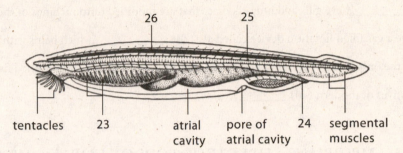

tentacles   23   atrial cavity   pore of atrial cavity   24   segmental muscles

Labeling [p.436]

Label the following groups on the Chordate family tree.

27. _____
28. _____
29. _____
30. _____
31. _____

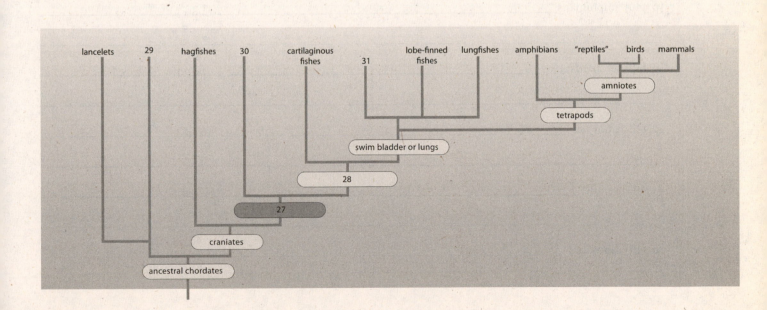

Fill-in-the-Blanks [p.438]

        Existing jawless fish include the scavenging (32) _____ and the often parasitic (33) _____.

Cartilaginous fishes include about 850 species of rays, skates, and (34) _____. They have a conspicuous fins

and five to seven (35) _____. (36) _____ fish are the most diverse vertebrates on the planet. (37)

_____ have gills and lung-like sacs that are modified outpouchings of the gut wall. Bony fish (osteichthyes)

have a gas filled flotation device called a(n) (38) _____ which helps with buoyancy in water. (39) _____

are the only living representatives of the lobe-finned fish. These lobe-finned fish are most likely the closest living

relatives of modern (40) _____.

26.5. AMPHIBIANS—FIRST TETRAPODS ON LAND [pp.440-441]

26.6. VANISHING ACTS [p.441]

26.7. THE RISE OF AMNIOTES [pp.442-443]

26.8. SO LONG, DINOSAURS [p.443]

26.9. DIVERSITY OF MODERN REPTILES [pp.444-445]

Selected Words: frog [p.441], salamander [p.441], caecilian [p.441], Carboniferous [p.442], Lystrosaurus [p.442], synapsids [p.442], Triassic [p.443], Jurassic [p.443], Cretaceous [p.443], turtles [p.444], lizards [p.444], tuataras [p.444], snakes [p.444], crocodilians [p.445]

Boldfaced, Page-Referenced Terms

[p.440] amphibians_____

_____

[p.440] tetrapods _____

_____

[p.442] amniotes _____

_____

[p.442] reptile _____

_____

[p.443] dinosaurs _____

_____

[p.443] K-T asteroid impact theory _____

_____

[p.444] ectotherms _____

_____

[p.444] cloaca _____

_____

Fill-in-the-Blanks [pp.440-443]

Natural selection acting on lobe-finned fishes during the Devonian period favored adaptations that enabled fish to survive on (1) _____. These adaptations include a (2) _____, (3) _____, and changes to the (4) _____ and (5) _____. Moving onto land allowed amphibians to escape predators as well as presented them with a new food source, (6) _____.

There are three groups of existing amphibians: (7) _____, frogs and toads, and caecilians, All of which are (8) _____ as adults. Amphibians require free-standing (9) _____ or at least moist habitats to (10) _____. Amphibian skin must also be kept moist since it acts as an additional (11) _____. Many amphibian species are declining worldwide due to (12) _____ activity.

(13) _____ were the first truly land animals. They produced water-conserving (14) _____ with four protective membranes. They also had dry, tough, or scaly skin, which helped prevent (15) _____ from the body. One group of amniotes, the (16) _____, gave rise to the mammals, the (17) _____ gave rise to the crocodilians and eventually modern (18) _____. At the end of the (19) _____, a huge asteroid slammed into Earth and obliterated many amniotes, including the (20) _____ that had been the dominate animals for almost 125 million years.

Matching [pp.444- 445]

Match each of the "reptiles" with its characteristics.

21. _____ crocodilians

22. _____ lizards

23. _____ snakes

24. _____ tuataras

25. _____ turtles

A. Posses a bony, scale-covered shell

B. The most diverse and numerous "reptiles"

C. With an amphibian-like brain, they are very primitive "reptiles"

D. Evolved from a short-legged, long bodied lizard

E. Closest living relatives of birds; posses a 4 chambered heart

26.10. BIRDS—THE FEATHERED ONES [pp.446-447]

26.11. THE RISE OF MAMMALS [pp.448-449]

26.12. MODERN MAMMALIAN DIVERSITY [pp.450-451]

Selected Words: Confuciusornis sanctus [p.446], *lift* [pp.446], biological clock [pp.447], egg-laying mammals [pp.448], pouched mammals [pp.448], teeth [p.448], morphological convergence [p.449]

Boldfaced, Page-Referenced Terms

[p.446] birds _____

_____

[p.446] endotherms _____

_____

[p.448] mammals _____

_____

[p.448] monotremes _____

_____

[p.448] marsupials _____

_____

[p.448] placental mammals _____

_____

[p.450] placenta _____

_____

Labeling [p.446]

Label the structures pictured at the right.

1. _____

2. _____

3. _____

Labeling [pp. 447]

Label the structures pictured at the right.

4. _____

5. _____

6. _____

7. _____

8. _____

9. _____

Fill-in-the-Blanks [pp.446-451]

Birds descended from (10) _____ that ran around on two legs during the (11) _____ era. Like their closest relatives, birds have (12) _____ on their legs. All birds have (13) _____ that insulate and help with flight. Generally, birds have a greatly enlarged (14) _____ to which flight muscles are attached. Bird bones contain (15) _____, which decrease weight. They also have lungs with attached airs sacs that increase gas exchange. Flight demands high (16) _____ rates, which requires an abundant supply of (17) _____ being pumped to all parts of the body by means of a large, durable (18) _____ chambered heart. Almost 9,000 species of birds show amazing variation in body structure. The smallest known species of bird weighs only (19) _____ grams. The largest existing bird is the (20) _____, which weighs about 150 kilograms and is a flightless sprinter. One complex behavior seen in birds is (21) _____, a seasonal movement from one region to another. The champion of this behavior is the artic tern, which travels from the Arctic to the (22) _____ and back each year.

Most mammals have (23) _____ as a means of insulation, and (24) _____ with which females will nurse their young. Unlike "reptiles," mammals have 4 different types of (25) _____ in their jaws. The ancestors of today's mammals diverged from small, hairless reptiles called (26) _____ at the same time that the dinosaurs were becoming the dominant species on Earth. Mammals coexisted with the diverse groups of (27) _____ through the Cretaceous. When the dinosaurs became extinct, diverse adaptive zones awaited exploitation by the three principal mammalian lineages: those that lay eggs, also called the (28) _____; those that use a pouch for reproduction called the (29) _____; and those that use a (30) _____ for reproduction. (31) _____ has occurred as evolutionary distinct, geographically isolated lineages evolved similar physical features.

Choice [pp.450-451]
Label each of the numbered mammals with one of the following lettered classifications.

    a. monotreme   b. marsupial   c. eutherian

32. _____ bat

33. _____ human

34. _____ kangaroo

35. _____ koala

36. _____ platypus

37. _____ spiny anteater

38. _____ Tasmanian devil

39. _____ whale

26.13. FROM EARLY PRIMATES TO HOMINIDS [pp.452-453]

26.14. EMERGENCE OF EARLY HUMANS [pp.454-455]

26.15. EMERGENCE OF MODERN HUMANS [pp.456-457]

Selected Words: prosimians [p.452], Hominoids [p.452], foramen magnum [p.452], prehensile [p.453], opposable [p.453], Eocene [p.453], Sahelanthropus tchadensis [p.454], "Lucy" [p.454], Australopithecus [p.454], Paranthropus [p.454], Homo habilis [p.455], Olduvai Gorge [p.455], Homo erectus [p.456], Homo sapiens [p.456], Homo floresiensis [p.456], cultural [p.457]

Boldfaced, Page-Referenced Terms

[p.452] primate _____

_____

[p.452] hominids _____

_____

[p.452] bipedal _____

_____

[p.453] cultural _____

_____

[p.454] australopiths _____

_____

[p.455] humans _____

_____

[p.456] multiregional model _____

_____

[p.456] replacement model _____

_____

Fill-in-the-Blanks [pp.452-457]

During the Cenozoic era a group of tree dwelling mammals arose called the (1) _____, the earliest of which were the (2) _____. The (3) _____ include (4) _____, apes, and (5) _____. The evolution from early prosimians to hominids required several adaptations. Excellent (6) _____ vision with (7) _____ facing eyes allowed them to judge depth and distances better. This shift to greater visual abilities lessened their dependence upon their sense of (8) _____. The (9) _____ and (10) _____ movements of the hands allowed hominids to (11) _____, which eventually led to the ability to make (12) _____. The foramen magnum was adapted for (13) _____ while their (14) _____ and jaws evolved to a mixed diet. As the (15) _____ expanded, more complex (16) _____ occurred and this eventually lead to the development of (17) _____.

In central (18) _____, 6 or 7 million years ago, hominids were becoming distinct from the apes. During the Miocene and Pliocene, many different "southern apes" or (19) _____ evolved. The earliest forms of genus (20) _____ evolved in the East African Rift Valley. About (21) _____ million years ago, hominids began making crude stone tools. *H. erectus* had a larger (22) _____ and was a more advanced (23) _____ than its ancestors. This group eventually migrated out of Africa into both (24) _____ and (25) _____. By (26) _____ years ago, modern humans, (27) _____ were seen in East Africa and around (28) _____ years ago in the Middle East. As they spread, earlier lineages, like the massively built, cold-adapted (29) _____ of Europe, disappeared.

One debate in human evolution concerns the way in which *H. sapiens* arose. According to the (30) _____ model, modern humans arose in sub-Saharan Africa and migrated out, replacing (31) _____ populations from prior migrations. This is supported by the discovery of the oldest (32) _____ fossils in Africa. The (33) _____ model proposes that the *H. erectus* populations which migrated earlier, independently evolved into (34) _____.

Choice [pp.452-453]

Label each of the numbered primates with one of the following lettered groups.

    a. Prosimians   b. Anthropoids   c. Hominoids

35. _____ Humans
36. _____ Lemur
37. _____ Gibbon
38. _____ Baboon
39. _____ Gorillas
40. _____ Tarsier
41. _____ Orangutan

SELF-TEST

____ 1. In true fishes, the gills serve primarily _____ function. [p.437]
   a. a gas exchange
   b. a feeding
   c. a water elimination
   d. both a feeding and a gas exchange

____ 2. Jawed fish have paired _____ that help provide lift while swimming. [p.438]
   a. fins
   b. scales
   c. gills
   d. kidneys

____ 3. Which of the following is an important adaptation to life on land? [p.440]
   a. amniote egg
   b. lungs
   c. 3 chambered heart
   d. all of the above

____ 4. What is believed to be the most likely reason for the extinction of the dinosaurs? [p.443]
   a. Increase in the Earth's temperature
   b. K-T asteroid impact
   c. Disease
   d. Competition with mammals
   e. Killed off by Neanderthals

____ 5. Primitive primates generally live _____. [p.453]
   a. in tropical forest canopies
   b. in temperate savanna and grassland habitats
   c. near rivers, lakes, and streams in the East African Rift Valley
   d. in caves where there are abundant supplies of insects

____ 6. Which group is the closest living relative of birds? [p.445]
   a. Turtles
   b. Snakes
   c. Dinosaurs
   d. Crocodiles

____ 7. A hominid in Europe and Near East that became extinct nearly 30,000 years ago was the _____. [p.456]
   a. dryopith
   b. *Australopithecus*
   c. *Homo erectus*
   d. Neanderthal

____ 8. Which of the following is NOT a key trend that led to human traits? [p.452]
   a. Enhanced daytime vision
   b. Upright walking
   c. Jaws and teeth adapted to eat meat
   d. Opposable thumbs

____ 9. The oldest, undisputed, bipedal hominid is _____. [p.454]
   a. Ardipithecus ramidus
   b. paranthropus
   c. Homo habilis
   d. Australopithecus afarensis

____ 10. Which hominid was the 1st to migrate out of Africa? [p.456]
   a. *Homo erectus*
   b. *Homo neanderthalensis*
   c. *Homo sapiens*
   d. *Homo floresiensis*

Matching [pp.452-457]

Choose the one most appropriate answer for each term.

11. _____ anthropoids
12. _____ australopiths
13. _____ *Homo habilis*
14. _____ hominids
15. _____ hominoids
16. _____ *Homo erectus*
17. _____ primates
18. _____ prosimians

A. A group that includes apes and humans

B. Organisms in a suborder that includes New World and Old World monkeys, apes, and humans

C. One of the earliest members of genus Homo; found in Africa

D. A group that includes humans and human-like species

E. The first member of the genus Homo to leave Africa and migrate to the far corners of Eurasia

F. Organisms in a suborder that includes tarsiers, lemurs, and others

G. A group that includes prosimians and anthropoids

H. Bipedal organisms living from about 4 million to 1 million years ago, petite with a narrow jaw and small teeth

Matching [pp.434-451]

Match the following groups and classes with the corresponding characteristics (a-i) and representatives (A-I).

19. ___ , ___ Amphibians
20. ___ , ___ Birds
21. ___ , ___ Bony fishes
22. ___ , ___ Cartilaginous fishes
23. ___ , ___ Jawless fishes
24. ___ , ___ Mammals
25. ___ , ___ "Reptiles"
26. ___ , ___ Urochordates

a. Hair + mammary glands

b. Feathers + hollow bones

c. Jawless + cartilaginous skeleton (in existing species)

d. Two pairs of limbs (usually) + glandular skin + "jelly" covered eggs

e. Amniote eggs + scaly skin + bony skeleton

f. Invertebrate + sessile adult (cannot swim)

g. Jaws + cartilaginous skeleton + vertebrae

h. In adult, notochord stretches from head to tail; mostly burrowed-in, adult can swim

i. Bony skeleton + skin covered with scales, adapted to aquatic environment

A. Lancelet

B. Loons, cardinals, and eagles

C. Tunicates and sea squirts

D. Sharks and manta rays

E. Lampreys

F. Sea horses and groupers

G. Lizards and turtles

H. Caecilians and salamanders

I. Platypuses and opossums

CHAPTER OBJECTIVES / REVIEW QUESTIONS

1. List four characteristics found only in chordates. [p.434]
2. State what sort of changes occurred in the primitive chordate body plan that could have promoted the emergence of vertebrates. [pp.436-437]
3. Describe the differences between primitive and advanced fishes in terms of skeleton, jaws, special senses, and brain. [pp.438-439]
4. Describe the changes that enabled aquatic fishes to give rise to amphibians. [pp.440-441]
5. What key features are present in birds that show their reptilian heritage? [pp.446-447]
6. List the five key adaptations of primate evolution. [pp.452-453]
7. State which anatomical features underwent the greatest changes along the evolutionary line from early anthropoids to humans. [pp.452-453]
8. Explain how you think *Homo sapiens* arose. Use the evidence presented throughout the chapter to support your explanation. [pp.456-457]

INTEGRATING AND APPLYING KEY CONCEPTS

1. Birds and mammals both have four-chambered hearts, high metabolic rates, and can efficiently regulate their body temperatures. Evidence indicates that both of these groups evolved from a reptilian ancestor. Data suggests that reptiles have a heart that is between three and four-chambered, a lower metabolic rate, and are not that efficient at internally regulating their body temperature. If the bird / mammalian traits had evolved in reptiles how might their evolution been different?
2. Suppose someone told you that between 12 and 6 million years ago various populations of dryopiths were forced by predators to flee from their native forests and take up refuge along the banks of rivers, estuarine, and costal habitats. Over time and through various mutations they eventually lost their body hair, became bipedal, developed subcutaneous fat deposits as insulation, and developed a bridged nose that provided an advantage in a watery environment. The dryopith populations that did not flee to the safety of the watery habitats remained isolated and never developed these particular features. Over time the dryopiths that remained inland were killed by predators and their population became extinct. The water dwelling populations continued to adapt and successfully expand their population by learning to access the variety of available food sources (shellfish, fish, wild rice, oats, and various nuts, fruits, and tubers. In these aquatic habitats the first food gathering tools were developed. In order to be effective at food collecting some early form of communication was also established. How does this story fit with current speculations about the evolution of early humans? What evidence could be used to support or negate this story?

27

PLANTS AND ANIMALS—COMMON CHALLENGES

INTRODUCTION

This chapter introduces the concept of homeostasis, a process that gives living organisms the ability to maintain a stable environment for their chemical reactions. A key component of homeostasis is the use of feedback mechanisms, which provide the fine-tuned regulation of the metabolic activities. Later chapters in the text will relate these mechanisms to the physiology of specific organ systems in animals and plants.

FOCAL POINT

* Figure 27.7 and Figure 27.8 [p.466] outline the mechanisms by which homeostasis occurs in many systems.

INTERACTIVE EXERCISES

Impacts, Issues: A Cautionary Tale [p.460]

27.1. LEVELS OF STRUCTURAL ORGANIZATION [pp.462-463]

27.2. RECURRING CHALLENGES TO SURVIVAL [pp.464-465]

Selected Words: anatomy [p.460], physiology [p.460], quantitative terms [p.462], vascular tissue [p.464], surface to volume ratio [p.464]

Boldfaced, Page-Referenced Terms

[p.462] tissue _____

_____

[p.462] organ _____

_____

[p.462] organ system _____

_____

[p.462] growth _____

_____

[p.462] development _____

_____

[p.463] extracellular fluid _____

_____

[p.463] homeostasis _____

_____

[p.464] diffusion _____

_____

[p.464] passive transport _____

_____

[p.464] active transport _____

_____

Matching [pp.460-465]

1. _____ physiology
2. _____ growth
3. _____ development
4. _____ anatomy
5. _____ homeostasis
6. _____ tissue
7. _____ organ
8. _____ organ system

A. Increase in the size of, number, and volume of cells

B. The form of an organism

C. The maintenance of internal body conditions

D. Successive stages in the formation of specialized tissues, organs, and organ systems

E. Study of how the body's parts are put to use

F. Two or more organs interacting physically or chemically

G. Consists of at least two tissues organized into certain proportions and patterns

H. Community of cells that interact in one or more tasks

Fill-in-the-Blanks [pp.463-465]

Plant and animals cells are constantly bathed in (9) _____. The body's systems must work in unison to keep the fluid (10) _____ and (11) _____ stable. If these levels shift, the body is no longer maintaining (12) _____. Molecules or ions tend to move from areas of higher concentration to areas of lower concentration by a process called (13) _____. This helps maintain concentration levels of various substances. When solutes are pumped from a region of low concentration to one of high concentration it is called (14)

_____.

Choice [pp.464-465]

For each of the numbered statements, choose the most appropriate terms from the lettered list. Some choices may be used more than once and some statements may have more than one answer.

a. active transport   b. habitat   c. surface to volume ratio   d. diffusion

15. _____   Movement of ions and molecules away from an area of concentration
16. _____   Responsible for adjusting the kinds, amounts, and directional movement of substances
17. _____   Movement against a concentration gradient with the use of a protein pump
18. _____   Place where specific species will live
19. _____   As organisms grow, this will decrease
20. _____   May be influenced by biotic factors such as predators and availability of producers
21. _____   The development of vascular tissues overcomes some limits imposed by this
22. _____   Most effective over small distances

27.3. HOMEOSTASIS IN ANIMALS [pp.466-467]

27.4. HEAT-RELATED ILLNESS [p.467]

27.5. DOES HOMEOSTASIS OCCUR IN PLANTS [pp.468-469]

27.6. HOW CELLS RECEIVE AND RESPOND TO SIGNALS [pp.470-471]

Selected Words: Metabolic reactions [p.467], "sleep" position [p.469], signaling molecules [p.470]

Boldfaced, Page-Referenced Terms

[p.466] receptor _____

_____

[p.466] integrator _____

_____

[p.466] effectors _____

_____

[p.466] negative feedback mechanisms _____

_____

[p.467] positive feedback mechanisms _____

_____

[p.468] system acquired resistance _____

_____

[p.468] compartmentalization _____

_____

[p.469] circadian rhythm _____

_____

[p.470] apoptosis _____

_____

Matching [pp.466-467]

Choose the most appropriate description for each of the following terms.

1. _____ integrator
2. _____ negative feedback mechanism
3. _____ positive feedback mechanism
4. _____ receptor
5. _____ effector
6. _____ stimulus

A. The location that processes information regarding stimuli and that signals the response

B. The specific form of energy detected by a receptor

C. Chain of events that intensify change from the original condition

D. The portion of the body that carries out the response to the stimulation

E. Change that leads to a response that reverses the change

F. Cell that changes in response to stimuli

Choice [pp.466-467]

Determine the type of feedback mechanism that each of the following cases represents.

   a. positive feedback   b. negative feedback

7. _____ Regulation of body temperature in mammals

8. _____ Less common feedback system of the two

9. _____ Secretion of oxytocin during childbirth

10. _____ The response of the system cancels or counteracts the effect of the original stimulus

11. _____ The response is an intensification of the effect of the original stimulus

12. _____ Prevents overheating during exercise

Choice [pp.466-467]

Choose whether each of the following represents a plant or an animal mechanism for regulating the internal environment.

   a. plant   b. animal

13. _____ Compartmentalization of infected cells

14. _____ Makes adjustments to the internal systems in order to maintain homeostasis

15. _____ Uses interactions among receptors and effectors to maintain homeostasis

16. _____ Structural adaptations for water retention in sand-dwelling organisms

17. _____ A system acquired resistance develops in response to an infection

Fill-in-the-Blanks [pp.470-417]

   In big (18) _____ organisms, one cell type signals others in response to cues from both the internal and (19) _____ environment. Signal molecules will often (20) _____ through (21) _____ in an animal cell and (22) _____ in plant cells.

   (23) _____ mechanisms by which cells "talk" to one another evolved early in the history of life. They often have three parts. First, a specific (24) _____ is activated, as by reversibly binding a signaling molecule. Second, the signal is (25) _____—it is converted into a form that can operate inside the cell. Third, the cell makes a functional (26) _____.

   Most receptors are membrane (27) _____. An activated receptor may activate a(n) (28) _____ that in turn activates many molecules of a different enzyme, which activates many molecules of another kind and so on. These chains of cascading (29) _____ inside the cell greatly (30) _____ the original signal. One example is that of programmed cell death called (31) _____.

SELF-TEST

___ 1. Which of the following terms is used to indicate the process by which an organism goes through successive stages in the formation of specialized tissues and organs? [p.462]
a. Anatomy
b. Physiology
c. Growth
d. Development

___ 2. The action of an organism to keep the operating conditions of the internal environment within specified limits is called _____. [p.463]
a. development
b. homeostasis
c. adaptation
d. diffusion

___ 3. In animals, a mechanism of controlling homeostasis in which the output of a metabolic reaction is used to slow or reverse a pathway is called _____. [p.466]
a. a positive feedback mechanism
b. adaptation
c. a negative feedback mechanism
d. physiology

___ 4. Which of the following processes moves molecules against their concentration gradient? [p.464]
a. Diffusion
b. Anatomy
c. Compartmentalization
d. Active transport

___ 5. The overall size of a cell is determined by its _____. [p.464]
a. evolutionary importance
b. surface to volume ratio
c. compartmentalization
d. all of the above

___ 6. Apoptosis is _____. [p.470]
a. the movement of molecules against a concentration gradient
b. a form of receptor
c. programmed cell death
d. a process by which cells grow in size

___ 7. Which of the following represents a community of cells interacting to perform a task? [p.462]
a. Organ system
b. Tissue
c. Organ
d. Integrator

___ 8. _____ describes a biological activity that is repeated in cycles of about 24 hours. [p.469]
a. Apoptosis
b. Physiology
c. Transduction
d. Compartmentalization
e. Circadian rhythm

___ 9. The internal environment in which body cells live is called the _____. [p.463]
a. extracellular fluid
b. sweat
c. plasma
d. phloem

___ 10. Movement of molecules from a high concentration to a low concentration is called _____. [p.464]
a. osmosis
b. active transport
c. diffusion
d. adaptation

CHAPTER OBJECTIVES / REVIEW QUESTIONS

1. Define *homeostasis*. [p.463]
2. Explain the difference in organization between a tissue, an organ, and an organ system. [p.462]
3. Distinguish between the terms *growth* and *development*. [p.462]
4. Define homeostasis in relation to the internal environment of an organism. [p.463]
5. Explain how a cell may use diffusion and active transport to maintain an internal environment. [p.464]
6. Explain how the surface to volume ration defines the physical size of a cell. [p.464]
7. Define *habitat*. [p.465]
8. Explain the action of sensory receptors, using the terms *stimulus, integrators,* and *effectors*. [p.466]
9. Explain the difference between a negative and positive feedback mechanism and give an example of each. [pp.466-467]
10. Give examples of how plants may maintain homeostasis. [pp.468-469]
11. Define the term *circadian rhythm*. [p.469]
12. Explain the process of apoptosis. [pp.470-471]

INTEGRATING AND APPLYING KEY CONCEPTS

1. Describe the warning signals that someone will experience as they are going into heat exhaustion. Indicate what should be done for someone that is entering heat exhaustion.
2. Explain how our knowledge of apoptosis benefits medicine. Give examples of when it would be beneficial to "turn on or turn off" the process of apoptosis.
3. Explain how traveling from one time zone to another can disrupt the circadian rhythms in humans.

28

PLANT TISSUES

INTRODUCTION

This chapter examines simple and complex plant tissues and the specific types of cells that comprise those tissues. It also addresses the way these tissues are arranged to form plant organs, how the structure and function of these tissues and organs are interrelated, and how the organs combine to form a unified, functional plant body.

FOCAL POINTS

- Figure 28.2 *(animated)* [p.476] illustrates the basic structure of a flowering plant.
- Figure 28.3 *(animated)* [p.477] compares structural differences between Eudicots and Monocots.
- Figure 28.4 [p.477] illustrates the apical and lateral meristems responsible for primary and secondary growth.
- Table 28.1 [p.478] summarizes the components of plant tissues and their functions.
- Figure 28.14 *(animated)* [p.483] is an excellent review of the tissue structure of the leaf and the movement of materials into and out of it.
- Figure 28.16 *(animated)* [p.484] details the production and organization of mature tissues in plant roots.
- Figures 28.19 and 28.20 *(animated)* [pp.486-487] provide a comprehension of how lateral meristems function in the secondary growth of woody plants.

INTERACTIVE EXERCISES

Impacts, Issues: Droughts Versus Civilization [pp.474]

28.1. COMPONENTS OF THE PLANT BODY [pp.476-477]

Selected Words: stems [p.476], leaves [p.476], flowers [p.476], epidermis [p.476], seed leaves [p.476], vascular cambium [p.477], cork cambium,[p.477], vascular bundle [p.477]

Boldfaced, Page-Referenced Terms

[p.476] shoots _____

_____

[p.476] eudicots _____

_____

[p.476] monocots _____

_____

[p.476] roots _____

_____

[p.476] ground tissue system _____

_____

[p.476] vascular tissue system _____

_____

[p.476] dermal tissue system _____

_____

[p.476] cotyledons _____

_____

[p.476] meristems _____

_____

[p.477] apical meristems _____

_____

[p.477] primary growth _____

_____

[p.477] secondary growth _____

_____

[p.477] lateral meristem _____

_____

Labeling [p.476]

Identify the numbered parts of the illustration. Choose from the following:

dermal tissues   root system   ground tissues   shoot system   vascular tissues

1. _____

2. _____

3. _____

4. _____

5. _____

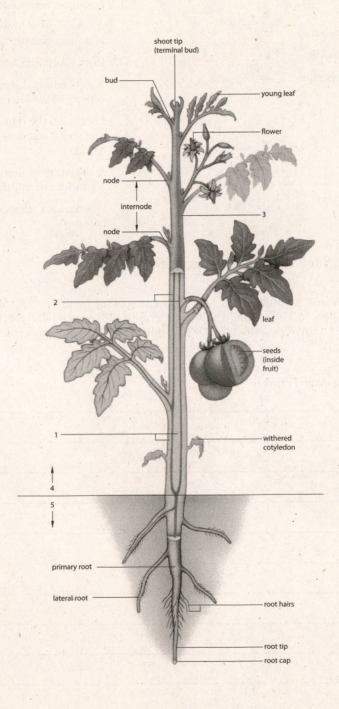

shoot tip
(terminal bud)

bud

young leaf

flower

node

internode

3

node

2

leaf

seeds
(inside
fruit)

1

withered
cotyledon

4

5

primary root

lateral root

root hairs

root tip

root cap

Matching [pp.476-477]

Choose the most appropriate answer for each term.

6. _____ secondary growth

7. _____ cotyledons

8. _____ meristems

9. _____ apical meristems

10. _____ primary growth

11. _____ lateral meristems

12. _____ Monocot

13. _____ Eudicot

A. A lengthening of stems and roots originating from cell divisions at apical meristems

B. Localized regions of dividing cells

C. A subgroup of flowering plants; flower parts in 3s and multiples of 3; parallel leaf veins

D. Cork Cambium and Vascular Cambium

E. A thickening of stems and roots caused by activity of the lateral meristems

F. Located in the dome-shaped tips of all shoots and roots; responsible for lengthening those organs

G. A subgroup of flowering plants; flower parts in 4's and 5's or multiples thereof; net-veined leaves

H. "Seed leaves" that contain food for the growing embryo

28.2. COMPONENTS OF PLANT TISSUES [pp.478-479]

Selected Words: growth [p.478], differentiate [p.478], pectin [p.478], lignin [p.478], fibers [p.478], sclereids [p.478], sieve-tube members [p.479], cutin [p.479], periderm [p.478], guard cells [p.479], radial section [p.478], tangential section [p.478], transverse section [p.478], simple tissues [p.478], complex tissues [p.478], sieve plates [p.479]

Boldfaced, Page-Referenced Terms

[p.478] parenchyma _____

_____

[p.478] mesophyll _____

_____

[p.478] collenchyma _____

_____

[p.478] sclerenchyma _____

_____

[p.478] xylem _____

_____

[p.478] tracheids _____

_____

[p.478] vessel members _____

_____

[p.479] phloem _____

_____

[p.479] sieve tubes _____

_____

[p.479] companion cells _____

_____

[p.479] epidermis _____

_____

Labeling [p.478]

It is important to understand the terms that identify the thin sections (slices) of plant organs and tissues that are prepared for study. Label each of the following three diagrams. Choose from radial section (cut along the radius of the organ), transverse or cross section (cut perpendicular to the long axis of the organ), or tangential section (cut made at a right angle to the radius of the organ). Note: The dark area indicates the slice.

1. _____ section
2. _____ section
3. _____ section

Choice [pp.478-479]

For questions 4-13, choose from the following Simple Tissues:

  a. parenchyma   b. collenchyma   c. sclerenchyma

4. _____ This tissue makes up the bulk of soft primary growth of plant organs

5. _____ The cells are dead at maturity and have thick, lignin-impregnated walls

6. _____ Cells are alive at maturity and can continue to divide

7. _____ Some types specialize in storage, secretion, and repair of wounds

8. _____ Cells are elongated, alive at maturity, and contain pectin in their cell walls

9. _____ Provides flexible support for primary tissues

10. _____ Supports mature vascular tissues and strengthens seed coats

11. _____ Cells are thin-walled, pliable, and many-sided

12. _____ This type of cell, when specialized for photosynthesis, is called mesophyll

13. _____ Fibers and sclereids belong to this type of simple tissue

Choice [pp.478-479]

(a) Choose the cell type that matches the numbered description: companion cell, vessel member, tracheid, sieve tube member

(b) State whether the cell is **dead** or **alive** at maturity

(c) List the cell as belonging to the complex tissue of **Xylem** or **Phloem**

14. Water-conducting tube; cells tapered at ends, have pits in sides

  (a) _____

  (b) _____

  (c) _____

15. Stacks of cells connect by perforated plates to move sugars throughout the plant

  (a) _____

  (b) _____

  (c) _____

16. Parenchyma cells that load sugars into #15

  (a) _____

  (b) _____

  (c) _____

17. Cylindrical cells stacked end to end, cell walls impregnated with lignin

  (a) _____

  (b) _____

  (c) _____

Short Answer [pp.478-479, 487]

18. Compare and contrast epidermis and periderm. _____

_____

_____

_____

28.3. PRIMARY STRUCTURE OF SHOOTS [pp.480-481]

Selected Words: axillary buds [p.480], bud scales [p.480], shoot apical meristem [p.480], node [p.480], internode [p.480], cortex [p.481], pith [p.481]

Boldfaced, Page-Referenced Terms

[p.480] terminal bud _____

_____

[p.480] lateral bud _____

_____

[p.481] vascular bundles _____

_____

Name the structures numbered in the following illustrations of the monocot stem. Choose from these terms:

Collenchyma, vessel, vascular bundle, sieve tube, pith, companion, cell epidermis

1. _____

2. _____

3. _____

4. _____ sheath cell

5. _____ in xylem

6. _____ in phloem

7. _____ in phloem

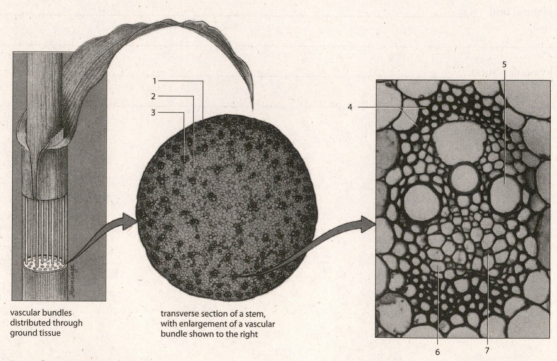

vascular bundles
distributed through
ground tissue

transverse section of a stem,
with enlargement of a vascular
bundle shown to the right

Stem fine structure for corn (*Zea mays*), a monocot

Labeling [p 481]

Name the structures numbered in the following illustrations of the herbaceous eudicot stem. Choose from these terms:

Phloem, vascular bundle, xylem, cortex, vascular cambium, pith, epidermis

8. _____

9. _____

10. _____

11. _____

12. _____

13. What specific layer of meristematic cells is seen here?

_____

14. _____

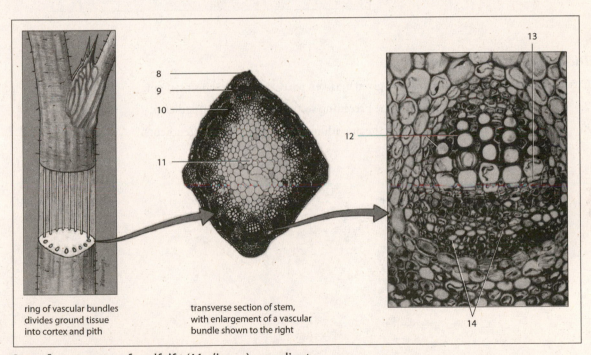

Stem fine structure for alfalfa (*Medicago*), a eudicot

ring of vascular bundles divides ground tissue into cortex and pith

transverse section of stem, with enlargement of a vascular bundle shown to the right

28.4. A CLOSER LOOK AT LEAVES [pp.482-483]

Selected Words: deciduous [p.482], evergreen [p.482], blade [p.482], petiole [p.482], simple leaves [p.482], compound leaves [p.482], leaflet [p.482], cuticle [p.482], coleoptile sheath [p.482], epidermal hairs [p.482], palisade mesophyll [p.483], spongy mesophyll [p.483], plasmodesmata [p.483]

Boldfaced, Page-Referenced Terms

[p.483] veins _____

_____

Labeling [p.482]

Name the structures numbered in the illustrations of leaf development and leaf forms.

1. _____

2. _____

3. _____

4. _____

5. _____

6. _____

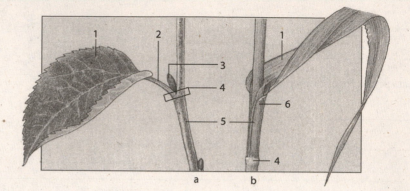

Dichotomous Choice [p.482]

Circle one of the two possible answers given between parentheses in each statement.

7. The leaf illustrated on the left in the preceding section is a (monocot/eudicot).

8. The leaf illustrated on the right in the preceding section is a (monocot/eudicot).

Labeling [p.483]

Identify each numbered part of the accompanying illustration.

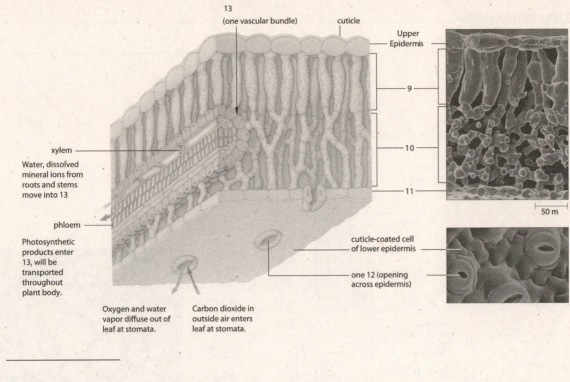

13
(one vascular bundle)

cuticle

Upper
Epidermis

9

10

11

50 m

xylem

Water, dissolved
mineral ions from
roots and stems
move into 13

phloem

Photosynthetic
products enter
13, will be
transported
throughout
plant body.

Oxygen and water
vapor diffuse out of
leaf at stomata.

Carbon dioxide in
outside air enters
leaf at stomata.

cuticle-coated cell
of lower epidermis

one 12 (opening
across epidermis)

9. _____

10. _____

11. _____

12. _____

13. _____

28.5. PRIMARY STRUCTURE OF ROOTS [pp.484-485]

Selected Words: primary root [p.485], root cap [p.485], root apical meristem [p.485], pericycle [p.485], lateral root [p.485], endodermis [p.485], adventitious roots [p.485]

Boldfaced, Page-Referenced Terms

[p.485] root hairs _____

_____

[p.485] vascular cylinder _____

_____

[p.485] taproot system _____

_____

[p.485] fibrous root system _____

_____

Labeling [p.484]

Identify each numbered part of the illustrations.

1. _____

2. _____

3. _____

4. _____

5. _____

6. _____

7. _____

8. _____

9. _____

10. _____

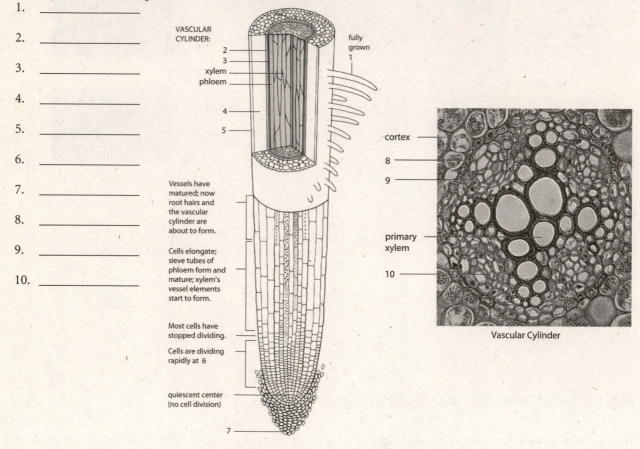

VASCULAR CYLINDER:

2
3
xylem
phloem

fully grown
1

4

5

Vessels have matured; now root hairs and the vascular cylinder are about to form.

Cells elongate; sieve tubes of phloem form and mature; xylem's vessel elements start to form.

Most cells have stopped dividing.

Cells are dividing rapidly at 6

quiescent center (no cell division)

7

cortex

8

9

primary xylem

10

Vascular Cylinder

Fill-in-the-Blanks [pp.484-485]

When a seed germinates, the first structure to emerge is usually the (11) _____. At its tip, a protective

layer, the (12) _____, forms. To increase absorption, some epidermal cells send out extensions called (13)

_____, which greatly increase the (14) _____ available for absorption. Water, minerals, and food are

transported through a large central (15) _____. Cells of one of its layers, the (16) _____, divide

perpendicular to the root axis to produce (17) _____ that branch off the original root. The root system of most

eudicots is a(n) (18) _____ system, whereas monocots usually have a(n) (19) _____ root system.

28.6. SECONDARY GROWTH [pp.486-487]

28.7. TREE RINGS AND OLD SECRETS [p.488]

Selected Words: secondary growth [p.486], lateral meristem [p.486], primary xylem [p.486], secondary xylem [p.486], primary phloem [p.486], secondary phloem [p.486], rays [p.486], suberin [p.487], early wood [p.487], late wood [p.487], growth ring [p.487], hardwood [p.487], softwood [p.487]

Boldfaced, Page-Referenced Terms

[p.486] vascular cambium _____

_____

[p.486] wood _____

_____

[p.487] cork cambium _____

_____

[p.487] periderm _____

_____

[p.487] bark _____

_____

[p.487] cork _____

_____

[p.487] heartwood _____

_____

[p.487] sapwood _____

_____

[p.487] sap _____

_____

Matching [pp.486-487]

Choose the most appropriate answer for each term.

1. _____ tree rings
2. _____ heartwood
3. _____ sapwood
4. _____ early wood
5. _____ suberin
6. _____ softwood
7. _____ bark
8. _____ cork
9. _____ late wood
10. _____ hardwood
11. _____ sap
12. _____ vascular cambium

A. Produced by the cork cambium
B. Includes all tissues external to the vascular cambium
C. Alternating bands of early and late wood; reveal secondary growth patterns of xylem in response to changes in climate
D. Responsible for the production of secondary phloem and xylem
E. The first xylem cells produced in response to moisture at the start of the growing season; tend to have large diameters and thin walls
F. Refers to conifers that possess tracheids but lack fibers and vessels in their xylem
G. Refers to the wood of eudicot trees that evolved in temperate and tropical regions; possess vessels, tracheids, and fibers in their xylem
H. Formed in response to dry seasons or summer, xylem cells with smaller diameters and thicker walls
I. A fluid containing sugars that travels through secondary xylem from roots to buds in spring
J. Compressed, non-conducting xylem cells at the center of older stems and roots; collects metabolic wastes such as resins, oils, gums, and tannins; often dark in color
K. Secondary growth located between hardwood and the vascular cambium; wet, usually pale; contains conducting xylem
L. A fatty substance in the cell walls of cork cells

Short Answers [p.487]

13. Compare and contrast heartwood and sapwood.

_____

_____

14. Which tissues make up the bark of a tree?

_____

_____

28.8. MODIFIED STEMS [p.489]

Selected Words: stolons [p.489], runners [p.489], rhizomes [p.489], bulbs [p.489], scales [p.489], corms [p.489], basal plate [p.489], tubers [p.489], cladodes [p.489]

Matching [p.489]

Choose the stem modification that matches each description.

1. _____ corm
2. _____ tuber
3. _____ bulb
4. _____ cladode
5. _____ stolon
6. _____ rhizome

A. A flattened photosynthetic stem that stores water

B. An underground stem with overlapping modified leaves called scales

C. A fleshy, scaly underground stem with nodes that form aboveground structures

D. Runners that branch from a main stem; roots and shoots sprout from the nodes

E. Solid thickened underground stem; roots grow from a basal plate

F. Thickened portion of underground stolon; has nodes; used for nutrient storage

---

SELF-TEST

___ 1. _____covers and protects the plant's surfaces. [p.476]
   a. Ground tissue
   b. Dermal tissue
   c. Vascular tissue
   d. Pericycle

___ 2. Which of the following is not considered a type of simple tissue? [p.478]
   a. Epidermis
   b. Parenchyma
   c. Collenchyma
   d. Sclerenchyma

___ 3. Of the following cell types, which one does not appear in vascular tissues? [pp.478-479]
   a. Vessel members
   b. Cork cells
   c. Tracheids
   d. Sieve-tube members
   e. Companion cells

___ 4. The _____is a leaflike structure that is part of the embryo; monocot embryos have one, eudicot embryos have two. [p.476]
   a. shoot tip
   b. root tip
   c. cotyledon
   d. apical meristem

___ 5. Each part of the stem where one or more leaves are attached is a(n) _____. [p.480]
   a. node
   b. internode
   c. vascular bundle
   d. cotyledon

___ 6. Which of the following structures is not considered meristematic? [pp.477, 486-487]
   a. Vascular cambium
   b. Apical meristem
   c. Cork cambium
   d. Endodermis

7. New plants grow and older plant parts lengthen through cell divisions at _____ meristems present at root and shoot tips; older roots and stems of woody plants increase in diameter through cell divisions at _____ meristems. [p. 477]
   a. lateral; lateral
   b. lateral; apical
   c. apical; apical
   d. apical; lateral

8. Vascular bundles called _____ form a network through a leaf blade. [p. 483]
   a. xylem
   b. phloem
   c. veins
   d. cuticles
   e. vessels

9. A primary root and its lateral branchings represent a(n) _____ system. [p. 485]
   a. lateral root
   b. adventitious root
   c. taproot
   d. prop root

10. The _____ layer of a root divides to produce lateral roots. [p.485]
    a. endodermis
    b. pericycle
    c. xylem
    d. cortex
    e. phloem

11. Primary growth is _____ to the main axis of the plant body; secondary growth is _____ to the main axis of the plant body. [p.477, 485]
    a. parallel; parallel
    b. perpendicular; perpendicular
    c. parallel; perpendicular
    d. perpendicular; parallel

12. Which of the following is not a Eudicot characteristic? [p.477]
    a. 2 cotyledons
    b. Flower parts in 3's or multiples
    c. Netlike veins
    d. Vascular bundles organized in a ring in the stem

13. Photosynthetic mesophyll cells are a type of [p.482]
    a. sclerenchyma
    b. collenchyma
    c. epidermis
    d. parenchyma

CHAPTER OBJECTIVES/REVIEW QUESTIONS

1. The aboveground parts of flowering plants are called _____; the plants descending parts are called _____. [p.476]
2. Distinguish among the ground tissue system, the vascular tissue system, and the dermal tissue system. [p.476]
3. Plants grow at localized regions of self-perpetuating embryonic cells called _____. [p.476]
4. Visually identify and generally describe the simple tissues called parenchyma, collenchyma, and sclerenchyma. [pp.478-479]
5. _____ tissue conducts soil water and dissolved minerals, and it mechanically supports the plant. [p.478]
6. _____ tissue transports sugars and other solutes. [p.479]
7. Name and describe the functions of the conducting cells in xylem and phloem. [pp.478-479]
8. What is the general function of the guard cells and stomata found within the epidermis of young stems and leaves? [p.479]
9. Visually distinguish between monocot stems and eudicot stems, as seen in cross section. [p.481]
10. How does the "simple" leaf type differ from "compound" leaves? [p.482]
11. Describe the structure (cells and layers) and major functions of leaf epidermis, mesophyll, and vein tissue. [pp.482-483]
12. How does a taproot system differ from a fibrous root system? [p.485]
13. Define the term adventitious root. [p.485]
14. Describe the origin and function of root hairs. [p.485]
15. Describe the relationship between apical and lateral meristems and primary and secondary growth. [pp.477, 486-487]
16. Explain the origin of the annual growth layers (tree rings) seen in a cross section of a tree trunk. [pp.486-487]

INTEGRATING AND APPLYING KEY CONCEPTS

1. Many plants can be asexually reproduced through stem cuttings. What plant processes must occur for this to work?
2. Consider how the following plant pathogens could affect the structure and function of plant organs and tissues. (1) a fungus that enters the leaf through the stoma and grows in the leaf air spaces, (2) a worm that burrows into a root and lives there, (3) a bug that lays its eggs beneath the epidermis of a stem where the larvae hatch and grow.

29

PLANT NUTRITION AND TRANSPORT

INTRODUCTION

This chapter introduces plant physiology. Plants, like other organisms, must regulate the exchange of materials with the external environment as well as within the organism itself. In this chapter you will examine how cells regulate the movement of both water and organic materials. The chapter focuses on the structure and function of the vascular tissue of plants—the xylem and phloem. It also introduces some important mutualistic relationships that enhance the plant's ability to exchange materials within the roots. In addition, you will learn about the adaptations plants have evolved to prevent water loss, inorganic nutrients they require, and how the soil in which a plant grows affects its life functions.

FOCAL POINTS

- Table 29.1 [p.494] provides a summary of plant nutrients and deficiency symptoms.
- Figure 29.4 [p.496] illustrates various root adaptations.
- Figure 29.5 [p.497] (animated) demonstrates how plants regulate water movement through root tissues and into the xylem.
- Figure 29.7 [p. 499] (animated) illustrates the cohesion-tension theory, essential to understanding the movement of water through the plant's xylem.
- Figure 29.8 [p.500] is a visual representation of the relationship between the stomatal opening and the physiological condition of the guard cells.
- Figure 29.10 [p.502] (animated) demonstrates the pressure-flow theory of sugar movement through the phloem of a plant from a source (site of photosynthesis) to a sink (area of storage or usage).

INTERACTIVE EXERCISES

Impacts, Issues: Leafy Clean-Up Crews [p.492]

29.1. PLANT NUTRIENTS AND AVAILABILITY IN SOIL [pp.494-495]

Selected Words: ground water [p.492], trichloroethylene [p.492], phytoremediation [p.492], plant physiology [p.492], macronutrients [p.494], micronutrients [p. 494], deficiency [p.494], sand [p.494], silt [p.494], clay [p.494], soil horizon [p.495], bedrock [p.495]

Boldfaced, Page-Referenced Terms

[p.494] nutrient _____

[p.494] soil _____

_____

[p.494] humus _____

_____

[p.494] loams _____

_____

[p.495] topsoil _____

_____

[p.495] leaching _____

_____

[p.495] soil erosion _____

_____

Matching [pp.494-495]

Choose the most appropriate statement for each term.

1. _____ soil
2. _____ humus
3. _____ loams
4. _____ phytoremediation
5. _____ topsoil
6. _____ nutrients
7. _____ macronutrients
8. _____ micronutrients
9. _____ leaching
10. _____ erosion

A. Element or molecule essential for an organism's growth and survival

B. The use of plants to clean contaminants from soil by degrading or concentrating them in their tissues

C. The decomposing organic material in soil

D. The loss of soil under the force of wind, running water, and ice

E. Chemical elements essential for plant growth necessary in only trace amounts

F. Consists of particles of minerals mixed with variable amounts of decomposing organic material

G. The removal of some of the nutrients in soil as water passes through it

H. Uppermost part of the soil, rich in organic matter (a horizon); zone of densest root growth

I. Soils having more or less equal proportions of sand, silt, and clay; best for plant growth

J. Essential elements that are required in amounts greater than 0.5% of the plant's dry weight

Short Answer

11. What role do clay, sand, and silt particles play in the nutrition of plants?

   _____

   _____

12. Compare and contrast the processes of leaching and erosion.

   _____

   _____

29.2. HOW DO ROOTS ABSORB WATER AND NUTRIENTS? [pp.496-497]

Selected Words: mutualism [p.496], fungal hyphae [p.496], legumes [p.496], nitrogen [p.496], nitrate [p.496], ammonia [p.496], Rhizobium [p.496], vascular cylinder [p.497], endodermis [p.497]

Boldfaced, Page-Referenced Terms

[p.496] root hairs _____

_____

[p.496] mycorrhiza (pl. mycorrhizae) _____

_____

[p.496] nitrogen fixation_____

_____

[p.496] root nodules_____

_____

[p.497] Casparian strip _____

_____

[p.497] exodermis_____

_____

Choice [p.496]

Choose the type of specialized absorptive structure best described by each statement.

    a. root hairs   b. mycorrhizae   c. root nodules   d. *Rhizobium*

1. _____ Fungi that live in a mutualistic relationship with roots

2. _____ Penetrate the soil for only a few millimeters; live only a few days

3. _____ Swollen masses of bacteria-infected root cells

4. _____ Bacteria that live in a mutualistic relationship with roots

5. _____ Thin extensions of root epidermal cells that increase surface area

6. _____ Responsible for nitrogen fixation in root nodules

7. _____ Organism that shares minerals it absorbs from the soil with plant roots

8. _____ Organisms that convert nitrogen gas to ammonia

Sequence [p.497]

Put the following steps of water absorption in correct sequential order by placing a 1 in the appropriate blank for the first step, a 2 for the second step, etc.

9. _____ Water passes through and around the parenchyma cells of the cortex

10. _____ Water is taken up by the xylem cells

11. _____ Water crosses the epidermal cell layer

12. _____ Transport proteins selectively move water through the endodermal cells

Labeling [p.497]

Label the parts of the vascular cylinder seen in the following diagram.

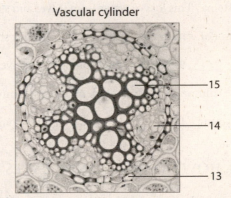

Vascular cylinder

13.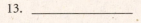

14. _____

15. _____

16. What is the name of the waterproof layer that seals the spaces between cells in the endodermis?

    _____

29.3. HOW DOES WATER MOVE THROUGH PLANTS? [pp.498-499]

Selected Words: evaporation [p.498], cohesion [p.498], tension [p.498], negative pressure [p.498], hydrogen bonds [p.498]

Boldfaced, Page-Referenced Terms

[p.498] tracheids _____

_____

[p.498] vessel members _____

_____

[p.498] cohesion–tension theory _____

_____

[p.498] transpiration _____

_____

Fill-in-the-Blanks [p. 498]

By the (1) _____ theory, water inside the (2) _____ is pulled upward by air's drying power, which creates a continuous negative pressure called (3) _____. The tension extends all the way from leaves to (4) _____. First, air's drying power causes (5) _____: the evaporation of water from aboveground plant parts. Most of the water a plant takes up is lost due to transpiration through (6) _____ on the plant's leaves and stems. This loss of water from leaves and stems causes a (7) _____ that exerts a pull on the continuous column of water in the xylem. The collective strength of many (8) _____ among water molecules imparts (9) _____ to liquid water. Thus, the negative pressure created by transpiration exerts (10) _____ on the entire column of water in a xylem tube extending from (11) _____ down through (12) _____ and on into young (13) _____ where water is pulled in from the (14) _____.

29.4. HOW DO STEMS AND LEAVES CONSERVE WATER? [pp.500-501]

Selected Words: cuticle [p.500], stomata [p.500], translucent [p.500], osmotic pressure [p.501], abscisic acid [p.501], CAM plant [p.501]

Boldfaced, Page-Referenced Terms

[p.500] guard cells _____

_____

Matching [pp.500-501]

Select the term that best matches the lettered description.

1. _____ their stomata open at night to conserve water loss

2. _____ a hormone released by root cells to signal a scarcity of soil water

3. _____ an ion pumped into cells to cause water to enter the cell by osmosis

4. _____ water availability, CO_2 level, light intensity affect stomatal opening

5. _____ a gap between guard cells in the epidermis

6. _____ a water impermeable layer covering the epidermis

7. _____ can temporarily or permanently block stomatal opening

8. _____ specialized epidermal cells that control size of stoma

A. stoma

B. cuticle

C. Guard cell

D. abscisic acid

E. environmental cues

F. potassium

G. CAM plant

H. airborne pollutants

29.5. HOW DO ORGANIC COMPOUNDS MOVE THROUGH PLANTS? [pp.502-503]

Selected Words: phloem [p.502], sieve tubes [p.502], companion cells [p.502], starch [p.502], sucrose [p.502], aphids [p.502], concentration gradients [p.502], turgor [p.503]

Boldfaced, Page-Referenced Terms

[p.502] translocation _____

_____

[p.502] source _____

_____

[p.502] sink _____

_____

[p.503] pressure-flow theory _____

_____

Labeling [p.503]

Provide the missing term for each of the numbered blanks in the following diagram.

1. _____

2. _____

3. _____

4. _____

5. _____

6. _____

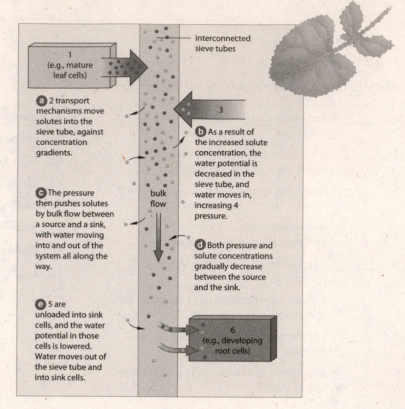

Steps in the translocation process

Matching [pp.502-503]

Choose the most appropriate statement for each term.

7. _____ translocation

8. _____ sieve tube

9. _____ companion cells

10. _____ source

11. _____ sink

12. _____ pressure-flow theory

A. Any region where organic compounds are being loaded into the sieve tubes

B. Cells adjacent to sieve-tube members that supply energy to load organic compounds at the source

C. Any region of the plant where organic compounds are being unloaded from the sieve-tube system to be used or stored.

D. Process that move sucrose and organic compounds through phloem of vascular plants

E. Internal pressure builds up at the source end of a sieve-tube system and pushes the solute-rich solution toward a sink, where solutes and water are removed

F. Living cells with porous ends; found in phloem

SELF-TEST

1. The use of plant species to remove toxic chemicals from soils is called_____ [p.492]
   a. transpiration
   b. nitrogen fixation
   c. phytoremediation
   d. translocation

2. Plant macronutrients are the nine dissolved mineral ions that _____. [p.494]
   a. are found in major concentrations in the earth's surface
   b. occur in only small traces in plant tissues
   c. are required in amounts above 0.5 percent of the plant's dry weight
   d. can function only without the presence of micronutrients

3. Soils that have the best oxygen and water penetration are called _____. [p.494]
   a. topsoils
   b. humus
   c. clays
   d. loams

4. The greatest amount of organic matter is found in which of the following soil layers? [pp.494-495]
   a. O Horizon
   b. A Horizon
   c. B Horizon
   d. C Horizon

5. Gaseous nitrogen is converted to a plant-usable form by _____. [p.496]
   a. root nodules
   b. mycorrhizae
   c. nitrogen-fixing bacteria
   d. sunlight and carbon dioxide

6. _____ prevent(s) inward-moving water from moving past the abutting walls of the root endodermal cells. [p.497]
   a. Cytoplasm
   b. Plasma membranes
   c. Osmosis
   d. Casparian strips

7. Most of the water moving into a leaf is lost through _____. [p.500]
   a. osmotic gradients being established
   b. transpiration
   c. pressure flow forces
   d. translocation

8. Mycorrhizae represent mutualistic relationships between roots and _____. [p.496]
   a. bacteria
   b. fungi
   c. animals
   d. all of the above

9. Without _____, plants would rapidly wilt and die during hot, dry spells. [p.500]
   a. a cuticle
   b. mycorrhizae
   c. phloem
   d. cotyledons

10. The _____ theory of water transport states that hydrogen bonding allows water molecules to maintain a continuous fluid column as water is pulled from roots to leaves. [p. 498]
    a. pressure-flow
    b. cohesion–tension
    c. evaporation
    d. nitrogen-fixation

11. In the pressure-flow theory, leaves are _____ regions while growing leaves, stems, fruits, seeds, and roots are _____ regions. [pp.502-503]
    a. source; source
    b. sink; source
    c. source; sink
    d. sink; sink

12. Which of the following is not involved in water conservation? [pp.500-501]
    a. Cuticle
    b. Guard cell
    c. Abscisic acid
    d. Companion cell

CHAPTER OBJECTIVES/REVIEW QUESTIONS

1. Explain the benefits of phytoremediation. [p.492]
2. Distinguish between macronutrients and micronutrients in relation to their role in plant nutrition. [p.494]
3. Define the terms soil, humus, and loam. [p.494]
4. Distinguish between leaching and erosion. [p.495]
5. Describe the roles of root nodules and mycorrhizae in plant nutrition. [p.496]
6. Define nitrogen fixation and explain what is involved in this process. [p.496]
7. Explain why root hairs are so valuable in root absorption. [p.496]
8. Explain the role of the Casparian strip in regulating water movement. [p.497]
9. Differentiate between the endodermis and the exodermis of the root cortex. [p.497]
10. Define transpiration. [p.498]
11. Use the cohesion–tension theory to describe how water moves in a plant. [pp.498-499]
12. Explain the role of the cuticle and stomata in regulating movements of materials into the plant. [pp.500-501]
13. Describe the role of phloem sieve-tube members and companion cells in the movement of organic compounds. [p.502]
14. Define the role of sinks and sources in translocation. [pp.502-503]
15. Explain the general principles of the pressure-flow theory. [p.503]

INTEGRATING AND APPLYING KEY CONCEPTS

1. How do you think maple syrup is made from maple trees? Which specific systems of the plant are involved, and why are maple trees tapped only at certain times of the year?
2. What might be the result of stripping vegetation from the soil and exposing it to elements of climate? What role do you think ground cover (plants that are short with thick root systems) plays in preventing soil erosion?
3. As noted in the chapter, cuticles can actually prevent the movement of gases in the direction that the plant prefers. What balance must the plant obtain between the amount of cuticle and gas transport? What factors are at work? In what types of environment would you expect to find plants with thick cuticles?
4. Many desert cacti have cladode stem modifications [refer to chapter 28.8, p.489] and leaves reduced to spines, and are CAM plants. How do you think these adaptations help cacti succeed in the extremes of desert climate?
5. Review the cohesion-tension theory and explain why the introduction of an air bubble into a cut rose stem will cause it to die quickly.

30

PLANT REPRODUCTION

INTRODUCTION

The ability of flowering plants to reproduce both asexually (vegetatively) and sexually has helped them become very successful inhabitants of this planet. Central to grasping the success of flowering plants is the understanding of floral and seed structure along with the various adaptations for pollination and seed dispersal that have coevolved in relation with members of the animal kingdom. This chapter follows the sexual reproductive process from alternation of gametophyte and sporophyte generations, through floral and pollinator co-evolution, pollination, the unique angiosperm process of double fertilization, to seed formation, fruiting, and fruit adaptations. It also discusses various types of vegetative reproduction.

FOCAL POINTS

- The introduction [p.506], Section 30.2 [pp. 510–511], and Table 30.1 [p. 511] focus on the important concept of coevolution of flowers and their pollinators.
- Figure 30.2 [p.508] (animated) illustrates the structure of the reproductive shoot known as a flower.
- Figure 30.3 [p.509] (animated) summarizes the life cycle of flowering plants.
- Figure 30.8 [pp.512-511] (animated) shows the life cycle of the cherry plant from the development of male and female gametophytes through pollination and fertilization.
- Figure 30.10 [p.515] (animated) demonstrates the transition of an ovule to a mature seed.
- Figures 30.12 and 30.13 [pp.516-517] illustrate the structure and function of various fruit adaptations.
- Table 30.2 [p.517] provides a synopsis of the classification of fruit.
- Section 30.7 describes the ability of flowering plants to populate their environment through asexual reproduction.

INTERACTIVE EXERCISES

Impacts, Issues: Plight of the Honeybee [p. 506]

30.1. REPRODUCTIVE STRUCTURES OF FLOWERING PLANTS [pp.508-509]

Selected Words: colony collapse disorder [p.506], hive [p.506], Israeli paralysis virus [p.506], insecticides [p.506], pesticides [p.506], coevolved [p.506], whorl [p.508], sepals [p.508], calyx [p.508], petals [p.508], corolla [p.508], pollinator [p.508], anther [p.508], filament [p.508], pollen sacs [p.508], stigma [p.508], style [p.508], pistil [p.508], modified branch [p.508], receptacle [p.508], perfect flowers [p.509], imperfect flowers [p.509], zygote [p.509], seed [p.509], regular flowers [p.509], irregular flowers [p.509], inflorescence [p.509], complete flower [p.509], incomplete flower [p.509], self-pollination [p.509], cross-pollination [p.509]

Boldfaced, Page-Referenced Terms

[p.508] sporophyte _____

_____

[p.508] gametophytes _____

_____

[p.508] stamens _____

_____

[p.508] pollen grains_____

_____

[p.508] carpels _____

_____

[p.509] ovary_____

_____

[p.509] ovule_____

_____

Labeling [p.509]

Identify each numbered part of the accompanying illustration.

1. _____

2. _____

3. _____

4. _____

5. _____

6. _____

7. _____

8. _____

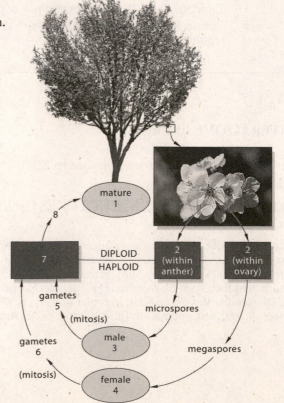

Short Answer [p.509]

9. Compare and contrast regular and irregular flowers.

_____

_____

10. Compare and contrast complete and incomplete flowers.

_____

_____

11. What is a perfect flower?

_____

_____

Labeling [p.508]

Identify each numbered part of the following diagram of the angiosperm flower.

12. _____

13. _____

14. _____

15. _____

16. _____

17. _____

18. _____

19. _____

20. _____

21. Taken together, #s 12 and 13 are the

_____

22. Taken together, #s 14, 15, and 16 are called the

_____ or the

_____

23. A whorl of #19 is called the

_____

24. A whorl of #20 is called the

_____

stamen
(male reproductive part)

carpel
(female reproductive part)

13
12
14
15
16
20
(all combined are
the flower's corolla)
19
(all combined are
the flower's calyx)
17
(forms
within
ovary)
18

30.2. FLOWERS AND THEIR POLLINATORS [pp.510-511]

Selected Words: wind pollination [p.510], animal pollinators [p.510], fragrance [p.510], UV-reflecting pigments [p.510], airborne chemicals [p.510], pollen [p.510], butterfly [p.510], hummingbird [p.510], moth [p.510], floral tube [p.510]

Boldfaced, Page-Referenced Terms

[p.510] pollination vector _____

_____

[p.510] pollinators _____

_____

[p.510] nectar _____

Fill-in-the-Blanks [pp.510-511]

In order to ensure pollination, many plants have adaptations to attract (1) _____ to transfer (2) _____ among individuals of the same species. (3) _____, (4) _____, and (5) _____ are common pollination vectors. About 90% of flowering plants have (6) _____ animal pollinators. Yellow, blue, and purple pigments attract (7) _____, as do pigments that reflect (8) _____ light.   Plants that release fruity or musky odors at night attract (9) _____ and (10) _____, and the odors of rotting flesh and dung attract (11) _____ and (12) _____.   Pollen and (13) _____ are nutritious rewards that attract pollinators. Honeybees convert nectar to (14) _____. Many flowers have specializations that exclude non-pollinators. An ample supply of nectar deeply hidden at the bottom of a long (15) _____ may be reached only by specific pollinators such as (16) _____, (17) _____, and (18) _____ that have matching feeding devices.

30.3. A NEW GENERATION BEGINS [pp. 512-513]

30.4. FLOWER SEX [p.514]

Selected Words: pollen sac [p.512], meiosis [p.512], pollen grain [p.512], mitosis [p.512], embryo sac [p.512], germinate [p.512], pollen tube [p.512], sperm cells [p.512], zygote [p.513], sperm nuclei [p.513], triploid [p.513], sporopollenin [p.514], cell signaling [p.514], furrows [p.514], pores [p.514]

Boldfaced, Page-Referenced Terms

[p.512] microspores _____

_____

[p.512] megaspores _____

_____

[p.512] endosperm mother cell _____

_____

[p.512] pollination _____

_____

[p.512] double fertilization _____

_____

[p.513] endosperm _____

_____

Sequence [p.513]

Place the numbers 1-6 in the space before each statement to indicate its proper place in the life cycle of the flowering plant from production of the female gametophyte to seed formation.

1. _____ Germination of the pollen tube

2. _____ Megaspores form by meiosis

3. _____ Seed consists of maternal integuments, a diploid embryo, and triploid endosperm

4. _____ One sperm nucleus fertilizes the egg, a second fuses with the endosperm mother cell

5. _____ An ovule forms within the ovary of the sporophyte

6. _____ Repeated mitoses of the megaspore form the embryo sac

Dichotomous Choice [pp.512-513]

Circle one of the two possible answers between the parentheses to complete each statement.

7. The gametophyte stage is (haploid/diploid).

8. The sporophyte stage is (haploid/diploid).

9. The megaspore is (haploid/diploid).

10. The microspore is (haploid/diploid).

11. Nuclei within a pollen grain are (haploid/diploid).

12. The embryo is (haploid/diploid).

Sequence [p.512]

Place numbers 1-6 in the space before each statement to indicate its proper place in the life cycle of the male gametophyte of a flowering plant.

13. _____   A pollen grain lands on the stigma

14. _____   Meiosis produces four haploid microspores

15. _____   A pollen tube germinates from the pollen grain

16. _____   Pollen sacs form in the anther of the sporophyte flower

17. _____   Mitosis of a microspore followed by cellular differentiation form a pollen grain

18. _____   A pollen grain is released from the anther

Fill-in-the-Blanks [p.514]

The outer layer of the coat of a pollen grain is made of (19) _____, a very durable substance. Sex in plants begins when (20) _____ on the epidermal cells of the stigma bind to molecules on the coat of the pollen grain. The pollen grain secretes lipids and proteins which bind to (21) _____ in the stigma cell membranes. Pollen is dry and its cells are (22) _____ until they receive nutrient-rich fluid from the stigma. The cells of the male gametophyte (pollen) then resume metabolism and grow a (23) _____ out of one of the (24) _____ or (25) _____ in the pollen's coat. Gradients of nutrients direct the growth of the pollen tube down through the (26) _____. The female gametophyte secretes (27) _____ to guide the pollen tube to the (28) _____. These signals are (29) _____ so that pollen tubes of different species may grow down the style but not be able to reach and (30) _____ the egg.

30.5. SEED FORMATION [p.515]

30.6. FRUITS [pp.516-517]

Selected Words: endosperm [p.515], embryo sporophyte [p.515], dormancy [p.515], grains [p.515], germ [p.515], bran [p.515], milling [p.515], seed dispersal [p.516], dispersal vectors [p.516], wings [p.516], air bladders [p.516], pods [p.516], spines [p.516], fleshy fruits [p.517], dry fruits [p.517], simple fruit [p.517], aggregate fruit [p.517], multiple fruit [p.517], true fruit [p.517], accessory fruit [p.517], dehiscent fruit [p.517], indehiscent fruit [p.517], drupe [p.517], berry [p.517], pome [p.517], pepo [p.517], hesperidium [p.517]

Boldfaced, Page-Referenced Terms

[p.515] seed _____

_____

[p.516] fruit _____

_____

Labeling [p.515]

Identify each numbered part of the accompanying illustration of a developing seed.

1. _____

2. _____

3. _____

4. _____

5. _____

embryo's 1
embryo's 2
5
3
4

Choice [p.515]

Choose the part of the seed that contains each substance.

a. seed coat   b. embryo   c. endosperm

6. _____ Vitamins

7. _____ Minerals

8. _____ Protein

9. _____ Fiber

10. _____ Starch

Short Answer [p.517]

11. Differentiate among simple fruits, aggregate fruits, and multiple fruits. Give an example of each.

_____

_____

_____

12. Differentiate between true fruits and accessory fruits. Give an example of each.

_____

_____

_____

Matching [p.517]

Match the fruit category with its description. Complete the exercise by inserting an example of each fruit type in the parentheses.

13. _____ dehiscent (_____)

14. _____ indehiscent (_____)

15. _____ drupe (_____)

16. _____ berry (_____)

17. _____ pome (_____)

A. Fleshy fruit with a hard pit surrounding the seed

B. Fleshy fruit; seeds in a core derived from the ovary; core is surrounded by fleshy tissue derived from the receptacle

C. Dry fruit wall splits along a seam to release the seeds

D. Fleshy fruit formed from a compound ovary; usually has many seeds; no pit

E. Dry fruit; wall does not split open; seeds are dispersed inside intact fruits

Choice [p.516]

Choose the type of seed dispersal that best fits the adaptation listed.

    a. wind-dispersed fruits   b. fruits dispersed by animals   c. water-dispersed fruits

18. _____ Pods that pop open ejecting the seeds

19. _____ Thick-husked fruits of coconut palms

20. _____ Seed coats abraded by digestive enzymes assist in seed germination

21. _____ Wings of maple fruits

22. _____ Air bladders of sedges

23. _____ Red fleshy fruits of crabapples

24. _____ Hairy modified sepals of dandelions, milkweed, and thistle

25. _____ Spines and hooks of cockleburs and bur clover

30.7. ASEXUAL REPRODUCTION OF FLOWERING PLANTS [pp.518-519]

Selected Words: clone [p.518], quaking aspen (Populus tremuloides) [p.518], root suckers [p.518], aneuploid [p.518], cuttings [p.518], somatic cell [p.519], grafting [pp.518-519], seedless fruit [p.519], triploid fruit [p.519], polyploidy [p.519], colchicine [p.519], tetraploid [p.519]

Boldfaced, Page-Referenced Terms

[p.518] vegetative reproduction _____

_____

[p.519] tissue culture propagation _____

_____

Matching [pp.518-519]

Match the following terms associated with asexual reproduction in flowering plants with their definitions.

1. _____ root suckers
2. _____ clone
3. _____ Tasmanian King's holly
4. _____ triploid
5. _____ sterile
6. _____ aneuploid
7. _____ tissue culture propagation
8. _____ cutting
9. _____ grafting
10. _____ polyploidy

A. In this laboratory technique a single somatic cell may be induced to form an embryo and then an entire new plant

B. Any organism that cannot reproduce sexually

C. Any multiple of the "n" chromosome number beyond diploid (2n)

D. A group of plants that have arisen from the vegetative reproduction of one ancestor and share genetic identity with that ancestor

E. Shoots that sprout from the lateral roots of a plant

F. A stem fragment of a parent plant that has the capability of forming shoots as a new individual

G. Having the '3n' number of chromosomes in each cell

H. The oldest known living plant, more than 43,000 years old

I. Any number of chromosomes in a cell that is not a perfect multiple of the 'n' number

J. Joining a cutting of one plant to the tissues of another plant, often a stem cutting to a rootstock

SELF-TEST

___ 1. The joint evolution of flowers and their pollinators is known as _____. [pp.510-511]
   a. adaptation
   b. coevolution
   c. joint evolution
   d. covert evolution

___ 2. A stamen is _____. [p.508]
   a. composed of a stigma
   b. the mature male gametophyte
   c. the site where microspores are produced
   d. part of the vegetative phase of an angiosperm

___ 3. The portion of the carpel that contains an ovule is the _____ . [p.508]
   a. stigma
   b. anther
   c. style
   d. ovary

___ 4. The phase in the life cycle of plants that gives rise to spores is known as the _____. [pp.512-513]
   a. gametophyte
   b. embryo
   c. sporophyte
   d. seed

___ 5. A gametophyte is _____. [pp. 512-513]
   a. a gamete-producing plant
   b. haploid
   c. both a and b
   d. the plant produced by the fusion of gametes

___ 6. A characteristic of a seed is that it _____. [p.515]
   a. contains an embryo sporophyte
   b. represents an arrested growth stage
   c. is covered by hardened and thickened integuments
   d. all of these

___ 7. Ovaries become _____ while ovules become _____ . [pp.515- 516]
   a. fruit; megaspores
   b. fruits; seeds
   c. megaspores; fruits
   d. seeds; fruits

___ 8. In flowering plants, one sperm nucleus fuses with that of an egg, and a zygote forms that develops into an embryo. Another sperm fuses with _____. [p.513]
   a. a primary endosperm cell to produce three cells, each with one nucleus
   b. a primary endosperm cell to produce one cell with one diploid nucleus
   c. both nuclei of the endosperm mother cell, forming a primary endosperm cell with a single triploid nucleus
   d. one of the smaller megaspores to produce what will eventually become the seed coat

___ 9. "Simple, aggregate, multiple, and accessory" refer to types of _____. [p.517]
   a. carpels
   b. seeds
   c. fruits
   d. ovaries

___ 10. Calyx is to corolla as _____ is (are) to _____ [p.508]
   a. ovary; ovule
   b. stamens; pistil
   c. petals; sepals
   d. sepals; petals

___ 11. A tulip flower consisting of symmetric green sepals, pink petals, yellow stamens, and orange pistils is _____. [p.509]
   a. complete
   b. perfect
   c. regular
   d. all of the above

___ 12. Which of the following is not considered a valuable pollinator? [pp.510-511]
   a. Monarch butterfly
   b. Hummingbird
   c. Honeybee
   d. Leaf-cutter ants

CHAPTER OBJECTIVES /REVIEW QUESTIONS

1. Describe the role of a pollinator. [pp.506, 510-511]
2. Distinguish between sporophytes and gametophytes. [p.508]
3. Identify the various parts of a typical flower and state their functions. [p.508]
4. Distinguish between a flower that is *perfect* and one that is *imperfect*. [p.509]
5. What structures represent the male gametophyte and the female gametophyte in flowering plants? List the contents of each. [pp.512-513]
6. Describe the double fertilization that occurs uniquely in the flowering plant life cycle. [pp.512-513]
7. Review seed structure and discuss how its design prepared it for survival in the harsh, dry, terrestrial climate. [p.515]
8. Review the general types of fruits produced by flowering plants. [pp.516-517]
9. Describe the various methods of seed dispersal by fruits. [p.516]
10. Discuss polyploidy in plants including the reasons for its existence and consequences to the plant. [pp.518-519]
11. Distinguish between vegetative reproduction and tissue culture propagation, including a discussion of the various types of vegetative reproduction; cite an example for each. [pp.518-519]

INTEGRATING AND APPLYING KEY CONCEPTS

1. In terms of botanical morphology, a flower is interpreted as "a shoot modified for reproductive functions." After studying floral structures in this chapter and shoot/leaf structure in text Chapter 29, can you think of any comparable structural evidence that might have led botanists to arrive at this conclusion?
2. This chapter revealed the intimate interrelationships between flowering plants and animals as pollinators and seed dispersal agents. Consider how circumstances like habitat destruction, the use of pesticides and animal diseases can have grave consequences on the continued success of flowering plants.

31
PLANT DEVELOPMENT

INTRODUCTION

Like all living organisms, plants have mechanisms to address such essential life functions as energy requirements, reproduction, growth, development, and response to stimuli. Previous chapters have discussed energy requirements and reproduction. This chapter introduces the major mechanisms by which plants control and regulate their growth and development and respond to stimuli. Unlike many animals, plants do not have nervous tissue and so must rely solely on the use of hormone messengers and other chemical signaling. From germination of a seed and growth of its embryo to sexual reproduction and death, plants use chemical signaling to control all aspects of their life cycles. In order to gain insight into plant physiology, you should familiarize yourself with the major classes of hormones, their functions, and how horticulturists use hormones to enhance agricultural production.

FOCAL POINTS

- Figure 31.4 (animated) [p.525] is an excellent representation of the germination and early growth of monocot and eudicot seedlings.
- Table 31.1 [p.526] summarizes the various effects, and target tissues, of the five major plant hormone categories.
- Table 31.2 [p.527] shows some of the commercial uses of the five major plant hormone categories.
- Figure 31.7 [p.528] demonstrates the action of gibberellin in seed germination.
- Figure 31.8 (animated) [p.529] demonstrates the effect of auxin on monocot seedling growth.
- Figure 31.11 (animated) [p.530] illustrates the function of statoliths in root tip response to gravity.
- Figure 31.12 (animated) [p.531] demonstrates the effect of auxins on stem growth in response to light.
- Figure 31.15 (animated) [p.532] shows the response of phytochrome to environmental light cues.
- Figure 31.17 (animated) [p.533] illustrates the role of photoperiodism in plant flowering.

INTERACTIVE EXERCISES

Impacts, Issues: Foolish Seedlings, Gorgeous Grapes [p.522]

31.1. PATTERNS OF DEVELOPMENT IN PLANTS [pp.524-525]

Selected Words: foolish seedling [p.522], environmental cues [p.522], desiccation [p.524], dormancy [p.524], radicle [p.524], development [p.524], differentiate [p.524], coleoptile [p.525], plumule [p.525], hypocotyl [p.525]

Boldfaced, Page-Referenced Terms

[p.524] germination _____

_____

[p.524] growth _____

_____

Fill-in-the-Blanks [p.524]

A mature seed contains an embryonic plant complete with a (1) _____ and a (2) _____. The embryo is in a state of temporarily suspended development called (3) _____. (4) _____ is the process by which an embryo resumes metabolic activity and growth. Water seeping into a dormant seed softens tissues and activates (5) _____ that change stored starches into (6) _____ that can be used by (7) _____ cells as energy for cell division. The embryonic (8) _____ is the first part of the plant to break out of the softened (9) _____. In a eudicot, the shoot and cotyledons emerge later. Germination is one of many patterns of development in plants. Development includes (10) _____, an increase in cell number and size, and (11) _____, the formation of specialized tissues behind areas of meristematic growth. Seed dormancy evolved as an adaptation to (12) _____. It allows seeds to germinate at a time favorable to their growth and survival. In addition to water, (13) _____, (14) _____, (15) _____, (16) _____, and (17) _____ are other environmental cues that trigger germination in certain species.

Labeling [p.525]

Identify each numbered part of the accompanying illustration of eudicot seed germination.

18. _____

19. _____

20. _____

21. _____

22. _____

23. _____

24. _____

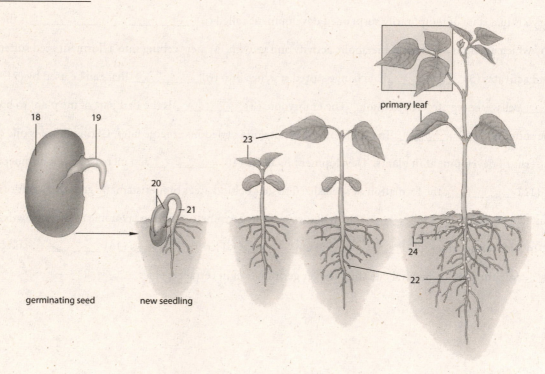

germinating seed   new seedling   primary leaf

31.2. PLANT HORMONES AND OTHER SIGNALING MOLECULES [pp.526-527]

31.3. EXAMPLES OF PLANT HORMONE EFFECTS [pp.528-529]

Selected Words: target cell [p.526], elongation [p.526], inhibitory [p.527], brassinosteroids [p.527], jasmonates [p.527], salicylic acid [p.527], nitric oxide [p.527], systemin [p.527], FT protein [p.527], aleurone [p.528], IAA (indole-3-acetic acid) [p.528], polarity [p.528], proton pumps [p.528], repressor proteins [p.528]

Boldfaced, Page-Referenced Terms

[p.526] plant hormones _____

_____

[p.526] gibberellins _____

_____

[p.527] auxins _____

_____

[p.527] apical dominance _____

_____

[p.527] abscisic acid _____

_____

[p.527] cytokinins _____

_____

[p.527] ethylene _____

_____

Choice [pp.526-527]

For each of the following, choose from the following classes of plant hormones.

    a. auxins   b. gibberellins   c. cytokinins   d. abscisic acid (ABA)   e. ethylene

1. _____ Used to prolong the shelf life of cut flowers
2. _____ Inhibits abscission of leaves and fruits
3. _____ Stimulates fruit ripening
4. _____ Induce cell division and elongation between stem nodes
5. _____ Responsible for apical dominance by inhibiting cell division and elongation in lateral buds
6. _____ Closes stomata; overrides the growth-stimulating effect of other plant hormones
7. _____ Breaks dormancy in seeds and stimulates germination
8. _____ Most common type in nature is indole-3-acetic acid (IAA)
9. _____ In autumn, induces abscission of leaves
10. _____ Used to increase fruit size; delay citrus fruit ripening
11. _____ Induce seedless fruit production
12. _____ Places plants into dormancy to prevent damage during shipping
13. _____ Allows fruit to be shipped green, then ripened at the store
14. _____ Promote root formation in cuttings
15. _____ Induces cell division and differentiation in vascular cambium

Fill-in-the-Blanks [pp.526-528]

Plant (16) _____ are signaling molecules that can stimulate or inhibit plant development. (17) _____ trigger the production and dispersal of plant hormones. Hormones bind to (18) _____ to produce a change in the plant's physiology. The five major types of plant hormones are (19) _____, (20) _____, (21) _____, (22) _____, and (23) _____. During germination, water absorbed by a seed causes the release of (24) _____ which induces transcription of the gene for (25) _____. Stored starch in the (26) _____ is then broken down into sugars that are used to make ATP to fuel rapid cell in the meristem cells. (27) _____ is a member of the auxin family of hormones that has many effects on plant development. It causes cells to expand by increasing the activity of (28) _____ thus softening the cell wall and allowing it to stretch irreversibly. It also causes the degradation of (29) _____ that block transcription of specific genes. IAA is made mainly in (30) _____ and (31) _____ where it is most highly concentrated, and is transported in (32) _____ over long distances from the site of its production. Many plants protect themselves from predatory (33) _____ with thorns or nasty tasting chemicals. Others enlist the help of parasitic (34) _____. Tissue damage caused by a caterpillar chewing on a leaf triggers the synthesis of (35) _____ that activate certain genes. The result is the release of herbivore-specific chemicals into the (36)

_____. These chemicals attract the correct species of wasp to the injured plant where they attack the leaf-munching caterpillar and deposit an (37) _____ inside it that will hatch into a caterpillar-munching (38) _____.

31.4. ADJUSTING THE DIRECTION AND RATES OF GROWTH [pp.530-531]

Selected Words: amyloplasts [p.530], actin microfilaments [p.530], cytoskeleton [p.530], efflux carriers [p.530], blue light [p.531], phototropins [p.531], TOUCH [p.531], mechanical stress [p.531], tendril [p.531]

Boldfaced, Page-Referenced Terms

[p.530] tropisms _____

_____

[p.530] gravitropism _____

_____

[p.530] statoliths _____

_____

[p.531] phototropism _____

_____

[p.531] thigmotropism _____

_____

Choice [pp.530-531]

For each description, choose the most appropriate tropic response.

   a. phototropism   b. gravitropism   c. thigmotropism   d. mechanical stress

1. _____ Growth response made possible by statoliths in plant cells
2. _____ TOUCH genes
3. _____ Growth response to the direction of light
4. _____ Auxin migrates to one side of a stem causing greater cell elongation on that side
5. _____ Growth response to the Earth's gravitational forces
6. _____ Growth response to contact with a solid object
7. _____ Growth response to factors such as prevailing winds and grazing animals
8. _____ Allows roots to grow around rocks in the soil

Fill-in-the-Blanks [pp.530-531]

A (9) _____ is a growth response of a plant to an environmental (10) _____. Tropisms are mediated by (11) _____. No matter how a seed is positioned in the soil, when a seed germinates, the radicle always grows (12) _____ and the primary shoot always grows (13) _____ due to a growth response known as (14) _____. Root cap cells contain starch-filled amyloplasts called (15) _____ that respond to gravity by occupying the side of the cell closest to gravity. Their placement at the bottom of cells causes a redistribution of (16) _____ which results in downward growth of root tips. The effects of (17) _____ cause stems to bend toward the source of light. The mechanism responsible for this involves the hormone (18) _____. Pigments called (19) _____ absorb (20) _____ and use its energy in a cascade of intracellular signals which ultimately redistributes auxin to the (21) _____ side of a shoot. The result is that cells on the shaded side of the stem (22) _____ faster than the cells on the illuminated side. When a vine's (23) _____ touch an object, they begin to curl around it due to the growth response called (24) _____. This is due to a decrease in cell (25) _____ at the point of contact while the cells on the opposite side of the shoot continue to elongate. (26) _____, such as shaking, wind shear, or the action of grazing animals, also inhibits stem elongation at the point of stress.

31.5. SENSING RECURRING ENVIRONMENTAL CHANGES [pp.532-533]

31.6. SENESCENCE AND DORMANCY [p.534]

Selected Words: photoreceptors [p.532], red light [p.532], far-red light [p.532], phototropin [p.532], short-day plants [p.532], long-day plants [p.532, day-neutral plants [p.532], activated phytochrome [p.533], deactivated phytochrome [p.533], biennials [p.533], perennials [p.533], deciduous [p.534], evergreen [p.534], growing season [p.534], abscission zone [p.534], middle lamella [p.534]

Boldfaced, Page-Referenced Terms

[p.532] biological clock _____

_____

[p.532] circadian rhythm _____

_____

[p.532] solar tracking _____

_____

[p.532] phytochromes _____

_____

[p.532] photoperiodism _____

_____

[p.533] vernalization_____

_____

[p.534] abscission _____

_____

[p.534] senescence _____

_____

[p.534] dormancy _____

_____

Matching [p.532-534]

1. _____ biological clock
2. _____ circadian rhythm
3. _____ solar tracking
4. _____ phytochromes
5. _____ photoperiodism
6. _____ short-day plant
7. _____ long-day plant
8. _____ day-neutral plant
9. _____ vernalization
10. _____ abscission
11. _____ dormancy
12. _____ senescence

A. A period of arrested growth that is triggered by, and ended by, environmental cues

B. A cycle of activity that has an approximate duration of 24 hours

C. Plants that flower only when the hours of darkness exceed a critical value

D. Blue green photoreceptor pigments sensitive to red and far-red light

E. The return to warmth following cold winter temperatures; necessary for flowering in some plants

F. Plants that flower regardless of night length

G. An internal mechanism that governs the timing of rhythmic cycles of activity

H. The phase of a plant's life cycle from maturity to the death of plant parts or the entire plant

I. A flower or leaf changing position in response to the changing direction of sunlight throughout the day

J. The process by which a plant sheds its parts in response to environmental cues

K. The plant's response to changes in the length of night relative to the length of day

L. Plants that flower only when the hours of darkness fall below a critical value

Fill-in-the-Blanks [pp.532-534]

Like a mechanical clock, a biological one can be reset. (13) _____ resets the biological clock in

plants by affecting light receptors called (14) _____. (15) _____ light changes phytochrome from an

inactive form (Pr) to an active form (Pfr). (16) _____ light causes phytochrome to revert to its inactive form.

Active phytochromes cause transcription of genes involved in (17) _____ and (18) _____. They also

affect flowering and germination. Many plants only flower at certain times of the year due to (19) _____. (20)

_____ plants flower when the nights are relatively long; (21) _____ plants flower when the nights are

relatively short. Flowering in (22) _____ plants is unaffected by photoperiod. Plants respond to the seasonal

change in photoperiod through the activation and deactivation of (23) _____ that affects many chemical

events in the plant finally leading to flowering. Seasonal variation in climates not near the equator affects other

events in the lives of plants. In a process called (24) _____ plants shed leaves during times of stress. (25)

_____ gas is the hormone that triggers leaf, flower and fruit drop.

SELF-TEST

For questions 1-4, choose from the following answers:
[pp.526-527]
    a. gibberellins
    b. ethylene
    c. abscisic acid
    d. auxins

___ 1. Promoting fruit ripening and abscission of leaves, flowers, and fruits is a function of _____.

___ 2. The effects of apical dominance are caused by _____.

___ 3. Stem lengthening is a result of cell division and elongation stimulated by _____.

___ 4. Closing of stomates is stimulated by _____.

___ 5. _____ is demonstrated by a germinating seed whose primary root always curves down while its primary shoot always curves up. [p.530]
    a. Phototropism
    b. Photoperiodism
    c. Gravitropism
    d. Thigmotropism

___ 6. Light in the _____ part of the visible spectrum is the main stimulus for phototropism. [p.531]
    a. blue
    b. yellow
    c. red
    d. green

___ 7. Plants whose flowers are closed during the day but open at night are exhibiting a _____. [p.532]
    a. growth movement
    b. circadian rhythm
    c. biological clock
    d. both b and c are correct

8. Vernalization is _____. [p.533]
   a. the death of plant parts in response to cold
   b. daily changes in plant orientation toward gravity
   c. flowering in response to warmth following a period of cold
   d. a protein-related disease

9. The phase of the plant life cycle from full maturity to death is called _____. [p.534]
   a. dormancy
   b. vernalization
   c. abscission
   d. senescence

10. Phytochrome is converted to an active form, _____, at sunrise and reverts to an inactive form, _____, at sunset or at night. [p.532]
   a. Pr; Pfr
   b. Pfr; Pfr
   c. Pr; Pr
   d. Pfr; Pr

11. Fruit ripening is induced by _____. [p.527]
   a. ethylene gas
   b. auxins
   c. gibberellins
   d. cytokinins

12. Plants that do not require a specific amount of darkness to flower are called day-_____ plants. [p.532]
   a. long
   b. short
   c. neutral
   d. indifferent

13. The embryonic root is called the _____. [p.524]
   a. hypocotyl
   b. radicle
   c. plumule
   d. coleoptile

14. A coleoptile _____. [pp.524-525]
   a. is found in eudicots
   b. protects the plumule
   c. is found in monocots
   d. both b and c

CHAPTER OBJECTIVES/REVIEW QUESTIONS

1. Define the term *hormone*. [p.526]
2. Describe what is happening inside of a seed during germination. [pp.524-525]
3. Explain the difference between growth and differentiation. [p.524]
4. Describe the general role of each class of plant hormone. [pp.526–527]
5. Understand how plant hormones can be manipulated to aid agriculture. [p.527]
6. Understand the general action of auxins. [p.527]
7. Define *phototropism*, *gravitropism*, and *thigmotropism*, and give examples of each. [pp.530-531]
8. Define the role of statoliths in gravitropism. [p.530]
9. Explain the role of phototropins in phototropism. [p.531]
10. Give an example of how mechanical stress can affect plants. [p.531]
11. Describe the process of photoperiodism as it relates to circadian cycles and biological clocks. [pp.532-533]
12. Describe the action of phytochromes. [p.532]
13. Describe the photoperiodic responses of "long-day," "short-day," and "day-neutral" plants. [p.532]
14. Define *vernalization*. [p.533]
15. Distinguish between abscission and senescence. [p.534]
16. Explain why solar tracking is an example of photoperiodism and not phototropism. [p.532]

INTEGRATING AND APPLYING KEY CONCEPTS

1. An oak tree has grown up in the middle of a forest. A lumber company has just cut down all the surrounding trees except for a narrow strip of woods that includes the oak. How will the oak likely adjust to its changed environment? To what new stresses will it be exposed? Which hormones will most probably be involved in the adjustment?

2. You have been hired by a company in Costa Rica to raise a rare breed of northern plant for distribution. This plant requires 12 hours of continuous dark over a three-week period to flower and form seeds. Explain what facilities you would need to make this possible.

3. As an agricultural consultant, your job is to maximize profits by growing plants as quickly as possible and reducing the amount of loss due to damage or rotting on the way to market. Describe the hormones and other chemicals that you would most likely use in your work and when they would be applied to the plant.

4. You just purchased a property with several greenhouses and want to supply florists with many flowers such as roses, chrysanthemums, carnations, daisies, irises, tulips, daffodils, sunflowers, poinsettias, and lilies. Considering the problem of photoperiodism and flowering, how would you go about designing special set-ups for short-day, long-day, and day-neutral plants? Which plants could be grouped together in each greenhouse?

32

ANIMAL TISSUES AND ORGAN SYSTEMS

INTRODUCTION

Chapter 32 starts with a description of how cells are organized and connected to form organs, organ systems, and multicellular animals. The four types of animal tissues are described: epithelial, connective, muscle, and nerve. Vertebrate organ systems are introduced along with a more detailed description of how tissues are integrated into a specific organ system—the integumentary system.

FOCAL POINTS

- Figure 32.2 [p.540] illustrates how cells are attached to each other to form coherent tissues.
- Figure 32.4 [p.541] shows various types of epithelium.
- Figure 32.5 and 32.6 [pp.542-543] illustrate connective tissues.
- Figure 32.8 [p.544] has images of muscle tissues.
- Figure 32.9 [p.545] shows a typical neuron.
- Figure 32.11 and 32.12 [pp.546-547] describe anatomical terms and outline the major human organ systems.
- Figure 32.13 [p.568] diagrams human skin structure.

INTERACTIVE EXERCISES

Impacts, Issues: Open or Close the Stem Cell Factories? [p.538]

32.1. ORGANIZATION OF ANIMAL BODIES [p.540]

Selected Words: embryonic stem cells [p.538], adult stem cells [p.538], anatomy [p.538], physiology [p.538], cell junctions [p.540], interstitial fluid [p.540], intermediate filaments [p.540]

Boldfaced, Page-Referenced Terms

[p.540] tissue _____

_____

[p.540] tight junction _____

_____

[p.540] adhering junction _____

_____

[p.540] gap junction _____

_____

[p.540] organ system _____

_____

[p.540] homeostasis _____

_____

Fill-in-the-Blanks [pp.538-540]

(1) _____ are cells that are undifferentiated and have the capability to produce any type of cell. You

started as a single call and all of the cells of your body came from (2) _____ stem cells. Repair and replacement

of cells in your body today starts from (3) _____ stem cells. (4) _____ is about how an animal is put

together while (5) _____ is about how animals function. (6) _____ junctions consist of strands of

proteins that help stop substances from leaking across a tissue. (7) _____ junctions cement cells together. (8)

_____ junctions help cells communicate by promoting the rapid transfer of ions and small molecules among

them. (9) _____ junctions in the epithelium of your stomach help prevent a condition called (10) _____

ulcer.

32.2. EPITHELIAL TISSUE [p.541]

Selected Words: squamous [p.541], secretion [p.541], simple epithelium [p.541], stratified epithelium [p.541]

Boldfaced, Page-Referenced Terms

[p.541] basement membrane _____

_____

[p.541] microvilli _____

_____

[p.541] glands _____

_____

[p.541] endocrine glands _____

_____

[p.541] exocrine glands _____

_____

Fill-in-the-Blanks

(1) _____ tissue has a free surface, which faces either a body fluid or the outside environment. (2) _____ has a single layer of cells and functions as a lining for body cavities, ducts, and tubes. (3) _____ has two or more layers and typically functions in protection, as it does in the skin. (4) _____ glands secrete mucus, saliva, earwax, milk, oil, digestive enzymes, and other cell products, These products are usually released onto a free (5) _____ surface through ducts or tubes. (6) _____ glands lack ducts; their products are (7) _____, which are secreted directly into the fluid bathing the gland. Typically, the (8) _____ picks up the hormone molecules and distributes them to target cells elsewhere in the body.

True/False [p.541]

If the statement is true, write a "T" in the blank. If the statement is false, make it correct by changing the underlined word(s) and writing the correct word(s) in the answer blank.

9. _____ Epithelial tissues are named based on the shape of the cells and the number of <u>nuclei</u> in each cell.

10. _____ Simple epithelial tissues have <u>one</u> cell layer.

11. _____ Because individual cells in stratified squamous epithelium are thin, the tissue is best suited for <u>diffusion and osmosis</u>.

12. _____ Columnar cells that are involved in absorption have microvilli on the free edge to increase <u>surface area</u>.

13. _____ Endocrine glands deliver their secretions through <u>ducts</u>.

32.3. CONNECTIVE TISSUES [pp.542-543]

32.4. MUSCLE TISSUES [pp.544-545]

32.5. NERVOUS TISSUES [p.545]

Selected Words: extracellular matrix [p.542], fibroblasts [p.542], collagen [p.542], elastin [p.542], ligaments [p.542], osteocyte [p.543], plasma [p.543], contract [p.544], muscle fibers, [p.544], sarcomere [p.544], striated [p.544], neuroglia [p.565], electrochemical signals [p.545]

Boldfaced, Page-Referenced Terms

[p.542] cartilage _____

_____

[p.543] bone tissue _____

_____

[p. 543] adipose tissue _____

_____

[p.543] blood _____

_____

[p.544] skeletal muscle _____

_____

[p.544] cardiac muscle _____

_____

[p.545] smooth muscle _____

_____

[p.545] neurons _____

_____

[p.545] neuroglial cells _____

_____

Choice [p.542]

For questions 1-8, choose from the following types of soft connective tissue.

    a. loose   b. fibrous, irregular   c. fibrous, regular

1. _____ Contains many fibers, mostly collagen-containing ones, in no particular orientation, and a few fibroblasts
2. _____ Rows of fibroblasts often intervene between the bundles of fibers
3. _____ Has its fibers and cells loosely arranged in a semifluid ground substance
4. _____ Has parallel bundles of many collagen fibers and resists being torn apart
5. _____ Forms protective capsules around organs that do not stretch much
6. _____ Often serves as a support framework for epithelium
7. _____ Found in tendons, which attach skeletal muscle to bones
8. _____ Found in elastic ligaments, which attach bones to each other

Matching [pp.542-543]

Match each of the specialized connective tissues to its definition.

9. _____ adipose tissue

10. _____ blood

11. _____ bone

12. _____ cartilage

A. Chock-full of large gat cells; stores excess energy as fats; richly supplied with blood

B. Intercellular material, solid yet pliable, resists compression; structural models for vertebrate embryo bones; maintains shape of nose, outer ear, and other body parts; cushions joints

C. Derived mainly from connective tissue, has transport functions; circulating within plasma are a great many red blood cells, white blood cells, and platelets

D. The weight-bearing tissue of vertebrate skeletons, which support or protect softer tissues and organs; mineral-hardened with calcium-salt-laden collagen fibers and ground substance; interact with skeletal muscles attached to them

Dichotomous Choice [pp.544-545]

Circle one of the two possible answers given between parentheses in each statement.

13. Contractile cells of (skeletal / smooth) muscle tissue taper at both ends.

14. Walls of the stomach and intestine contain (smooth / skeletal) muscle tissue.

15. The only muscle tissue attached to bones is (skeletal / smooth).

16. (Smooth / Skeletal) muscle cells are bundled together in parallel.

17. "Voluntary" muscle action is associated with (smooth / skeletal) muscle tissue.

18. The term *striated* means (bundled / striped).

19. The function of smooth muscle tissue is to (pump blood / cause movement in the internal organs).

20. Cell junctions fuse together the plasma membranes of (smooth / cardiac) muscle cells.

21. Excitable cells are the (neuroglia / neurons).

22. (Neuroglia / Muscle) cells protect the neurons and support them structurally and metabolically.

23. Different types of (neuroglia / neurons) detect specific stimuli, integrate information, and issue or relay commands for response.

Labeling and Matching [pp.541-545]

Label each of the following illustrations on the next page with one of the following terms for tissue types:

connective, epithelial, muscle, nervous

Complete the exercise by writing all appropriate letters from each of the following groups in the parentheses after each label.

24. _____ (   )

25. _____ (   )

26. _____ (   )

27. _____ (   )

28. _____ (   )

29. _____ (   )

30. _____ (   )

31. _____ (   )

32. _____ (   )

33. _____ (   )

34. _____ (   )

35. _____ (   )

A. Adipose tissue
B. Bone
C. Cardiac muscle
D. Fibrous, regular connective tissue
E. Loose connective tissue
F. Simple columnar epithelium
G. Simple cubodial epithelium
H. Simple squamous epithelium
I. Smooth muscle
J. Skeletal muscle
K. Blood
L. Neurons

a. Absorption
b. Communication by means of electrical signals
c. Energy reserve
d. Contraction for voluntary movements
e. Diffusion
f. Padding
g. Contracts to propel substances along internal passageways; not striated
h. Attaches muscle to bone and bone to bone
i. In vertebrates, provides the strongest internal framework of the organism
j. Elasticity
k. Secretion
l. Pumps circulatory fluid; striated
m. Insulation
n. Transport of nutrients and waste products to and from body cells

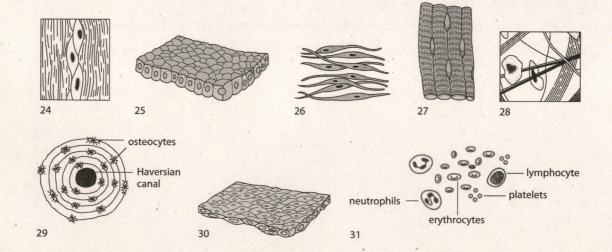

24   25   26   27   28

osteocytes

Haversian canal

29   30   31

lymphocyte
platelets
neutrophils
erythrocytes

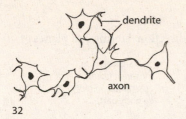

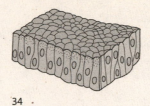
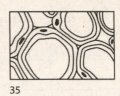

32   33   34   35

32.6. OVERVIEW OF MAJOR ORGAN SYSTEMS [p.546-547]

Selected Words: primary tissue layers [p.546], midsagittal [p.546], dorsal [p.546], ventral [p.546], frontal [p.546], transverse [p.546], posterior [p.546], superior [p.546], inferior [p.546], distal [p.546], proximal [p.546]

Boldfaced, Page-Referenced Terms

[p.546] germ layers _____

_____

[p.546] ectoderm _____

_____

[p.546] mesoderm _____

_____

[p.546] endoderm _____

_____

Matching [p.546]

Match each of the primary germ layers with the adult tissues it forms.

1. _____ ectoderm

2. _____ endoderm

3. _____ mesoderm

A. Forms internal skeleton and muscle, circulatory, reproductive, and urinary systems

B. Forms inner lining of gut and linings of major organs formed from the embryonic gut

C. Forms outer layer of skin and the tissues of the nervous system

Labeling [p.546]

Identify each numbered part of the accompanying illustration that review the directional terms and planes of symmetry for the human body.

4. _____

5. _____

6. _____

7. _____

8. _____

9. _____

10. _____

11. _____

4 (of two body parts, the one closer to head)

5 (farthest from trunk or from point of origin of a body part)

6 (closest to trunk or to point of origin of a body part)

11 plane

midsagittal plane

10 (at or near front of body)

7 (at or near back of body)

8 plane

9 (of two body parts, the one farthest from head)

Labeling and Matching [pp.547-548]

Write the name of each organ system described. Complete the exercise by entering the proper letter form the following illustration in the parentheses after each label.

12. _____ (   ) Rapidly transports many materials to and from cells; helps stabilize internal pH and temperature

13. _____ (   ) Rapidly delivers oxygen tot eh tissue fluid that bathes all living cells; removes carbon dioxide wastes of cells; helps regulate pH

14. _____ (   ) Maintains the volume and composition of internal environment; excretes excess fluid and blood-borne wastes

15. _____ (   ) Supports and protects body parts; provides muscle attachment sites; produces red blood cells; stores calcium, phosphorus

16. _____ (   ) Hormonally controls body function; works with nervous system to integrate short-term and long-term activities

17. _____ (   ) *Female:* produces eggs; after fertilization, affords a protected, nutritive environment for the development of new individual. *Male:* produces and transfers sperm to the female. Hormones of both systems also influence other organ systems

18. _____ (   ) Ingests food and water; mechanically and chemically breaks down food and absorbs small molecules into internal environment; eliminates food residues

19. _____ (   ) Moves body and its internal parts; maintains posture; generates heat (by increases in metabolic activity)

20. _____ (   ) Detects both external and internal stimuli; controls and coordinates responses to stimuli; integrates all organ system activities

21. _____ (   ) Protects body from injury, dehydration, and some pathogens; controls its temperature; excretes some wastes; receives some external stimuli

22. _____ (   ) Collects and returns some tissue fluid to the bloodstream; defends the body against infection and tissue damage

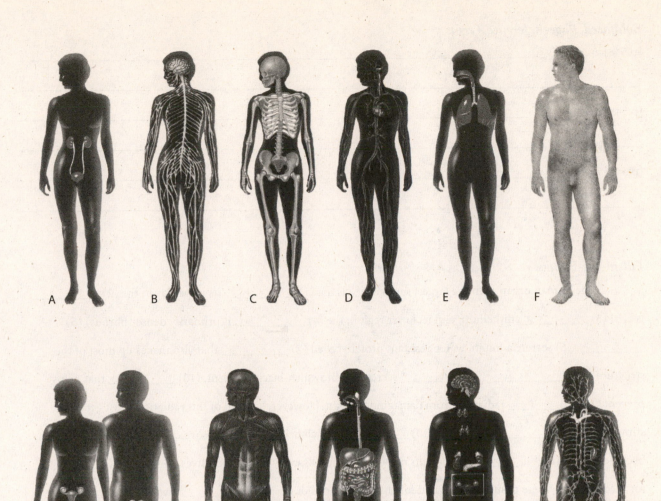

A   B   C   D   E   F

G   H   I   J   K

32.7. VERTEBRATE SKIN—EXAMPLE OF AN ORGAN SYSTEM [pp.548-549]

32.8. FARMING SKIN [p.549]

Selected Words: epidermis [p.548], hypodermis [p.548], ketatinocytes [p.548], keratin [p.548], dendritic cells [p.548], sebaceous glands [p.548], vitamin D [p.549], epidermolysis bollosa [p.549]

Boldfaced, Page-Reference Terms

[p.548] epidermis _____

_____

[p.548] dermis _____

_____

[p.548] melanin _____

_____

Fill-in-the-Blanks

The human organ with the largest surface area is the (1) _____. Its upper layer, the (2) _____, is a(n) (3) _____ epithelium, and its lower region, the (4) _____, is primarily dense (fibrous) (5) _____. (6) _____ secrete a tough, water-resistant protein called (7) _____ that also makes up most of the structure of (8) _____. (9) _____ secrete the brownish-black pigment, (10) _____ that helps prevent (11) _____ radiation from damaging the skin. However, enough of this radiation must get in to stimulate (12) _____ to produce (13) _____, which helps the body absorb calcium. As we age, skin becomes less (14) _____. Since adults make new skin cells every day of their lives, that can be (15) _____ for medical uses. These tissues can be used for chronic wounds or deep (16) _____ where all skin cells have been destroyed.

Short Answer [pp.548-549]

17. Describe how the skin serves as the first line of defense of the immune system. _____

_____

_____

_____

18. Skin tissues grown in a culture lack what structures normally found in the skin? _____

_____

_____

_____

SELF-TEST

____ 1. Which of the following is not one of the connective tissues? [pp.542-543]
   a. bone
   b. blood
   c. cartilage
   d. skeletal muscle

____ 2. Gland cells are contained in _____ tissues. [p.541]
   a. muscular
   b. epithelial
   c. connective
   d. nervous

____ 3. Blood is considered to be a(n) _____ tissue. [p.543]
   a. epithelial
   b. muscular
   c. connective
   d. none of these

____ 4. _____ are abundant in cardiac muscles where they promote diffusion of ions and small molecules from cell to cell. [pp.544-545]
   a. Adhesion junctions
   b. Filter junctions
   c. Gap junctions
   d. Tight junctions

____ 5. Muscle that is not striped and is involuntary is _____. [p.545]
   a. cardia
   b. skeletal
   c. striated
   d. smooth

____ 6. A(n) _____ is a group of cells and intercellular substances, all interacting in one or more tasks. [p.546]
   a. organ
   b. organ system
   c. tissue
   d. cuticle

____ 7. A graduate student in developmental biology accidentally stabbed a fish embryo. Later, the embryo developed into a creature that could not move and had no supportive or circulatory systems. Which embryonic tissue had suffered the damage? [p.546]
   a. ectoderm
   b. endoderm
   c. mesoderm
   d. protoderm

____ 8. A tissue whose cells are striated and fused at the ends be cell junctions so that the cells hold together during forceful contraction is called _____ tissue. [pp.544-545]
   a. smooth muscle
   b. dense fibrous connective
   c. supportive connective
   d. cardiac muscle

____ 9. The secretion of tears, milk, sweat, and oil are functions of _____ tissues. [p.541]
   a. epithelial
   b. loose connective
   c. lymphoid
   d. nervous

____ 10. Memory, decision making, and issuing commands to effectors are functions of _____ tissue. [p.545]
   a. connective
   b. epithelial
   c. muscle
   d. nervous

Matching [p.547]

Choose the most appropriate answer for each term.

11. _____ circulatory system

12. _____ digestive system

13. _____ endocrine system

14. _____ integumentary system

15. _____ muscular system

16. _____ nervous system

17. _____ reproductive system

18. _____ respiratory system

19. _____ skeletal system

20. _____ urinary system

A. Picks up nutrients absorbed from gut and transports them to cells throughout the body

B. Helps cells use nutrients by supplying them with oxygen and relieving them of CO_2 wastes

C. Helps maintain the volume and composition of body fluids that bathe the body's cells

D. Provided basic framework for the animal and supports other organs of the body

E. Uses chemical messengers to control and guide body functions

F. Produces younger, temporarily smaller versions of the animal

G. Breaks down larger food molecules into smaller nutrient molecules that can be absorbed by body fluids and transported to body cells

H. Consists of contractile parts that move the body through the environment and propel substances about in the animal

I. Serves as an electrochemical communications system in the animal's body

J. In the meerkat, serves as a heat catcher in the morning and protective insulation at night

CHAPTER OBJECTIVES/REVIEW QUESTIONS

1. Distinguish simple epithelium from stratified epithelium. [p.541]
2. Name and describe three kinds of cell junctions that occur in epithelia and other tissues. [p.540]
3. Name and describe the various types of epithelial tissues as well as their location and general functions. [p.541]
4. Describe the glands that usually secrete their products onto a free epithelial surface through ducts or tubes; cite examples of their products. [p.541]
5. Name the glands that lack ducts; describe their products which are secreted directly into the fluid bathing the gland. [p.541]
6. Distinguish between loose connective tissue; fibrous, irregular connective tissue; and fibrous, regular connective tissue on the basis of their structures and functions. [p.542]
7. Cartilage, bone, adipose tissue, and blood are known as the specialized connective tissues; describe their structures and various functions. [pp.542-543]
8. Distinguish among skeletal, smooth, and cardiac muscle tissues in terms of location, structure, and function. [pp.544-545]
9. Explain the differences between neurons and neuroglia. [p.545]
10. List each of the eleven principal organ systems in humans and list the main task of each. [p.547]
11. Name the directional terms and planes of symmetry used for description of the human body. [p.546]
12. Describe which germ layer gives rise to the skin's outer layer and tissues of the nervous system; gives rise to muscles, bones, and most of the circulatory, reproductive, and urinary systems; gives rise to the lining of the digestive tract and to organs derived from it. [p.546]

INTEGRATING AND APPLYING KEY CONCEPTS

1. What are some of the uses for adult and embryonic stem cells? Are both equally good for these uses?
2. Explain why, of all places in the body, marrow is located on the interior of long bones. Explain why your bones are remodeled after you reach maturity. Why does your body not keep the same mature skeleton throughout life?
3. You observe a tissue under the microscope and see multiple layers of cells along a free edge (outside environment). What other observations would you need to know to know exactly what type of tissue this is?
4. Explain why dense, regular connective tissue is the best type of connective tissue for ligaments.
5. Select any three organ systems and explain how they interact with each other.

33

NEURAL CONTROL

INTRODUCTION

This chapter provides an overview of how nervous systems, specifically vertebrate nervous systems, function. It describes how neurons work by developing a membrane and then transmitting action potentials. The chapter builds on this information by discussing the role of the major neurotransmitters and how signal integration occurs. The chapter then explores the anatomy and physiology of the human nervous system. It concludes with a discussion of how drugs influence the activity of the nervous system. One challenge in this chapter is that each section presents a number of new terms you need to understand before proceeding to the next section. The other complex process is how the neuron functions on a cell/molecular level—pay attention to the placement and movement of ions. Finally, focus on how the different parts of the nervous system interact with each other.

FOCAL POINTS

- Figure 33.4 [p.555] diagrams the overall structure of vertebrate nervous systems and the interaction of the various divisions.
- Sections 33.2 and 33.3 [pp.556-559] describe the role of a neuron and how it conducts messages using action potentials. Understanding this information is necessary in order to understand the remainder of the chapter.

INTERACTIVE EXERCISES

Impacts, Issues: In Pursuit of Ecstasy [p.552]

33.1. EVOLUTION OF NERVOUS SYSTEMS [pp.554-555]

Boldfaced, Page-Referenced Terms

[p.554] neuron _____

_____

[p.554] neuroglia _____

_____

[p.554] sensory neuron _____

_____

[p.554] interneuron _____

_____

[p.554] motor neuron _____

_____

[p.554] nerve net _____

_____

[p.554] cephalization _____

_____

[p.554] ganglion _____

_____

[p.555] central nervous system _____

_____

[p.555] peripheral nervous system _____

_____

Matching [pp.554-555]

Match each of the following terms with its correct definition.

1. _____ nerve net
2. _____ ganglion
3. _____ cephalization
4. _____ motor neuron
5. _____ interneurons
6. _____ sensory neurons
7. _____ nerves
8. _____ neuroglia

A. An asymmetrical mesh of neurons found in organisms with radial symmetry

B. Detect information about stimuli

C. Cells that structurally and metabolically support neurons

D. Accept and process sensory input

E. Relay signals to effectors that carry out responses

F. A cluster of nerve cell bodies

G. Having many sense organs concentrated at one end of the animal

H. Long-distance cable of the nervous system

Labeling [pp.554-555]

Provide the missing label for each numbered item in the accompanying figure.

9. _____

10. _____

11. _____

12. _____

13. _____

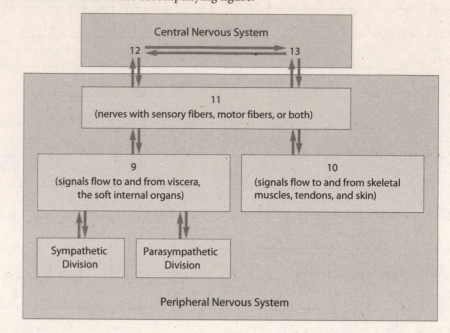

33.2. NEURONS—THE GREAT COMMUNICATORS [p.556]

Selected Words: cell body [p.556], axon terminal [p.556]

Boldfaced, Page-Referenced Terms

[p.556] dendrite _____

_____

[p.556] axon _____

_____

Labeling and Matching [p.556]

First, label each of the indicated structures in the accompanying diagram of the neuron. Then choose the correct function for each structure and place the appropriate letter in the parentheses.

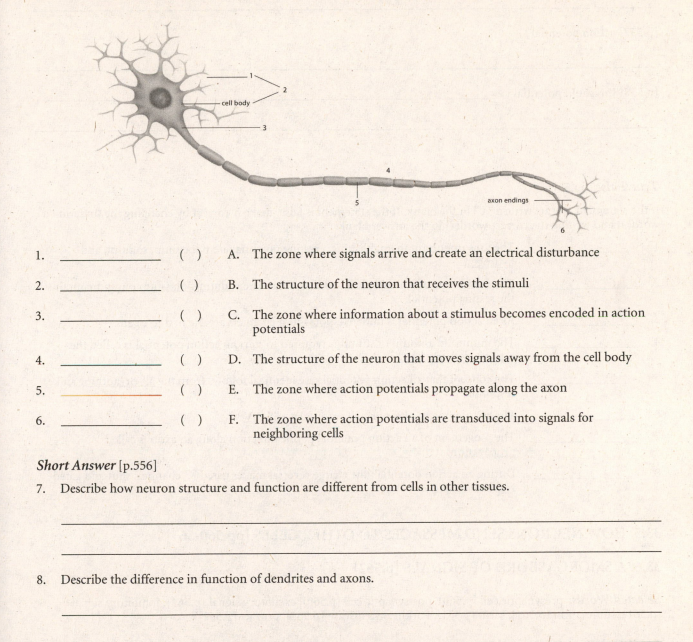

1. _____ ( ) A. The zone where signals arrive and create an electrical disturbance

2. _____ ( ) B. The structure of the neuron that receives the stimuli

3. _____ ( ) C. The zone where information about a stimulus becomes encoded in action potentials

4. _____ ( ) D. The structure of the neuron that moves signals away from the cell body

5. _____ ( ) E. The zone where action potentials propagate along the axon

6. _____ ( ) F. The zone where action potentials are transduced into signals for neighboring cells

Short Answer [p.556]

7. Describe how neuron structure and function are different from cells in other tissues.

_____

_____

8. Describe the difference in function of dendrites and axons.

_____

_____

33.3. MEMBRANE POTENTIALS [p.557]

33.4. A CLOSER LOOK AT ACTION POTENTIALS [pp.558-559]

Selected Words: membrane potential [p.557], threshold [p.558], all-or-nothing [p.558], propagation [p.559]

Boldfaced, Page-Referenced Terms

[p.557] resting membrane potential _____

_____

[p.557] action potential

_____

[p.558] threshold potential _____

_____

True/False [pp.557]

If the statement is true, write a "T" in the blank. If the statement is false, make it correct by changing the underlined word(s) and writing the correct word(s) in the answer blank.

1. _____ The ions important in establishing a resting potential are potassium, sodium, and <u>hydrogen</u>.

2. _____ The membrane of a neuron leaks sodium ions. The <u>sodium-potassium pump</u> maintains the resting potential.

3. _____ As an action potential is initiated, <u>potassium</u> channels are the first to open..

4. _____ The minimum amount of stimulus required to start an action potential is called the <u>refractory potential</u>.

5. _____ The concept that all action potentials are identical follows from the <u>all-or-nothing</u> spike response.

6. _____ The resting potential is normally around a <u>+55 mV</u>.

7. _____ The movement of an action potential in one direction along an axon is called <u>propagation</u>.

8. _____ During an action potential, the charge reversal makes gated <u>K^+</u> channels shut and gated <u>Na^+</u> channels open.

33.5. HOW NEURONS SEND MESSAGES TO OTHER CELLS [pp.560-561]

33.6. A SMORGASBORD OF SIGNALS [p.562]

Selected Words: presynaptic cell [p.560], postsynaptic cell [p.560], excitatory signal [p.561], inhibitory signal [p.561], summation [p.561], sarin [p.561], Parkinson's disease [p.562], prozac [p.562]

Boldfaced, Page-Referenced Terms

[p.560] synapse _____

_____

[p.560] neuromuscular junction _____

_____

[p.560] neurotransmitter _____

_____

[p.561] synaptic integration _____

_____

[p.562] neuromodulator _____

_____

Matching [pp.560-561]

Choose the most appropriate statement for each of the following terms.

1. _____ excitatory effect
2. _____ chemical synapse
3. _____ neurotransmitter
4. _____ inhibitory effect
5. _____ presynaptic neuron
6. _____ neuromuscular junction
7. _____ postsynaptic cell
8. _____ acetylcholine

A. Contains gated channels for calcium ions
B. Contains receptors for neurotransmitters
C. Moving a membrane away from the threshold of an action potential
D. A type of neurotransmitter used in neuromuscular junctions
E. A form of synapse between a motor neuron and skeletal muscle
F. Driving a membrane toward the threshold of an action potential
G. Signaling molecules made by neurons
H. A functional bridge between a neuron and another cell

True/False [pp.560-561]

If the statement is true, write a "T" in the blank. If the statement is false, make it correct by changing the underlined word(s) and writing the correct word(s) in the answer blank.

9. _____ The process of summing all the inputs to a neuron is called neuron differentiation.

10. _____ If an inhibitory signal and an excitatory signal arrive at the same time, they normally cancel each other.

11. _____ The action potential is passed to the next neuron through the output zone.

12. _____ Neurotransmitters diffuse across the synapse and attach to sodium ions on the postsynaptic neuron.

13. _____ Neurotransmitters must be cleaned out of the synapse after the signal transfer is complete.

Matching [p.562]

Match each of the following neurotransmitters and neuropeptides to the correct statement.

14. _____ serotonin

15. _____ GABA

16. _____ neuromodulators

17. _____ enkephalins and endorphins

18. _____ norepinephrine and epinephrine

19. _____ dopamine

20. _____ substance P

A. The major neurotransmitter inhibitor in the brain; derived from glutamate

B. Influences mood and memory; derived from tryptophan

C. Prime the body to respond to stress

D. A neuromodulator that enhances pain reception

E. Natural painkillers

F. Influences fine motor control and pleasure-seeking behaviors

G. A class of chemicals that magnify or reduce the effect of a neurotransmitter

33.7. DRUGS DISRUPT SIGNALING [p.583]

Selected Words: psychoactive [p.563], cocaine [p. 563], amphetamine [p.563], alcohol [p.563], barbituates [p.563], analgesic [p.563], LSD [p.563], marijuana [p.563]

Boldfaced, Page-Referenced Terms

[p.563] drug addiction _____

_____

[p.563] stimulant _____

_____

[p.563] depressant _____

_____

[p.563] analgesic _____

_____

[p.563] hallucinogen _____

_____

Complete the Table [p.563]

1. Drugs disrupt signaling at synapses. Complete the following table, which summarizes information about these molecules.

Drug	Category	Description/Function
a. nicotine		
b. cocaine		
c. amphetamine		
d. alcohol		
e. barbituates		
f. morphine		
g. oxycodone		
h. LSD		
i. marijuana		

Short Answer [p.563]

2. Describe what drug addiction is. _____

_____

_____

3. Explain five of the warning signs of drug addiction. _____

_____

_____

_____

33.8. THE PERIPHERAL NERVOUS SYSTEM [pp.564-565]

Selected Words: peripheral nerve [p.564], Schwann cell [p.564], myelin [p.564], autonomic nervous system [p.564]

Boldfaced, Page-Referenced Terms

[p.564] myelin sheath _____

_____

[p.564] somatic nervous system _____

_____

[p.564] autonomic nervous system _____

_____

[p.565] sympathetic neurons _____

_____

[p.565] parasympathetic neurons _____

_____

[p.565] fight-flight response _____

_____

Matching [pp.584-585]

Match each of the following terms with the appropriate statement.

1. _____ myelin sheath

2. _____ Schwann cell

3. _____ Autonomic Nervous System

4. _____ Peripheral Nervous System

5. _____ Fight-Flight Response

A. An electric insulator formed by Schwann cells

B. When nerve signals put you in a state of intense arousal

C. Contains all nerves outside of the central nervous system

D. A part of the nervous system concerned with signals to and from internal organs

E. A glial cell that produces the myelin around peripheral nerves

Short Answer [pp.564-565]

6. Explain the benefit of having a myelin sheath on a peripheral nerve.

_____

_____

7. Explain the interaction between the sympathetic and parasympathetic divisions of the nervous system.

_____

_____

Labeling [p.565]

Label each numbered part of the accompanying illustration.

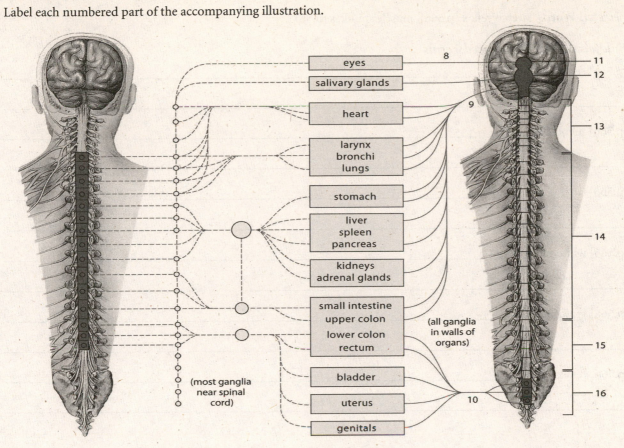

8. _____

9. _____

10. _____

11. _____

12. _____

13. _____

14. _____

15. _____

16. _____

33.9. THE SPINAL CORD [pp. 566-567]

Selected Words: stretch reflex [p.566], muscle spindles [p.567]

Boldfaced, Page-Referenced Terms

[p.566] spinal cord _____

_____

[p.566] meninges _____

_____

[p.566] cerebrospinal fluid _____

_____

[p.566] white matter _____

_____

[p.566] gray matter _____

_____

[p.566] reflex _____

_____

Labeling [p.566]

Identify the numbered parts of the accompanying illustration.

1. _____

2. _____

3. _____

4. _____

5. _____

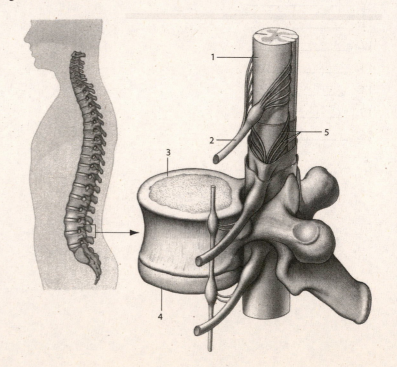

Matching

Match each of the following statements to the appropriate letter in the accompanying diagram.

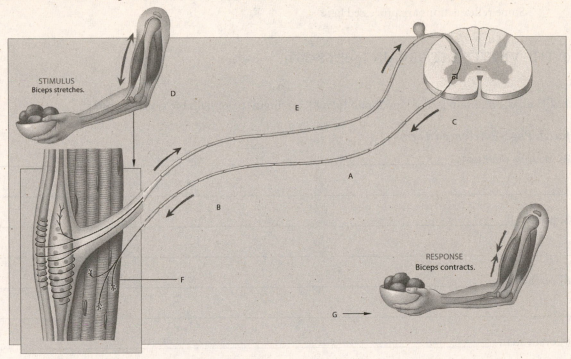

6. _____ Axons of the motor neurons synapse with muscle cells

7. _____ Neurotransmitter is released from the sensory neuron and stimulates the motor neuron

8. _____ Neurotransmitter released from the motor neuron stimulates the plasma membrane of muscle cells

9. _____ The action potential is propagated along the axon of the motor neuron

10. _____ The load is placed on the muscle tissue

11. _____ The stretching of the muscle tissue in response to the load stimulates receptors, generating an action potential

12. _____ The muscle is stimulated and contracts

Choice [pp.564-567]

For each of the numbered statements, choose the most appropriate group of nerves from the lettered list:

    a. peripheral—somatic nerves   b. peripheral—autonomic sympathetic nerves

    c. peripheral—autonomic parasympathetic nerves   d. spinal cord nerves

13. _____ Dominate when the body is in a state of relaxation.

14. _____ Dominate in times of stress, excitement, and danger.

15. _____ Can be attacked by meningitis.

16. _____ The sensory part of these nerves delivers information from receptors in the skin, skeletal muscles, and tendons to the central nervous system.

17. _____ Tend to slow down the body overall and divert energy to basic "housekeeping" tasks, such as digestion.

18. _____ The meninges belong to this division.

19. _____ An expressway for signals between the peripheral nervous system and the brain.
20. _____ The fight–flight response is directed by this division.
21. _____ Some reflex actions are processed here.

33.10. THE VERTEBRATE BRAIN [pp.568-569]

Selected Words: hindbrain [p.568], midbrain [p.568], forebrain [p.568], limbic system [p.568]

Boldfaced, Page-Referenced Terms

[p.568] medulla oblongata _____

_____

[p.568] pons _____

_____

[p.568] cerebellum _____

_____

[p.568] brain stem _____

_____

[p.568] cerebrum _____

_____

[p.568] thalamus _____

_____

[p.568] hypothalamus _____

_____

[p.569] blood-brain barrier _____

_____

[p.569] corpus callosum _____

_____

Matching [p.568]

Choose the most appropriate statement for each term.

1. _____ medulla oblongata

2. _____ cerebellum

3. _____ pons

4. _____ brain stem

5. _____ cerebrum

6. _____ thalamus

7. _____ hypothalamus

8. _____ blood–brain barrier

A. Main center for homeostatic control of the internal environment

B. Center for sorting out sensory information and sending it to the cerebrum

C. In vertebrates, site where olfactory information is integrated and responded to

D. Houses reflex centers for respiration and circulation

E. Protects the spinal cord and brain from harmful substances

F. Uses sensory input to control motor skills and posture

G. Controls signal traffic between the cerebellum and forebrain

H. The most ancient nervous tissue; continuous with the spinal cord

Choice [p.568]

Indicate the area of the brain to which each structure belongs.

    a. forebrain   b. hindbrain   c. midbrain

9. _____ cerebrum

10. _____ thalamus

11. _____ hypothalamus

12. _____ cerebellum

13. _____ medulla oblongata

14. _____ pons

15. _____ tectum

33.11. THE HUMAN CEREBRUM [pp.570-571]

33.12. THE SPLIT BRAIN [p.572]

Selected Words: motor cortex [p.570], premotor corex [p.570], association areas [p.571], Wernicke's area [p.570], Broca's area [p.770], declarative memory [p.571], skill memories [p.571], "gut reactions" [p.571], "split-brain" [p.572], epilepsy [p.572]

Boldfaced, Page-Referenced Terms

[p.570] cerebral cortex _____

_____

[p.571] limbic system _____

_____

Choice [pp.570-571]

For each statement, choose the appropriate area of the brain from the list provided. Some answers may be used more than once.

1. _____ Controls emotions and functions in memory

2. _____ Promotes chemical changes that affect states of consciousness

3. _____ Receiving center for information from skin and joints

4. _____ Governs learned patterns of motor skills

5. _____ Controls and coordinates movements of skeletal muscles

6. _____ Involved in interpreting social cues

7. _____ Contains the primary visual cortex

8. _____ Translates thoughts into speech

9. _____ Coordinates organ behavior for self-gratifying behaviors

10. _____ Perceptions of sound and odor arise here

11. _____ Personality and intellect start here

A. Broca's area

B. primary somato sensory cortex

C. temporal lobe

D. primary motor cortex

E. premotor cortex

F. occipital lobe

G. prefrontal cortex

H. limbic system

I. reticular formation

Short Answer [p.592]

12. In an effort to relieve the frequent seizures of severe epilepsy, neural surgeon Roger Sperry cut the neural bridge of the corpus callosum of several of these patients. The seizures did subside in frequency and intensity. Summarize the subsequent findings of Sperry regarding the function of the corpus callosum.

_____

_____

_____

Fill-in-the-Blanks [p.571]

(13) _____ term memory lasts just a few seconds or hours. This stage holds a few bits of information—a set of numbers, the words of a sentence, and so on. In (14) _____ -term memory, a seemingly unlimited quantity of larger bits is stored more or less (15) _____. Different forms of input are stored and called up by different mechanisms. Retention is best for (16) _____ memories, which are created when you consciously repeat an activity over and over. As the skill is being learned, the (17) _____ cortex signals (18) _____ areas of the cortex. Signals flow into the sensory cortex, the (19) _____, and the corpus striatum—a part of the (20) _____. Once the skill is mastered, the corpus striatum is able to call for appropriate movements, which frees you from having to consciously think about how to make the movements. (21) _____ memory allows you to remember how a lemon smells. It starts with signals from the (22) _____ cortex to the amygdala, which acts as the (23) _____ for those memories. The amygdala connects to the (24) _____, which serves as a(n) (25) _____ center. Signals must loop repeatedly through the hippocampus, cortex (26) _____, and (27) _____ for a memory to be retained.

33.13. NEUROGLIA—THE NEURON'S SUPPORT STAFF [p.573]

Selected Words: oligodendrites [p.573], microglia [p.573], astrocytes [p. 573], ependymal cells [p. 573], Multiple sclerosis [p. 573]

Complete the Table

1. There are four types of neuroglial cells. Complete the following table, which summarizes information about these cells.

RNA Molecule	Description/Function
a. Ependymal cell	
b. Microglia	
c. Astrocyte	
d. Oligodendrocyte	

___ 1. Which of the following is not true of an action potential? [pp.557-558]
a. It is a short-range message that can vary in size
b. It is an all-or-none brief reversal in membrane potential
c. It doesn't decay with distance
d. It is self-propagating

___ 2. The conducting zone of a neuron is the _____. [p.556]
a. axon
b. axon endings
c. cell body
d. dendrite

___ 3. The output zone of a neuron is the _____. [p.556]
a. axon
b. axon endings
c. cell body
d. dendrite

___ 4. An action potential is brought about by _____. [p.557]
a. a sudden membrane impermeability
b. the movement of negatively charged proteins through the neuronal membrane
c. the movement of lipoproteins to the outer membrane
d. a local change in membrane permeability caused by a greater-than-threshold stimulus

___ 5. The resting membrane potential _____. [p.557]
a. exists as long as a voltage difference sufficient to do work exists across a membrane
b. occurs because there are more potassium ions outside the neuronal membrane than inside
c. occurs because of the unique distribution of receptor proteins located on the dendrite exterior
d. is brought about by a local change in membrane permeability caused by a greater-than-threshold stimulus

___ 6. The phrase "all-or-nothing" used in conjunction with discussion about an action potential means that _____. [p.558]
a. a resting membrane potential has been received by the cell
b. an impulse does not diminish or dissipate as it travels away from the trigger zone
c. the membrane either achieves total equilibrium or remains as far from equilibrium as possible
d. propagation along the neuron is much faster than in other neurons

___ 7. An action potential passes from neuron to neuron across a synaptic cleft by _____. [p.560]
a. myelin bridges
b. the resting membrane potential
c. neurotransmitter substances
d. neuromodulator substances

___ 8. _____ nerves dominate when the body is not receiving much outside stimulation. [p.565]
a. Ganglia
b. Pacemaker
c. Sympathetic
d. Parasympathetic
e. All of the above

___ 9. Oligodendrocytes and astrocytes are examples of _____. [p.573]
a. neurons
b. neuroglial cells
c. nerves
d. neuropeptides

___ 10. The _____ are the protective coverings of the brain and spinal cord.[p.566]
a. ventricles
b. meninges
c. tectums
d. olfactory bulbs
e. pineal gland

11. The _____ regulates body temperature, thirst, and hunger and serves as an endocrine gland. [p.568]
    a. medulla
    b. pons
    c. thalamus
    d. hypothalamus

12. The part of the brain that controls the basic responses necessary to maintain life processes (respiration, blood circulation) is the _____. [p.568]
    a. cerebral cortex
    b. cerebellum
    c. corpus callosum
    d. medulla oblongata

13. The _____ integrates sensory input from the eyes, ears, and muscle spindles with motor signal from the fore-brain; it also helps control motor dexterity. [p.568]
    a. cerebrum
    b. pons
    c. cerebellum
    d. hypothalamus
    e. thalamus

14. The _____ evolved as a coordinating center for sensory input and as a relay station for signals to the cerebrum.[p.568]
    a. medulla
    b. pons
    c. reticular formation
    d. hypothalamus
    e. thalamus

15. Which neurotransmitter affects fine motor control and pleasure-seeking behaviors? [p.562]
    a. Epinephrine
    b. Dopamine
    c. GABA
    d. Serotonin

CHAPTER OBJECTIVES/REVIEW QUESTIONS

1. Define the terms *sensory neuron, interneurons, motor neurons,* and *neuroglia.* [p.554]
2. Explain the major stages in the evolution of nervous systems. [pp.554-555]
3. Draw a neuron and label it according to its three general zones, its specific structures, and the specific function(s) of each structure. [p.556]
4. Define *resting membrane potential*; explain what establishes it and how it is used by the cell neuron. [p.557]
5. Define *action potential* and state how sodium and potassium ions are used to generate an action potential. [pp.557-558]
6. Understand the importance of the sodium–potassium pump in maintaining the resting membrane potential. [p.557]
7. Explain how graded signals differ from action potentials. [p.557]
8. Explain the importance of a threshold level in generating an action potential. [p.558]
9. Understand the relationship between neurotransmitters and chemical synapses. [p.560]
10. Understand the difference between pre- and postsynaptic cells. [p.560]
11. Understand the process of synaptic integration. [p.561]
12. Understand the three mechanisms by which neurotransmitters are removed from the synaptic cleft. [p.561]
13. Recognize the general role of the major neurotransmitters and neuropeptides. [pp.560-562]
14. Explain the roles of the various types of neuroglial cells. [p.573]
15. Explain what the stretch reflex is and tell how it helps an animal survive. [pp.566-567]
16. Describe the function of each major division of the peripheral nervous system. [pp.564-565]
17. Describe the structure and function of the spinal cord. [pp.566-567]
18. List the parts of the brain found in the hindbrain, midbrain, and forebrain, and tell the basic functions of each. [pp.568-569]
19. In terms of structure and function, explain how the mechanism called the blood–brain barrier protects the brain and spinal cord. [p.569]
20. Recognize the general function of each area of the cerebrum. [pp.570-571]
21. Explain the function of the limbic system. [p.571]

22. State what the results of the "split-brain" experiments suggest about the functioning of the cerebral hemispheres. [p.572]
26. Distinguish between short-term and long-term information storage. [p.571]
27. List the major classes of psychoactive drugs and provide an example of each class. [p.563]

INTEGRATING AND APPLYING KEY CONCEPTS

1. What do you think might happen to human behavior if inhibitory postsynaptic potentials did not exist and if the threshold stimulus necessary to provoke an action potential were much higher?
2. Suppose that anger is eventually determined to be caused by excessive amounts of specific transmitter substances in the brains of angry people. Also suppose that an inexpensive antidote to anger that neutralizes these anger-producing transmitter substances is readily available. Can violent murderers now argue that they have been wrongfully punished because they were victimized by their brain's transmitter substances and could not have acted in any other way? Suppose an antidote is prescribed to curb violent tempers in an easily angered person. Suppose also that the person forgets to take the pill and subsequently murders a family member. Can the murderer still claim to be victimized by transmitter substances?

34
SENSORY PERCEPTION

INTRODUCTION

This chapter describes the different kinds of sensory receptors and shows how they are used in the special senses.

FOCAL POINTS

- Figures 34.8 and 34.9 [p.582] depict the structures involved in the chemical senses.
- Figure 34.12 [pp.584-585] shows the anatomy of the human ear.
- Figure 34.17 [p.588] diagrams the human eye.
- Figures 34.21 and 34.22 [p.590] look at rods and cones and how they are arranged in the retina.
- Section 34.10 [pp.592-593] describes visual disorders and eye diseases.

INTERACTIVE EXERCISES

Impacts, Issues: A Whale of a Dilemma [p.576]

34.1. OVERVIEW OF SENSORY PATHWAYS [pp.578-579]

Selected Words: sensory receptors [p.578], action potentials [p.578], mechanical energy [p.578], nociceptors [p.578], sensory signals [p.579], sensations [p.579], sensory nerve pathways [p.579]

Boldfaced, Page-Referenced Terms

[p.578] mechanoreceptors _____

_____

[p.578] pain receptors _____

_____

[p.578] thermoreceptors _____

_____

[p.578] chemoreceptors _____

_____

[p.578] osmoreceptors _____

_____

[p.579] photoreceptors _____

_____

[p. 579] sensory adaptation _____

_____

True/False [pp.578-579]

If the statement is true, write a "T" in the blank. If the statement is false, make it correct by changing the underlined word(s) and writing the correct word(s) in the answer blank.

1. _____ <u>Perception</u> is the conscious awareness of change in internal or external conditions.

2. _____ A generic term for any energy that activates a specific receptor is a <u>stimulus</u>.

3. _____ <u>Nociceptors</u> detect chemicals dissolved in the fluid around them.

4. _____ Changes in concentration of solutes in the body are detected by <u>osmoreceptors</u>.

5. _____ Signals coming over the optic nerve are always interpreted as <u>vision</u> by the brain regardless of the original stimulus.

6. _____ A weak stimulus recruits the <u>same number of</u> sensory receptors as a stronger stimulus.

7. _____ The fact that you don't feel a hat on your head a few minutes after putting it on is an example of <u>sensory integration</u>.

8. _____ Sensory receptors in the skin and skeletal muscles and near joints produce <u>visceral</u> sensations.

Choice [pp.578-579]

Select the best term for each of the following descriptions.

  a. chemoreceptors b. mechanoreceptors c. nociceptors d. photoreceptors   e. thermoreceptors

9. _____ Associated with vision

10. _____ Associated with pain

11. _____ Detect odors

12. _____ Detect sounds

13. _____ Detect CO_2 concentration in the blood

14. _____ Detect environmental temperature

15. _____ Detect internal body temperature

16. _____ Detect touch

17. _____ Rods and cones

18. _____ Hair cells in the ear's organ of Corti

19. _____ Pacinian corpuscles in the skin

20. _____ Olfactory receptors in nose

21. _____ Associated with the movement of fluid in the inner ear

34.2. SOMATIC AND VISCERAL SENSATIONS [pp.580-581]

34.3. SAMPLING THE CHEMICAL WORLD [p.582]

Selected Words: somatosensory cortex [p.580], Meissner's corpuscle [p.580], bulb of Krause [p.580], Ruffini endings [p.580], Pacinian corpuscle [p.580], somatic pain [p.580], visceral pain [p.580], substance P [p.581], endorphins [p.581], enkephalins [p.581], referred pain [p.581], olfactory bulbs [p.582], taste buds [p.582], umami [p.582]

Boldfaced, Page-Referenced Terms

[p.580] somatosensory cortex _____

_____

[p.581] referred pain _____

_____

[p.582] olfaction _____

_____

[p.582] pheromone _____

_____

[p.582] vomeronasal organ _____

_____

[p.582] taste receptors _____

_____

Fill-in-the-Blanks

Somatic sensations arise in the primary (1) _____. The two largest parts of this brain area correspond

to body parts with the most sensory activity, the (2) _____ and the (3) _____. This region responds to

input from receptors of various types. (4) _____ are simple, usually unmyelinated receptors that respond to

pressure, temperature, and (5) _____. Encapsulated receptors, like the (6) _____ abundant in the lips,

fingertips, and genitalia, also detect sensory input. Tissue injury triggers the release of chemicals that stimulate (7)

_____ receptors. Sometimes, this stimulation leads to the release of the natural opiates (8) _____ and

(9) _____. Chemical receptors detect molecules that become dissolved in fluid next to them. Receptors are the

modified endings of (10) _____ neurons. Animals smell substances by means of (11) _____ receptors,

such as the ones in your (12) _____; a human has about (13) _____ million of these. Sensory nerve

pathways—the (14) _____ bulb and nerve tract—lead from the nasal cavity to the region of the brain where odors are identified and associated with their sources. (15) _____ are signaling molecules secreted by one

individual that influence the behavior of another. These molecules also target (16) _____ receptors in the (17)

_____ organ. In the case of taste, receptors are located on animal tongues, often as part of sensory organs

called (18) _____, which are enclosed by circular papillae.

34.4. SENSE OF BALANCE [p.583]

34.5. SENSE OF HEARING [pp.584-585]

34.6. NOISE POLLUTION [p.586]

Selected Words: equilibrium [p.583], utricle [p.583], saccule [p.583], semicircular canals [p.583], dynamic equilibrium [p.583], static equilibrium [p.583], vertigo [p.583], motion sickness [p.583], amplitude [p.584], frequency [p.584], pitch [p.585], outer ear [p.584], pinna [p.584], auditory canal [p.584], middle ear [p.584], eardrum [p.584], hammer, anvil, stirrup [p.584], oval window [p.584], inner ear [p.584], basilar membrane [p.585], organ of Corti [p.585], tectorial membrane [p.585]

Boldfaced, Page-Referenced Terms

[p.583] organs of equilibrium _____

_____

[p.583] vestibular apparatus _____

_____

[p.583] hair cells _____

_____

[p.584] hearing _____

_____

[p.584] outer ear _____

_____

[p.584] middle ear _____

_____

[p.584] eardrum _____

_____

[p.584] inner ear _____

_____

[p.584] cochlea _____

_____

[p.585] organ of Corti _____

_____

True/False [pp.583-585]

If the statement is true, write a "T" in the blank. If the statement is false, make it correct by changing the underlined word(s) and writing the correct word(s) in the answer blank.

1. _____ Semicircular canals sense changes in rotation when hair cells are deformed by the movement of <u>calcium grains</u>.

2. _____ <u>Vertigo</u> is the sensation that the world is spinning around you.

3. _____ The saccule and utricle sense <u>dynamic equilibrium</u>.

4. _____ The <u>eardrum</u> transforms sound waves into vibrations of a thin membrane.

5. _____ The <u>middle ear bones</u> amplify the sound waves and transfer them to the fluid of the inner ear.

6. _____ The sound waves are converted to action potentials when movement of the <u>tectorial membrane</u> causes bending of hair cells.

7. _____ The inner ear contains both the vestibular apparatus and the <u>hammer, anvil, and stirrup</u>.

8. _____ Differences in pitch of a sound are determined by which part of the <u>basilar membrane</u> vibrates.

Labeling [pp.584-585]

Identify each indicated part of the accompanying illustrations.

9. _____

10. _____

11. _____

12. _____

13. _____

14. _____

15. _____

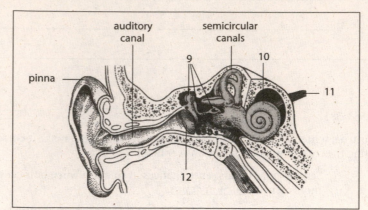

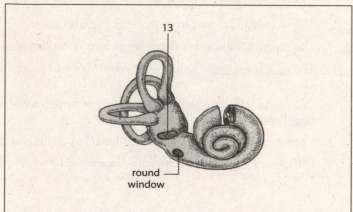

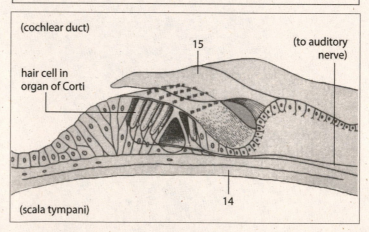

34.7. SENSE OF VISION [pp.586-587]

34.8. A CLOSER LOOK AT THE HUMAN EYE [pp.588-589]

34.9. FROM RETINA TO THE VISUAL CORTEX [590-591]

34.10. VISUAL DISORDERS [p.592]

Selected Words: ommatidia [p.587], ciliary body [p.588], aqueous humor [p.588], vitreous body [p.588], fovea [p.590], optic nerve [p.590], visual cortex [p.590], rhodopsin, [p. 591], red–green color blindness [p.592], astigmatism [p.592], nearsightedness (myopia) [p.592], farsightedness (hyperopia) [p.592], LASIK [p.592], macular degeneration [p.592], glaucoma [p.592], cataract [p.592], trachoma [p.592], onchocerciasis [p.592]

Boldfaced, Page-Referenced Terms

[p.586] vision _____

_____

[p.586] eyes _____

_____

[p.587] compound eyes _____

_____

[p.587] camera eyes _____

_____

[p.587] retina _____

_____

[p.588] conjunctiva _____

_____

[p.588] cornea _____

_____

[p.588] sclera _____

_____

[p.588] choroid _____

_____

[p.588] iris _____

_____

[p.588] pupil _____

_____

[p.589] visual accommodation _____

_____

[p.589] ciliary muscle _____

_____

[p.590] fovea _____

_____

[p.590] rod cells _____

_____

[p.590] cone cells _____

_____

[p.591] blind spot _____

_____

Fill-in-the-Blanks [pp.586-591]

Complex eyes have a(n) (1) _____ to focus light rays onto the retina. Even more complex eyes have a(n) (2) _____ for additional focusing and a ring of contractile tissue, the (3) _____, which helps to regulate the amount of light entering the eye. The (4) _____, a dense fibrous layer, protects the eyeball. In humans, muscles can change the shape of the (5) _____ in a process called (6) _____. Two types of receptors are found in the retina, (7) _____, which respond to dim light, and (8) _____, which respond to bright light and color. (8) _____ are most concentrated in a retinal area called the (9) _____. Rods contain (10) _____, which is stimulated by photons of (11) _____ wavelengths. Sensory input from the receptors is carried to the brain via the (12) _____. The region where the optic nerve exits the eye has no rods or cones and is called the (13) _____.

Labeling [p.588]

Identify each indicated part of the accompanying illustration.

14. _____

15. _____

16. _____

17. _____

18. _____

19. _____

20. _____

21. _____

22. _____

23. _____

24. _____

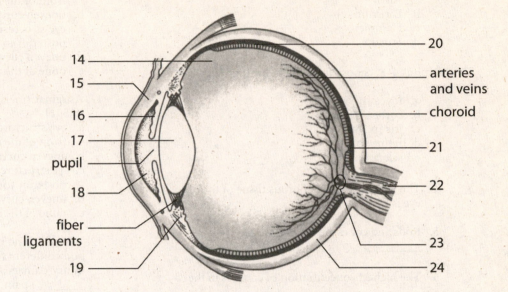

True/False [p.592]

If the statement is true, write a "T" in the blank. If the statement is false, make it correct by changing the underlined word(s) and writing the correct word(s) in the answer blank.

25. _____ <u>Nearsightedness</u> occurs when images are focused behind the retina.

26. _____ Macular degeneration occurs when there a loss of cones in the <u>fovea</u>.

27. _____ Increased pressure of the aqueous humor is called <u>glaucoma</u>.

28. _____ Cataracts are defects in the <u>cornea</u> that cloud vision.

29. _____ Astigmatism is caused by defects in the <u>lens</u> that prevent light from being focused properly.

30. _____ Nutritional blindness can be caused by a lack of <u>vitamin A</u> in the diet.

31. _____ The most common form of color blindness is the inability to distinguish <u>blue</u> from green.

32. _____ Onochocerciasis is a form of blindness caused by the <u>bacterium *Chlamydia trachomatis*.</u>

SELF-TEST

___ 1. Mechanoreceptors are important to
_____. [p.578]
   a. hearing
   b. regulating blood pressure
   c. both a and b
   d. none of the above

___ 2. Chemoreceptors are important to
_____. [p.578]
   a. taste
   b. smell
   c. both a and b
   d. none of the above

3. Which part of the middle ear directly transfers vibrations to the oval window? [p.584]
   a. Anvil
   b. Eardrum
   c. Hammer
   d. Stirrup

For questions 4–8, choose from the following answers:
   a. fovea [p.590]
   b. cornea [p.588]
   c. iris [p.588]
   d. retina [p.588]
   e. sclera [p.588]

4. The white protective fibrous tissue of the eye is the _____.

5. Rods and cones are located in the _____.

6. The highest concentration of cones is in the _____.

7. The adjustable ring of contractile and connective tissues that controls the amount of light entering the eye is the _____.

8. The outer transparent protective covering of part of the eyeball is the _____.

9. Visual accommodation in mammals involves the ability to _____. [p.589]
   a. change the sensitivity of the rods and cones by means of transmitters
   b. change the thickness of the lens by relaxing or contracting certain muscles
   c. change the curvature of the cornea
   d. adapt to large changes in light intensity
   e. all of the above

10. Nearsightedness is caused by _____. [p.592]
   a. eye structure that focuses an image in front of the retina
   b. uneven curvature of the lens
   c. eye structure that focuses an image posterior to the retina
   d. uneven curvature of the cornea
   e. none of the above

11. Astigmatism is caused by _____. [p.592]
   a. eye structure that focuses an image in front of the retina
   b. uneven curvature of the lens
   c. eye structure that focuses an image posterior to the retina
   d. uneven curvature of the cornea
   e. none of the above

12. When visceral pain is felt in an area of the body different than where the injury/damaged tissues are, that is called _____ pain. [p.581]
   a. phantom
   b. continuous
   c. referred
   d. chronic

CHAPTER OBJECTIVES/REVIEW QUESTIONS

1. Define and distinguish among chemoreceptors, mechanoreceptors, photoreceptors, and thermoreceptors. Name at least one example of each type that appears in an animal. [p.578]
2. Explain how a taste bud works. [p.582]
3. Explain how the three semicircular canals of the human ear detect changes of position and acceleration in a variety of directions. [p.583]

4. Follow a sound wave from the pinna to the organ of Corti; mention the name of each structure it passes and state where the sound wave is amplified and where the pattern of pressure waves is translated into electrochemical impulses. [pp.583-585]
5. Contrast the structure of compound eyes with the structures of invertebrate ommatidia and of the human eye. [pp.587-589]
6. Describe how the human eye perceives color and black-and-white. [pp.590-591]
7. Define *nearsightedness* and *farsightedness* and relate each to eyeball structure. [p.592]
8. Indicate the causes of the following disorders of the human eye: (a) red-green color blindness, (b) astigmatism, (c) cataracts, (d) glaucoma, and (e) retinal detachment. [p.592]

INTEGRATING AND APPLYING KEY CONCEPTS

1. Explain why adaptation is a positive process for animals.
2. Discuss the function of pain in an animal's behavior.
3. How might human behavior be changed if human eyes were compound eyes composed of ommatidia and if humans perceived only vibrations—as fish do—rather than sounds?
4. Human and squid eyes are quite similar. Is it more likely that humans and squid shared a common ancestor with camera-type eyes or that these eyes are a product of convergent evolution?

35

ENDOCRINE CONTROL

INTRODUCTION

This chapter looks at the glands that make up the endocrine system, the hormones they produce, and the effects of those hormones. It will also focus on the important interaction between the endocrine and nervous systems that essentially allows them to operate as a single system. Types of hormones will be described as well as how the endocrine glands are controlled to ensure the appropriate level of each hormone is produced and circulated. Disorders caused by incorrect amounts of some hormones are also discussed.

FOCAL POINT

- Figure 35.2 [p.599] shows the main endocrine glands and describes what they do.
- Figure 35.3 [p.601] diagrams the differences between the mechanisms of steroid and protein hormone action.
- Figures 35.4 and 35.5 [pp.602-603] show the function of the pituitary and its relationship to other endocrine glands.
- Figure 35.12 [p.608] describes the function of the pancreas in blood sugar regulation.
- Tables 35.2 and 35.3 [pp.602-605] review the functions of the endocrine glands.

INTERACTIVE EXERCISES

Impacts, Issues: Hormones in the Balance [p.596]

35.1. INTRODUCING THE VERTEBRATE ENDOCRINE SYSTEM [pp.598-599]

Selected Words: atrazine [p.596], endocrine disruptors [p.596], signaling molecules [p.598], "target" cells [p.598], glandular epithelium [p.598], pituitary gland [p.599], adrenal glands [p.599], pancreas [p.599], thyroid gland [p.599], parathyroid glands [p.599], pineal gland [p.599], thymus [p.599], gonads [p.599], hypothalamus [p.620]

Boldfaced, Page-Referenced Terms

[p.598] neurotransmitter _____

_____

[p.598] signaling molecules _____

_____

[p.598] animal hormones _____

_____

Fill-in-the-Blanks

(1) _____ is an insecticide that is effective and widely used in the U.S. Unfortunately,

studies have shown that it is also a(n) (2) _____. Male tadpoles exposed to (1) in the lab often became (3)

_____. Field-collected male frogs in agricultural areas where (1) was in use showed (4) _____ sex

organs. Humans may also be affected. Two pesticides banned in the U.S., (5) _____ and (6) _____, may

also act as endocrine disruptors.

Matching [p.598]
Choose the most appropriate answer for each term.

7. _____ hormones
8. _____ neurotransmitters
9. _____ Gap Junctions
10. _____ target cells
11. _____ local signaling molecules
12. _____ pheromones

A. Signaling molecules released from axon endings of neurons; act swiftly on target cells molecule

B. Cell connections that allow signals to move directly from cytoplasm of one cell to another

C. Released by many types of body cells; alter conditions within localized regions of tissues

D. Signaling molecules released by one animal that act on cells of other animals of the same species and help integrate social behavior

E. Secretions from endocrine glands, endocrine cells, and some neurons; distributed by the bloodstream to nonadjacent target cells

F. Cells that have receptors for a given type of signaling

Complete the Table [p.599]

13. Complete the following table. First, identify the numbered components of the endocrine system shown in the accompanying illustration. Then list the hormones produced by each.

Gland Name	Number	Hormones Produced
a. Hypothalamus		
b. Pituitary, Anterior lobe		
c. Pituitary, posterior lobe		
d. Adrenal gland (cortex)		
e. Adrenal gland (medulla)		
f. Ovaries		
g. Testes		
h. Pineal		
i. Thyroid		
j. Parathyroid		
k. Thymus		
l. Pancreas		

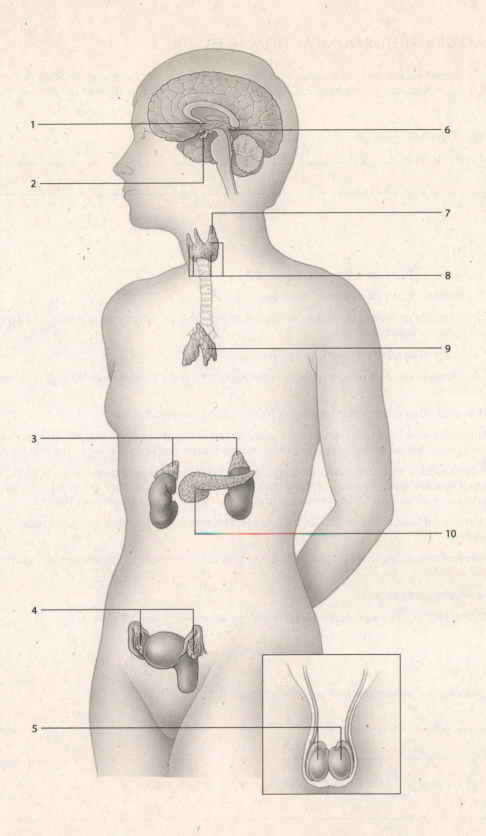

1

2

6

7

8

9

3

10

4

5

35.2. THE NATURE OF HORMONE ACTION [pp.600-601]

Selected Words: steroid hormones [p.600], amine hormones [p.600], peptide hormones [p.600], protein hormones [p.600], cAMP (cyclic adenosine monophosphate) [p.600], ADH (antidieuetic hormone) [p.600], vasopressin [p.601]

Boldfaced, Page-Referenced Terms

[p.600] second messenger _____

_____

Choice [pp.600-601]

For questions 1-10, choose from the following:

   a. steroid hormones   b. peptide/protein hormones

1. _____ Lipid-soluble molecules derived from cholesterol; can diffuse directly across the lipid bilayer of a target cell's plasma membrane

2. _____ Various peptides, polypeptides, and glycoproteins

3. _____ One example involves testosterone, defective receptors, and a condition called androgen insensitivity syndrome

4. _____ Hormones that often require assistance from second messengers

5. _____ Hormones that bind to receptors at the plasma membrane of a cell; the receptor then activates specific membrane-bound enzyme systems, which in turn initiate reactions leading to the cellular response

6. _____ A lipid-soluble molecule moves through the target cell's plasma membrane to the nucleus, where it binds to some type of protein receptor; this hormone-receptor complex moves into the nucleus and interacts with specific DNA regions to stimulate or inhibit transcription of mRNA

7. _____ Water-soluble signaling molecules that may incorporate anywhere from 2 to more than 150 amino acids

8. _____ Involves molecules such as cAMP, which activates many enzymes in cytoplasm that in turn alter some cell activity.

9. _____ Glucagon is an example

10. _____ Cyclic AMP relays a signal into the cell's interior to activate protein kinase A

Short Answer [p.600]

11. Explain what a second messenger does and why it is needed.

_____

_____

12. Explain what happens when the target cell lacks the receptor for a hormone.

_____

_____

35.3. THE HYPOTHALAMUS AND PITUITARY GLAND [pp.602-603]

Selected Words: posterior lobe [p.602], anterior lobe [p.602], positive feedback [p.603], negative feedback [p.603]

Boldfaced, Page-Referenced Terms

[p.602] hypothalamus _____

_____

[p.602] pituitary gland _____

_____

[p.603] releasers _____

_____

[p.603] inhibitors _____

_____

Labeling and Matching [p.602; also see summary table on p.605]

Label each hormone listed below with an A if it is secreted by the anterior lobe of the pituitary, a P if it is released from the posterior pituitary, or an I if it is released from intermediate tissue. Complete the exercise by entering the letter of the corresponding action in the parentheses after each label.

1. _____ ( ) ACTH
2. _____ ( ) ADH
3. _____ ( ) FSH
4. _____ ( ) STH (GH)
5. _____ ( ) LH
6. _____ ( ) MSH
7. _____ ( ) oxytocin (OCT)
8. _____ ( ) PRL
9. _____ ( ) TSH

A. Stimulates egg and sperm formation in ovaries and testes

B. Targets pigmented cells in skin and other surface coverings; induces color changes in response to external stimuli and affects some behaviors

C. Stimulates and sustains milk production in mammary glands

D. Stimulates progesterone secretion, ovulation, and corpus luteum formation in females; promotes testosterone secretion and sperm release in males

E. Induces uterine contractions and milk movement into secretory ducts of the mammary glands

F. Stimulates release of thyroid hormones from the thyroid gland

G. Acts on the kidneys to conserve water; required in control of extracellular fluid volume

H. Stimulates release of adrenal steroid hormones from the adrenal cortex

I. Promotes growth in young; induces protein synthesis and cell division; roles in adult glucose and protein metabolism

If the statement is true, write a "T" in the blank. If the statement is false, make it correct by changing the underlined word(s) and writing the correct word(s) in the answer blank.

10. _____ The <u>posterior pituitary</u> gland contains nerve endings that originate in the hypothalamus and secrete hormones.

11. _____ The anterior pituitary secretions are regulated by the <u>posterior pituitary</u>.

12. _____ Releasers are hormones released from the <u>anterior pituitary</u>.

13. _____ Anterior pituitary hormones are produced in the <u>hypothalamus</u>.

14. _____ Oxytocin and antidiuetic hormones are synthesized in the <u>hypothalamus</u>.

15. _____ The adrenocorticotropic hormone and thyroid stimulating hormone are synthesized in the <u>hypothalamus</u>.

16. _____ The anterior pituitary consists of <u>nerve</u> tissue and the posterior pituitary consists of <u>glandular</u> tissue.

17. _____ The production and subsequent release of milk require two hormones: <u>prolactin and luteinizing hormone.</u>

35.4. GROWTH HORMONE FUNCTION AND DISORDERS [p.604]

35.5. SOURCES AND EFFECTS OF OTHER VERTEBRATE HORMONES [p.605]

Selected Words: gigantism [p.604], acromegaly [p.604], pituitary dwarfism [p.604], secretin [p.605], leptin [p.605], erythropoietin [p.605], atrial natriuretic peptide [p.605]

Matching [pp.604-605]

Match each of the following terms with its correct definition/hormone.

1. _____ Pituitary dwarfism

2. _____ Acromegaly

3. _____ Pituitary gigantism

4. _____ Small intestine

5. _____ Kidney

6. _____ Heart

A. Excess growth hormone as an adult

B. Erythropoietin

C. Secretin

D. Atrial natriuretic peptide

E. Excess growth hormone as a child

F. Lack of growth hormone as a child

35.6. THYROID AND PARATHYROID GLANDS [pp.606-607]

35.7. TWISTED TADPOLES [p.607]

Selected Words: triiodothyronine [p.606], thyroxine [p.606], "thyroid hormone" [p.606], TRH [p.606], thyroid-stimulating hormone (TSH) [p.606], goiter [p.606], hypothyroidism [p.606], hyperthyroidism [p.606], calcitonin [p.606], Graves' disease [p.607], parathyroid hormone (PTH) [p.607], rickets [p.607], metamorphosis [p.607]

[p.606] thyroid gland _____

_____

[p.607] parathyroid glands _____

_____

Fill-in-the-Blanks

In a(n) (1) _____ loop, an increase in a hormone's level can inhibit the secretion of the hormone. This type of loop is in effect when TRH from the (2) _____ prevents the pituitary from secreting (3) _____, which in turn stops the (4) _____ from secreting its hormones. The mineral (5) _____ is required for the production of thyroid hormone. Deficiencies of (5) in the diet can lead to (6) _____. Excess thyroid hormone, or (7) _____, can cause many symptoms including anxiety and tremors. The four (8) _____ glands, located on the posterior surface of the thyroid, control the levels of (9) _____ in the blood. Low levels of (9) increase PTH secretion and can lead to (10) _____ in young children. Thyroid hormone is also important in the (11) _____ of a tadpole to a frog. Some pesticides are (12) _____ and can lead to developmental defects including lack of (11) in frogs living in contaminated ponds.

35.8. PANCREATIC HORMONES [p.608]

35.9. BLOOD SUGAR DISORDERS [p.609]

35.10. THE ADRENAL GLANDS [p.610]

35.11. TOO MUCH OR TOO LITTLE CORTISOL [p.611]

35.12. OTHER ENDOCRINE GLANDS [p.612]

Selected Words: alpha cells [p.608], beta cells [p.608], insulin [p.608], diabetes mellitus [p.609], ketoacidosis [p.609], type 1 diabetes [p.609], type 2 diabetes [p.609], hypoglycemia [p.609], cortisol [p.610], CRH (corticotropin-releasing hormone) [p.610], ACTH (adrenocorticotropin) [p.610], long-term stress [p.610], epinephrine [p.610], norepinephrine [p.610], fight–flight response [p.610], stress response [p.611], Cushing's syndrome [p.611], gonads [p.612], estrogens [p.612], progesterone [p.612], testosterone [p.612], libido [p.612], seasonal affective disorder [p.632]

Boldfaced, Page-Referenced Terms

[p.608] pancreas _____

_____

[p.608] glucagon _____

_____

[p.610] adrenal glands _____

_____

[p.610] adrenal cortex _____

_____

[p.610] adrenal medulla _____

_____

[p.610] aldosterone _____

_____

[p.612] gonads _____

_____

[p.612] testosterone _____

_____

[p.612] estrogen _____

_____

[p.612] progesterone _____

_____

[p.612] puberty _____

_____

[p.612] pineal gland _____

_____

[p.612] melatonin _____

_____

[p.612] thymus _____

_____

Matching

First match the gland/organ or cell group to the hormone(s) it produces. Then choose the best description of the action of the hormone(s).

a. glucagon b. testosterone c. cortisol d. estrogens and progesterone

e. epinephrine f. melatonin g. insulin

1. ___,___ adrenal cortex [p.610]

2. ___,___ adrenal medulla [p.610]

3. ___,___ testes [p.612]

4. ___,___ ovaries [p.612]

5. ___,___ pancreas (alpha cells) [p.608]

6. ___,___ pancreas (beta cells) [p.608]

7. ___,___ pineal [p.612]

A. Required in egg maturation and release, and in preparation of uterine lining for pregnancy and its maintenance in pregnancy; influence growth, development, and female genital development; maintains sexual traits

B. Promotes protein and fatty acid breakdown to provide fuel to most cells other than in the brain

C. Increases glucose uptake in liver, muscle, and adipose tissue, resulting in lower blood sugar levels

D. Required in sperm formation, male genital development and maintenance of sexual traits; influences growth and development

E. Influences daily biorhythms, gonad development, and reproductive cycles

F. Raises metabolism; increases heart rate and force of contraction

G. Raises blood sugar level, an effect opposite to insulin

Matching [pp.629-632]

Match each of the following diseases/disorders with the hormone and gland involved. Some letters may be used more than once.

a. too little insulin b. too little testosterone c. too much cortisol

d. too much insulin e. too much melatonin

8. ___,___ diabetes mellitus

9. ___,___ hypoglycemia

10. ___,___ Cushing's syndrome

11. ___,___ seasonal affective disorder

12. ___,___ decreased libido

A. Pancreas

B. Adrenal cortex

C. Pineal gland

D. Gonads

True/False [pp.608-611]

If the statement is true, write a "T" in the blank. If the statement is false, make it correct by changing the underlined word(s) and writing the correct word(s) in the answer blank.

13. _____ Type I diabetes usually begins when the person is an <u>adult</u>.

14. _____ Hypoglycemia occurs when blood sugar levels are too <u>high</u>.

15. _____ Cortisol levels are controlled by a <u>negative</u> feedback loop.

16. _____ Cushing's disease is characterized by <u>depressed</u> levels of cortisol.

17. _____ Hypercotisolism results in <u>lowered</u> white blood cell counts.

18. _____ The adrenal cortex hormones are involved in body responses to <u>short-term</u> stress.

35.13. A COMPARATIVE LOOK AT A FEW INVERTEBRATES [p.613]

Selected Words: molting [p.613], cuticle [p.613], X-organ [p.613], Y-organ [p.613], molt-inhibiting hormone [p.613]

Boldfaced, Page-Referenced Terms

[p.613] ecdysone _____

_____

Fill-in-the-Blanks [p.613]

Many invertebrate organisms (1) _____, periodically shedding their body covering as they grow. This process is controlled by hormones, such as the arthropod hormone (2) _____, which is structurally a(n) (3) _____ hormone. In crustaceans, the (4) _____ organ at the base of the (5) _____ inhibits the secretion of (2). The process is similar in (6) _____, but they do not have a(n) (7) _____ hormone.

SELF-TEST

___ 1. The _____ region of the forebrain monitors internal organs, influences certain forms of behavior, and secretes some hormones. [p.602]
   a. hypothalamus
   b. pancreas
   c. thyroid
   d. pituitary
   e. thalamus

___ 2. If you were lost in the desert and had no fresh water to drink, the level of _____in your blood would increase as a means to conserve water. [p.602]
   a. insulin
   b. corticotropin
   c. oxytocin
   d. antidiuretic hormone
   e. salt

For questions 3-5, choose from the following answers:
   a. estrogen
   b. PTH
   c. FSH
   d. growth hormone (GH)
   e. prolactin

___ 3. Stimulates bone cells to release calcium and phosphate and the kidneys to conserve it. [p.607]

___ 4. Stimulates and sustains milk production in mammary glands. [p.603]

___ 5. Is the hormone associated with pituitary dwarfism, gigantism, and acromegaly. [p.604]

For questions 6-8, choose from the following answers:
   a. adrenal medulla
   b. adrenal cortex
   c. thyroid
   d. anterior pituitary
   e. posterior pituitary

___ 6. Produces cortisol and other glucocorticoids that help increase the level of glucose in blood. [p.610]

___ 7. The gland that is most closely associated with emergency situations is the _____. [p.610]

___ 8. The overall metabolic rates of warm-blooded animals, including humans, depend on hormones secreted by the _____ gland. [p.606]

Matching

Choose the most appropriate answer for each term.

9. _____ ADH [p.602]

10. _____ ACTH and TSH [pp.602-603, 606]

11. _____ FSH and LH [pp.602-603, 612]

12. _____ GH [pp.602-603, 604]

13. _____ cortisol [pp.605, 610-611]

14. _____ epinephrine and norepinephrine [pp.605, 610]

15. _____ thyroxine and triiodothyronine [pp.605, 606]

16. _____ estrogens and progesterone [pp.605, 612]

17. _____ PTH [pp.605, 607]

18. _____ glucagon [pp.605, 608]

19. _____ testosterone [pp.605, 612]

20. _____ insulin [pp.605, 608]

21. _____ pheromones [p.598]

22. _____ melatonin [pp.605, 612]

23. _____ ecdysone [p.613]

A. In times of excitement or stress, these adrenal medulla hormones help adjust blood circulation and fat and carbohydrate metabolism

B. Hormones secreted by the ovaries; influence secondary sexual traits

C. Hormone secreted by beta pancreatic cells; lowers blood glucose level

D. Abnormal amount of this anterior pituitary lobe hormone has different effects on human growth during childhood and adulthood

E. Anterior pituitary lobe hormones that orchestrate secretions from the adrenal gland and thyroid gland, respectively

F. Adrenal cortex hormone that helps increase the level of glucose in blood

G. Hormone secreted by alpha pancreatic cells; raises the blood glucose level

H. The hormone that largely controls molting in insects and crustaceans

I. Major thyroid hormones having widespread effects such as controlling the overall metabolic rates of warm-blooded animals

J. Hormone secreted by the testes; influences secondary sexual traits

K. Hormone secreted by the pineal gland; influences the growth and development of gonads

L. Antidiuretic hormone released by the posterior pituitary lobe; promotes water reabsorption when the body must conserve water

M. Hormone-like secretions of certain exocrine glands; they diffuse through water or air to cellular targets outside the animal body

N. Anterior pituitary lobe hormones that act through the gonads to influence gamete formation and secretion of the sex hormones required in sexual reproduction

O. Hormone secreted by the parathyroid glands in response to low blood calcium levels

CHAPTER OBJECTIVES/REVIEW QUESTIONS

1. Define the terms *neurotransmitters*, *local signaling molecules*, and *pheromones*. [p.598]
2. Collectively, the body's sources of hormones came to be called the _____ system. [p.598]
3. Contrast the proposed mechanisms of hormonal action on target cell activities by (a) steroid hormones and (b) peptide and protein hormones. [pp.600-601]
4. The _____ and the pituitary gland interact closely as a major neural-endocrine control center. [p.602]
5. Identify the hormones released from the posterior lobe of the pituitary, and state their target tissues. [p.602]
6. Identify the hormones produced by the anterior lobe of the pituitary, and tell which target tissues or organs are affected. [pp.602-603]
7. Pituitary dwarfism, gigantism, and acromegaly are all associated with abnormal secretion of _____ by the pituitary gland. [p.604]
8. Outline the major human hormone sources, their secretions, main targets, and primary actions as shown in text Tables 35.2 and 35.3. [pp.602, 605]
9. Describe the role of cortisol in a stress response. [pp.610-611]
10. List the features of the fight–flight response. [p.610]
11. Describe the characteristics of hypothyroidism and hyperthyroidism. [pp.606-607]
12. Name the glands that secrete PTH, and state the function of this hormone. [p.607]
13. Name the hormones secreted by alpha and beta pancreatic cells; list the effect of each. [p.608]
14. Describe the symptoms of diabetes mellitus, and distinguish between type 1 and type 2 diabetes. [p.609]
15. The pineal gland secretes the hormone _____; relate two examples of the action of this hormone. [p.612]

INTEGRATING AND APPLYING KEY CONCEPTS

1. Suppose you suddenly quadruple your already high daily consumption of calcium. State which body organs would be affected, and tell how they would be affected. Name two hormones whose levels would most probably be affected, and tell whether your body's production of them would increase or decrease. Suppose you continue this high rate of calcium consumption for 10 years. Predict which organs would be subject to the most stress as a result.
2. What would happen to hormonal control if the nerves between the hypothalamus and pituitary were severed?
3. Explain how events that are perceived through the nervous system can affect hormone levels in the endocrine system. Give one example of this occurring in humans.

36

STRUCTURAL SUPPORT AND MOVEMENT

INTRODUCTION

This chapter explores how muscles and bones interact to allow movement. It starts with a discussion of skeletal systems in the invertebrates and introduces the evolutionary reasons for the development of an endoskeleton. However, the majority of the chapter details how muscles work at the cellular level. A key point is the sliding-filament model of how muscle contractions are powered.

FOCAL POINTS

- Figure 36.8 [p.621] gives an overview of the human skeleton; it is the framework the muscles are attached to.
- Figure 36.18 [p.629] illustrates the sliding-filament model of muscle contraction and focuses on the interaction of the proteins myosin and actin.

INTERACTIVE EXERCISES

Impacts, Issues: Pumping up Muscles [p.616]

36.1. INVERTEBRATE SKELETONS [pp.618-619]

36.2. THE VERTEBRATE ENDOSKELETON [pp.620-621]

Selected Words: androstenedione [p.616], creatine [p.616], gastrovascular cavity [p.618], circular muscles [p.618], longitudinal muscles [p.618], cartilaginous skeleton [p.620]

Boldfaced, Page-Referenced Terms

[p.618] hydrostatic skeleton _____

_____

[p.619] endoskeleton _____

_____

[p.620] vertebral column _____

_____

[p.620] vertebrae _____

_____

[p.620] intervertebral disk _____

_____

[p.620] axial skeleton _____

_____

[p.620] appendicular skeleton _____

_____

[p.620] foramen magnum _____

_____

Choice [pp.618-619]

For each of the following statements, choose the most appropriate category of skeleton.

    a. hydrostatic skeleton   b. endoskeleton   c. exoskeleton

1. _____ Uses external body parts to receive the applied force of muscle contractions
2. _____ Earthworms and soft-bodied invertebrates use this form of skeleton
3. _____ Echinoderms use this type of skeleton
4. _____ Uses internal body parts to receive the applied force of muscle contractions
5. _____ In this case muscles work against an internal body fluid
6. _____ This skeleton type is typical of the arthropods

Matching [pp.620-621]

Match each of the following structures with its correct definition/description.

7. _____ axial skeleton

8. _____ appendicular skeleton

9. _____ pectoral girdle

10. _____ pelvic girdle

11. _____ vertebrae

A. Consists of two sets of fused bones that support weight of upper body when standing
B. The central supporting column of the trunk and head
C. Bony segments that are stacked to make up the backbone
D. Consists of the pectoral girdle, pelvic girdle and limbs
E. The set of bones in the upper trunk to which the arms are attached

Labeling [p.621]

Identify each indicated part of the accompanying illustration.

12. _____

13. _____

14. _____

15. _____

16. _____

17. _____

18. _____

19. _____

20. _____

21. _____

22. _____

23. _____

24. _____

25. _____

26. _____

27. _____

28. _____

29. _____

30. _____

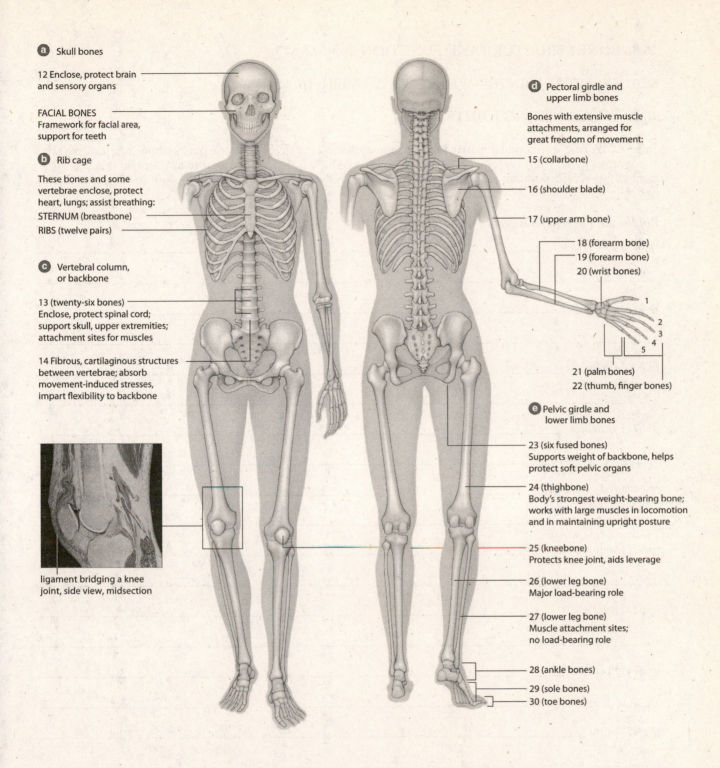

a Skull bones

12 Enclose, protect brain and sensory organs

FACIAL BONES
Framework for facial area, support for teeth

b Rib cage

These bones and some vertebrae enclose, protect heart, lungs; assist breathing:

STERNUM (breastbone)

RIBS (twelve pairs)

c Vertebral column, or backbone

13 (twenty-six bones)
Enclose, protect spinal cord; support skull, upper extremities; attachment sites for muscles

14 Fibrous, cartilaginous structures between vertebrae; absorb movement-induced stresses, impart flexibility to backbone

ligament bridging a knee joint, side view, midsection

d Pectoral girdle and upper limb bones

Bones with extensive muscle attachments, arranged for great freedom of movement:

15 (collarbone)

16 (shoulder blade)

17 (upper arm bone)

18 (forearm bone)

19 (forearm bone)

20 (wrist bones)

1
2
3
4
5

21 (palm bones)

22 (thumb, finger bones)

e Pelvic girdle and lower limb bones

23 (six fused bones)
Supports weight of backbone, helps protect soft pelvic organs

24 (thighbone)
Body's strongest weight-bearing bone; works with large muscles in locomotion and in maintaining upright posture

25 (kneebone)
Protects knee joint, aids leverage

26 (lower leg bone)
Major load-bearing role

27 (lower leg bone)
Muscle attachment sites; no load-bearing role

28 (ankle bones)

29 (sole bones)

30 (toe bones)

36.3. BONE STRUCTURE AND FUNCTION [pp.622-623]

36.4. SKELETAL JOINTS—WHERE BONES MEET [p. 624]

36.5. THOSE ACHING JOINTS [p.625]

Selected Words: extracellular matrix [p.622], compact bone [p.622], spongy bone [p.622], bone marrow [p.622], osteoporosis [p.623], fibrous joint [p.624], cartilaginous joints [p.624], synovial joint [p.624], sprain [p.625], meniscus [p.625], dislocation [p.625], osteoarthritis [p.625], rheumatoid arthritis [p.625], gout [p.625], bursitis [p.625]

Boldfaced, Page-Referenced Terms

[p.622] osteoblast _____

_____

[p.622] osteocyte _____

_____

[p.620] osteoclast _____

_____

[p.621] vertebrae _____

_____

[p.622] red marrow _____

_____

[p.622] yellow marrow _____

_____

[p.624] joint _____

_____

[p.624] ligament _____

_____

[p.625] bursa _____

_____

Matching [pp.622-623]

Choose the most appropriate statement for each of the following terms.

1. _____ yellow marrow
2. _____ osteoclasts
3. _____ osteocytes
4. _____ ligaments
5. _____ joints
6. _____ red marrow
7. _____ osteoblasts
8. _____ bone remodeling

A. Bone-forming cells

B. An ongoing task between osteoblasts and osteoclasts

C. The cells that break down bone using acids and enzymes

D. Straps of dense connective tissue

E. Mature bone cells

F. The fatty region of most mature bones in adults

G. The location, within bone, of red blood cell formation

H. Areas of contact, or near-contact, between bones

Choice [pp.624-625]

For each of the following statements, choose the most appropriate type of joint from the list below.

a. fibrous joint   b. cartilaginous joint   c. synovial joint

9. _____ Contains fluid to help lubricate joint
10. _____ Connect vertebrae to each other
11. _____ Joint between pelvic girdle and femur
12. _____ Bones held securely together—little flex
13. _____ Contain ligaments
14. _____ Pads or disks connect bones
15. _____ Has widest range of movement
16. _____ Hold teeth in sockets

Short Answer [p.625]

17. Describe in general what happens in a sprain. _____

_____

_____

18. Describe what happens when a cruciate ligament tears. Include what joint this would be in. _____

_____

_____

19. Describe the difference between arthritis and bursitis. _____

_____

_____

36.6. SKELETAL-MUSCULAR SYSTEMS [pp.626-627]

36.7. HOW DOES SKELETAL MUSCLE CONTRACT? [pp.628-629]

Selected Words: skeletal muscle [p.626], muscle fiber [p.628], Z band [p.629], ATP [p.629]

Boldfaced, Page-Referenced Terms

[p.626] muscle fiber _____

_____

[p.626] tendon _____

_____

[p.628] myofibril _____

_____

[p.628] sarcomere _____

_____

[p.628] actin _____

_____

[p.629] myosin _____

_____

[p.629] sliding-filament model _____

_____

Fill-in-the-Blanks [p.626]

Skeletal muscle cells are not your typical cells. Groups of them fuse together into one multinucleated (1)

_____. A cordlike or strap-like (2) _____ attaches to (3) _____. They act as a(n) (4) _____

system, in which a rigid rod is attached to a(n) (5) _____ point and moves about it. Muscles connect to bones

near a(n) (6) _____. When they contract, they transmit (7) _____ that makes the bones move. Bear in

mind, only (8) _____ muscle is the functional partner of bone. Often, two muscles work in (9) _____ to

each other; the action of one muscle (10) _____ the action of the other. (11) _____ muscle is mainly a

component of soft internal organs, such as the (12) _____. Cardiac muscle forms only in the (13) _____

wall. A strap-like (14) _____ connects skeletal muscles to bones.

Labeling and Matching [p.627]

On the next page, provide the correct name for each indicated muscle in this diagram. Then, in the parentheses, match the muscle to its correct function.

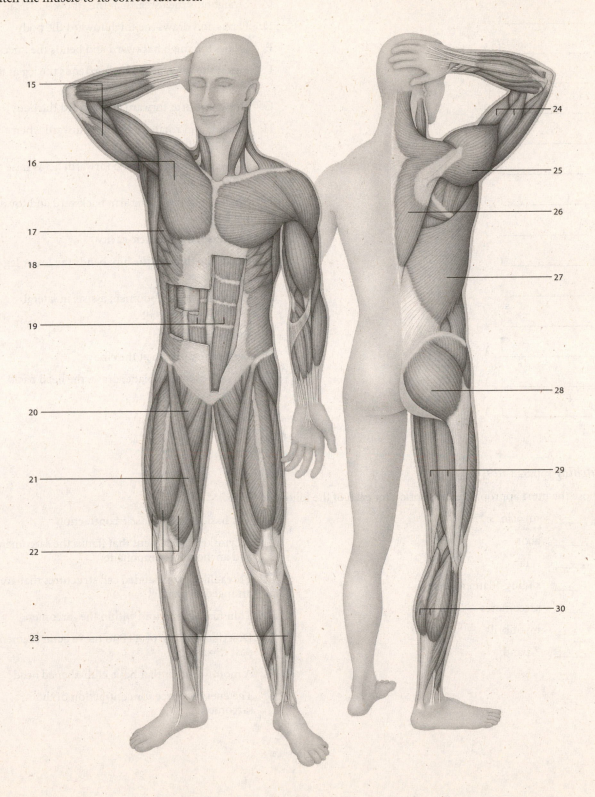

15. _____ ( )

16. _____ ( )

17. _____ ( )

18. _____ ( )

19. _____ ( )

20. _____ ( )

21. _____ ( )

22. _____ ( )

23. _____ ( )

24. _____ ( )

25. _____ ( )

26. _____ ( )

27. _____ ( )

28. _____ ( )

29. _____ (_)

30. _____ ( )

A.   Flexes the foot toward the shin

B.   Bends the lower leg at the knee while walking

C.   Straightens the forearm at the elbow

D.   Flexes and draws the thigh toward the body

E.   Draws the thigh backward and bends the knee

F.   Flexes the thigh at the hips; extends the leg at the knee

G.   Draws the arm forward and toward the body

H.   Extends and rotates the thigh outward when walking and running

I.   Draws the shoulder blade forward; helps raise the arm

J.   Rotates and draws the arm backward and toward the body

K.   Depresses the thoracic cavity

L.   Bends the thigh at the hip; bends the lower leg at the knee

M.   Compresses the abdomen; assists in lateral rotation of the torso

N.   Raises the arm

O.   Bends the forearm at the elbow

P.   Lifts the shoulder blade; draws the head back

Matching [pp.628-629]

Choose the most appropriate description for each of the following terms.

31. _____ myosin

32. _____ actin

33. _____ ATP

34. _____ sliding-filament model

35. _____ sarcomeres

36. _____ myofibrils

37. _____ Z band

A.   The basic unit of muscle contraction

B.   A cytoskeletal element that flanks the sarcomere and anchors its components

C.   Threadlike, crossbanded cell structures that are arranged in parallel

D.   Thin filaments found within the sarcomere

E.   Describes the use of ATP to move actin filaments past myosin

F.   A motor protein that has a club-shaped head

G.   The energy source for contraction of the sarcomere

True/False [pp.628-629]

If the statement is true, write a "T" in the blank. If the statement is false, make it correct by changing the underlined word(s) and writing the correct word(s) in the answer blank.

38. _____ <u>Myosin</u> proteins are attached to the Z band.

39. _____ Small, golf club shaped heads on myosin bind to <u>actin</u> filaments.

40. _____ When a myofibril contracts, the Z bands move <u>farther apart</u>.

41. _____ When a myofibril contracts, the A band <u>gets smaller</u>.

42. _____ When a myofibril contracts, the I band gets <u>larger</u>.

43. _____ Cross bridges refer to connections between the <u>actin and myosin</u> proteins.

44. _____ The thin fibers in a myofibril are made up of <u>myosin</u>.

45. _____ Movement of myosin heads is powered by <u>ATP</u>.

36.8. FROM SIGNAL TO RESPONSE: A CLOSER LOOK AT CONTRACTION [pp.630-631]

36.9. ENERGY FOR CONTRACTION [p.631]

36.10. PROPERTIES OF WHOLE MUSCLES [p.632]

36.11 DISRUPTION OF MUSCLE CONTRACTION [p.633]

Selected Words: T tubules [p.630], troponin [p.630], tropomyosin [p.630], creatine phosphate [p.631], aerobic respiration [p.631], lactate fermentation [p.631], isotonic contraction [p.632], isometric contraction [p.632], aerobic exercise [p.632], muscular dystrophies [p.633], polio [p.633], amyotrophic lateral sclerosis [p.633], botulism [p.633], tetanus [p.633]

Boldfaced, Page-Referenced Terms

[p.630] sarcoplasmic reticulum _____

_____

[p.632] motor unit _____

_____

[p.632] muscle twitch _____

_____

[p.632] tetanus _____

_____

[p.632] muscle tension _____

_____

[p.632] muscle fatigue _____

_____

Choose the most appropriate statement for each of the following terms.

1. _____ muscle cramp
2. _____ muscle fatigue
3. _____ muscular dystrophies
4. _____ tetanus (disease)
5. _____ botulism
6. _____ tetanus (muscle state)
7. _____ motor unit
8. _____ muscle twitch
9. _____ creatine phosphate
10. _____ excitable cells
11. _____ sarcoplasmic reticulum
12. _____ action potential

A. A disease in which over-stimulated muscles stiffen and contract; sometimes called lockjaw

B. A disease caused by bacteria that affects the neurons that synapse with muscle cells

C. The result of a sustained state of contraction due to high-frequency stimulation

D. A decrease in a muscle's capacity to generate force

E. Involuntary, painful contractions of muscles, probably the result of dehydration

F. Genetic disorders in which the muscles progressively weaken and degenerate

G. The brief interval in which a motor unit contracts

H. A motor neuron and all of the muscle cells that form junctions with its endings

I. Restores muscle fuel for about the rest 15 seconds of muscle contraction

J. The cellular structure that takes up and stores calcium ions

K. A reversal of the voltage difference across a membrane

L. Have the ability to reverse the voltage difference across their membrane as a result of stimulation

Short Answer [pp.630-632]

13. Explain the roles of troponin and tropomyosin in muscle contraction. _____

_____

_____

14. Explain the difference between isotonically and isometrically contracting muscles. _____

_____

_____

SELF-TEST

1. Insects have what form of skeleton? [pp.618-619]
   a. Endoskeleton
   b. Exoskeleton
   c. Hydroskeleton
   d. No skeleton

2. The major site of blood cell formation in the human body is the _____ . [p.622]
   a. ligaments
   b. osteoclasts
   c. yellow marrow
   d. red marrow

3. Bone-forming cells are called _____. [p.622]
   a. osteocytes
   b. osteoclasts
   c. osteoblasts
   d. none of the above

4. Which of the following is made from dense connective tissue? [p.624]
   a. Sarcomeres
   b. Joints
   c. Cartilage
   d. Ligaments

5. The protein responsible for contraction in skeletal muscle is called _____. [p.629]
   a. actin
   b. myosin
   c. Z band
   d. myofibril

6. The sarcoplasmic reticulum does which of the following? [p.630]
   a. Is involved in the sliding-filament model
   b. Restores voltage differences following an action potential
   c. Stores and releases calcium ions
   d. Restores ATP levels following contraction
   e. None of the above

7. _____ is a decrease in a muscle's capacity to generate force. [p.632]
   a. Tetanus
   b. Muscle fatigue
   c. Muscle tension
   d. Muscle twitch

8. Which of the following has a genetic component? [p.633]
   a. muscular dystrophies
   b. muscle fatigue
   c. muscle cramps
   d. tetanus
   e. botulism

CHAPTER OBJECTIVES/REVIEW QUESTIONS

1. Define and give examples of the three types of skeletons found in animals. [pp.618-619]
2. Give examples of how the vertebrate skeleton allows life on land. [p.620-621]
3. Distinguish between the axial and appendicular skeleton. [pp.620-621]
4. Explain the importance of intervertebral disks for upright locomotion. [p.620]
5. Recognize the major bones of the human body and their roles. [p.621]
6. Explain the various roles of osteoblasts, osteoclasts, osteocytes, red marrow, and yellow marrow. [p.622]
7. Distinguish between joints and ligaments. [p.624]
8. Understand the relationship between the human skeleton and sprains, osteoarthritis, rheumatoid arthritis, and osteoporosis. [p.625]
9. Understand the interaction of skeletal muscle and bones. [p.626]
10. Recognize the locations and functions of the major muscles. [p.627]
11. Describe the fine structure of a muscle fiber; use terms such as *myofibril, sarcomere, motor unit, actin,* and *myosin.* [pp.628-629]
12. Understand the sliding-filament model for muscle contraction. [p.629]
13. Understand the importance of calcium in muscle contraction. [pp.630-631]

14. Identify the major sources of energy for muscle contraction. [p.631]
15. Distinguish between isotonically and isometrically contracting muscles. [p.632]
16. Recognize how the terms *tetanus, muscle fatigue, muscle tension,* and *muscle twitch* relate to the properties of whole muscles. [p.632]
17. Identify how the diseases muscular dystrophy, tetanus, and botulism interact with the muscle system. [p.633]

INTEGRATING AND APPLYING KEY CONCEPTS

1. If humans had an exoskeleton rather than an endoskeleton, would they move differently from the way they do now? Name any advantages or disadvantages that having an exoskeleton instead of an endoskeleton would present in human locomotion.
2. Why should pregnant mothers increase their calcium intake? Where would the fetus get its calcium supply from naturally? Why should mothers continue increased calcium consumption following pregnancy?
3. What would be the effect of a chemical that prevented the interaction of myosin and actin?
4. What are the benefits of hydrostatic skeletons and exoskeletons to invertebrates? Why are endoskeletons so rare in these animals?

37

CIRCULATION

INTRODUCTION

This chapter looks at the structure and function of the circulatory system and its components, concentrating on humans. You will learn about the evolutionary background of our circulatory system as well as its parts—blood, blood vessels, and the heart. The interaction of these parts with the rest of the body will be covered.

FOCAL POINTS

- Figure 37.2 [p.638] compares open and closed systems.
- Figure 37.3 [p.639] compares the circulatory systems of fish, amphibians, and birds and mammals.
- Figure 37.4 [p.640] analyzes the components of blood.
- Figure 37.5 [p.641] shows how the various blood cells are derived.
- Figure 37.8 [p.643] illustrates blood types and cross reactions between them.
- Figure 37.10 [p.644] compares the systemic and pulmonary circuits.
- Figure 37.11 [p.645] shows the major vessels of the human circulatory system.
- Figure 37.13 [p.646] illustrates the human heart.
- Figure 37.17 [p.648] compares various blood vessels.
- Figure 37.26 [p.654] diagrams the human lymphatic system.

INTERACTIVE EXERCISES

Impacts, Issues: And Then My Heart Stood Still [p.656]

37.1. THE NATURE OF BLOOD CIRCULATION [pp.638-639]

Selected Words: sudden cardiac arrest [p.636], cardiopulmonary resuscitation (CPR) [p.636], automated external defibrillator (AED) [p.636], internal environment [p.658], gills [p.639], lungs [p.639], single circuit [p.639], two circuits [p.639]

Boldfaced, Page-Referenced Terms

[p.638] circulatory system _____

_____

[p.638] blood _____

_____

[p.638] heart _____

_____

[p.638] interstitial fluid _____

_____

[p.638] open circulatory system _____

_____

[p.638] closed circulatory system _____

_____

[p.638] capillaries _____

_____

[p.639] pulmonary circuit _____

_____

[p.639] systemic circuit _____

_____

Fill-in-the-Blanks [pp.636-639]

(1) _____ occurs when the heart abruptly stops beating. If oxygen is not delivered to the brain, irreversible damage occurs.(2) _____ is used to keep blood and oxygen flowing until the heart can be shocked into beating with a(n) (3) _____. In humans and many other animals, substances move rapidly to and from living cells by way of a(n) (4) _____ circulatory system. (5) _____, a fluid connective tissue within the (6) _____ and blood vessels, is the transport medium. Most of the cells of animals are bathed in a(n) (7) _____; blood is constantly delivering nutrients and removing wastes from that fluid. The (8) _____ generates the pressure that keeps blood flowing. Blood flows (9) _____ [choose one] (rapidly, slowly) through large-diameter vessels to and from the heart, but where the exchange of nutrients and wastes occurs, in the (10) _____ beds, the blood is divided up into vast numbers of smaller-diameter vessels with tremendous surface area that enables the exchange to occur by diffusion. The speed of the blood (11) _____ in these vessels. Fish have a(n) (12) _____ circuit of blood flow, whereas in birds and mammals, blood flows through the (13) _____ circuit to the lungs and through the (14) _____ circuit to the rest of the body. One of the advantages of having two circuits is that blood (15) _____ can be regulated independently in each.

Labeling and Short Answer [p.638]

Label the numbered parts in the following illustrations.

16. _____

17. _____

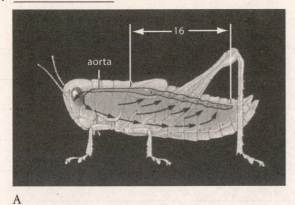

A

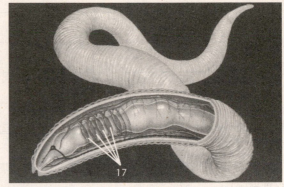

B

Describe the kind of circulatory system in:

18. Creature A _____

_____

19. Creature B _____

_____

37.2. CHARACTERISTICS OF BLOOD [pp.640-641]

37.3. HEMOSTASIS [p.642]

37.4. BLOOD TYPING [pp.642-643]

Selected Words: plasma proteins [p.640], erythrocytes [p.640], hemoglobin [p.640], leukocytes [p.641], neutrophils [p.641], basophils [p.641], eosinophils [p.641], phagocytes [p.641], inflammation [p.641], monocyte [p.641], macrophage [p.641], lymphocyte [p.641], B cell [p.641], T cell [p.641], megakaryocyte [p.641], coagulation [p.642], clot [p.642], fibrinogen [p.642], fibrin [p.642], prothrombin [p.642], hemophilia [p.642], antibodies [p.642]

Boldfaced, Page-Referenced Terms

[p.640] plasma _____

_____

[p.640] red blood cells _____

_____

[p.641] cell count _____

_____

[p.641] white blood cells _____

_____

[p.641] platelets _____

_____

[p.642] hemostasis _____

_____

[p.642] agglutination _____

_____

[p.642] ABO Blood Typing _____

_____

[p.643] Rh Blood Typing _____

_____

Complete the Table [p.640]

1. Fill in items a–g in the following table, which describes the components of blood.

Components	Relative Amounts	Functions
Plasma Portion (50%-60% of total volume)		
Water	91%-92% of plasma volume	a.
b.	7%-8%	Defense, clotting, lipid transport, roles in extracellular fluid volume, and so forth
Ions, sugars, lipids, amino acids, hormones, vitamins, dissolved gasses	1%-2%	Roles in extracellular fluid volume, pH, and so on
Cellular Portion (40%-50% of total volume)		
c.	4,800,000-5,400,000 per microliter	O2, CO2 transport
White Blood Cells		
d.	3,000-6,750	Phagocytosis
e.	1,000-2,700	Immunity
Monocytes (macrophages)	150-720	Phagocytosis
Eosinophils	100-360	f.
Basophils	25-90	Roles in inflammatory response, fat removal
g.	250,000-300,000	Roles in clotting

Short Answer [p.640]

2. Describe the functions of blood. _____

_____

_____

True/False [pp.640-641]

If the statement is true, write a "T" in the blank. If the statement is false, make it correct by changing the underlined word(s) and writing the correct word(s) in the answer blank.

3. _____ Blood is categorized as a <u>connective</u> tissue.

4. _____ Blood makes up approximately <u>15 percent</u> of the total body weight.

5. _____ By volume, 50-60 percent of the blood is made up of actual <u>cells</u>.

6. _____ All blood cells are produced in the <u>thymus gland</u>.

7. _____ Red blood cells are very thin to <u>facilitate exchange of oxygen and carbon dioxide</u>.

8. _____ Red blood cells are <u>very long lived</u> cells.

9. _____ <u>Eosinophil</u>s are the primary phagocytes on the blood.

10. _____ <u>Monocytes</u> are essential for effective clotting.

Sequence [p.642]

Arrange the following hemostatic events in the correct time sequence. Write the letter of the first event next to 11. The letter of the last event is written next to 16.

11. _____   A. Thrombin converts fibrinogen to fibrin

12. _____   B. Platelets stick together and plug the damaged vessel

13. _____   C. A blood vessel is damaged

14. _____   D. Enzyme cascade activates Factor X

15. _____   E. Fibrin forms a net that collects cells and platelets, forming a clot

16. _____   F. A vascular spasm constricts the vessel

Labeling and Matching [pp.640-641]

Identify the numbered cell types in the accompanying illustration. Complete the exercise by matching and entering the letter of the appropriate function in the parentheses (if any) after the given cell types. A letter may be used more than once.

17. _____ ( )

18. _____ ( )

19. _____ ( )

20. _____ ( )

21. _____ ( )

22. _____ ( )

23. _____ ( )

24. _____ ( )

A. Phagocytosis

B. A role in the inflammatory response

C. A role in clotting

D. Immunity

E. O$_2$, CO$_2$ transport

F. Immature, unspecialized blood cells

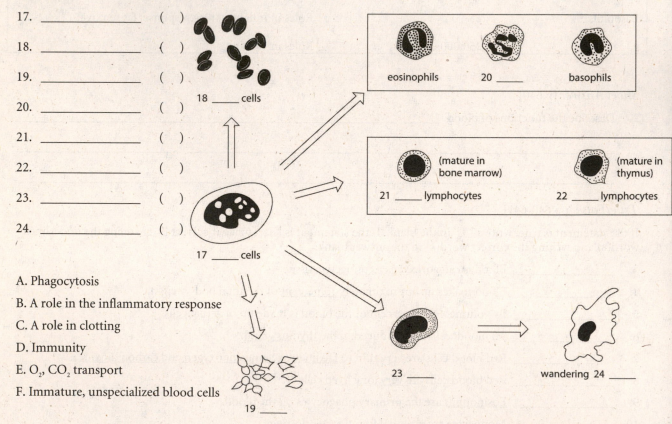

Matching [pp.642-643]

Match each of the following blood types, with the blood type(s) the individual could safely receive in a whole blood (includes plasma and cells) transfusion.

25. _____ A Rh$^+$

26. _____ AB Rh$^+$

27. _____ O Rh$^+$

28. _____ O Rh$^-$

29. _____ B Rh$^-$

A. A$^+$

B. B$^+$

C. O$^+$

D. AB$^+$

E. A$^-$

F. B$^-$

G. O$^-$

H. AB$^-$

Short Answer [pp.642-643]

30. Explain what can happen during a second pregnancy when the mother is Rh$^-$ and the fetus is Rh$^+$.

_____

_____

_____

_____

37.5. HUMAN CARDIOVASCULAR SYSTEM [pp.644-645]

37.6. THE HUMAN HEART [pp.646-647]

Selected Words: cardiovascular [p.644], pulmonary [p.644], systemic [p.644], capillary bed [p.644], hepatic portal vein [p.644], artery [p.645], arteriole [p.645], venule [p.645], pericardium [p.646], myocardium [p.646], endothelium [p.646], atrioventricular valve [p.646], semilunar valve [p.646], diastole [p.646], systole [p.646], atrium [p.646], ventricle [p.646], sinoatrial (SA) node [p.647], atrioventricular (AV) node [p.647]

Boldfaced, Page-Referenced Terms

[p.644] aorta _____

_____

[p.646] atrium _____

_____

[p.646] ventricle _____

_____

[p.646] cardiac cycle _____

_____

[p.646] cardiac muscle _____

_____

[p.647] cardiac conduction system _____

_____

[p.647] cardiac pacemaker _____

_____

True/False [pp.644-645]

If the statement is true, write a "T" in the blank. If the statement is false, make it correct by changing the underlined word(s) and writing the correct word(s) in the answer blank.

1. _____ The systemic circuit moves blood to the lungs and back to the heart.

2.. _____ When blood moves from one capillary bed through a vein to a second capillary bed before returning to the heart, the vessel between the two capillary beds is called a transfer vein.

3. _____ The pulmonary circuit takes blood to the lungs to pick up oxygen and get rid of carbon dioxide.

4. _____ Arteries always contain oxygenated blood.

5. _____ The aorta is the largest artery in the body.

Sequence [p.647]

Arrange the following events in the cardiac cycle in the correct time sequence. Write the letter of the first event next to 6. The letter of the last event is written next to 11.

6. _____   A. AV node receives the signal

7. _____   B. Ventricles contract

8. _____   C. Atria contract

9. _____   D. SA node initiates signal

10. _____   E. Signal spreads across the atria

11. _____   F. Signal sent down bundle fibers to apex of the heart

Fill-in-the-Blanks [p.645]

Fill in the missing words for the labels indicated in the figure. To complete the exercise, color in red on the figure all vessels that contain oxygen-enriched blood, and color in blue all vessels that contain oxygen-poor blood.

12. _____VEINS (from brain, neck, head tissues)

13. _____ _____ _____(from neck, shoulder, arms—

        VEINS of upper body)

14. _____VEINS (from lungs to heart)

15. _____VEIN (from kidneys back to heart)

16. _____ _____ _____(receives blood from all veins

    below the diaphragm)

17. _____VEINS (carry blood from pelvic organs and lower abdominal wall)

18. _____VEIN (from thigh and inner knee)

19. _____ARTERY (to thigh and inner knees)

20. _____ARTERIES (to pelvic organs, lower abdominal wall)

21. _____AORTA (to digestive tract, pelvic organs)

22. _____ARTERY (to kidney)

23. _____ARTERY (to arm, hand)

24. _____ARTERIES (to cardiac muscle)

25. _____ARTERIES (to lungs)

26. _____ARTERIES

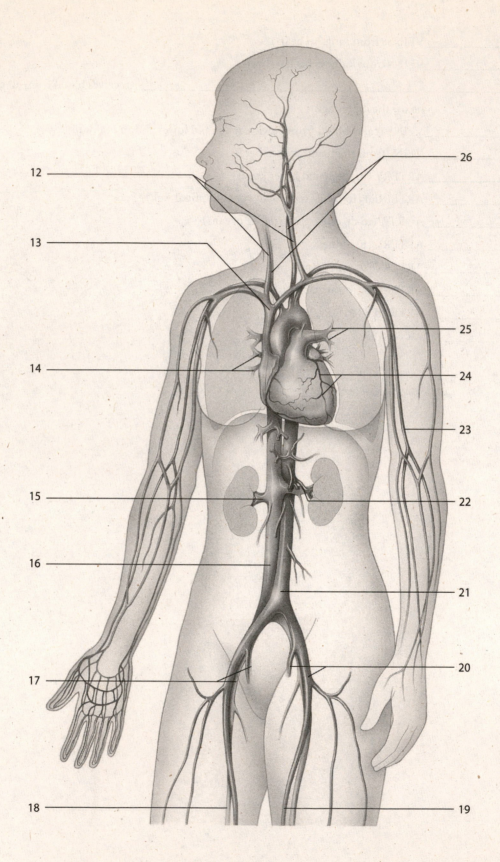

12

13

14

15

16

17

18

26

25

24

23

22

21

20

19

Labeling [p.646]

Identify each indicated part of the accompanying illustrations. Color in red all vessels and parts of the heart that contain oxygen-rich blood. Color in blue all vessels and parts of the heart that contain oxygen-poor blood.

27. _____

28. _____ _____ _____

29. _____ _____ _____

30. _____ _____ _____

31. _____ _____ _____

32. _____ _____ _____

33. _____ _____ _____

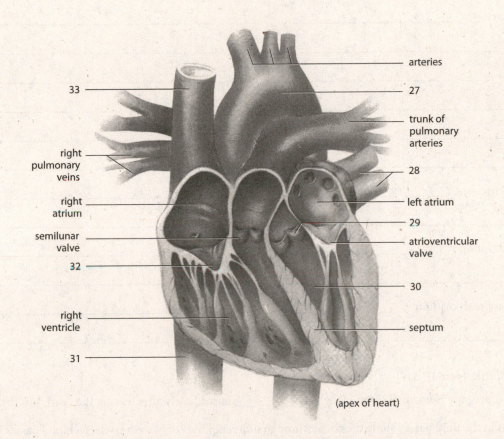

(apex of heart)

37.7. PRESSURE, TRANSPORT, AND FLOW DISTRIBUTION [pp.648-649]

37.8. DIFFUSION AT CAPILLARIES, THEN BACK TO THE HEART [pp.650-651]

37.9. BLOOD AND CARDIOVASCULAR DISORDERS [pp.652-653]

Selected Words: dilation [p.649], systolic pressure [p.649], diastolic pressure [p.649], endothelial cell [p.650], diffuse [p.650], endocytosis [p.650], blood-brain barrier [p.650], hydrostatic pressure [p.650], osmosis [p.651], lymph vessels [p.651], edema [p.651], varicose veins [p.651], hemorrhoids [p.651], anemia [p.652], hemolytic anemia [p.652], beta thalassemia [p.652], polycythemia [p.652], mononucleosis [p.652], leukemia [p.652], lymphoma [p.652], hemophilia [p.652], thrombus [p.652], embolus [p.652], atherosclerosis [p.652], plaque [p.652], hypertension [p.653], arrthymia [p.653], tachycardia [p.653], atrial fibrillation [p.653]

Boldfaced, Page-Referenced Terms

[p.648] artery _____

_____

[p.648] arteriole _____

_____

[p.648] capillary _____

_____

[p.648] venule _____

_____

[p.648] vein _____

_____

[p.648] blood pressure _____

_____

[p.651] ultrafiltration _____

_____

[p.651] capillary reabsorption _____

_____

Fill-in-the-Blanks [pp.648-649]

Blood pressure is normally high in the (1) _____ immediately after leaving the heart, but then the

pressure drops as the fluid passes along the circuit through different kinds of blood vessels. As blood passes into

smaller-diameter vessels, flow (2) _____ mainly because of increases in (3) _____. (4) _____

guide the flow of blood into various organs. Some signals can make the vessels relax, which causes (5) _____,

and more blood flows into the organ. Other signals cause (6) _____, which decreases blood flow into tissues.

(7) _____ in the walls of some arteries keep the brain apprised of blood flow. Long-term control is

exerted by hormones that act on the (8) _____.

Labeling [p.648]

Identify each indicated part of the accompanying illustrations.

9. _____

10. _____

11. _____

12. _____

13. _____ , _____

14. _____

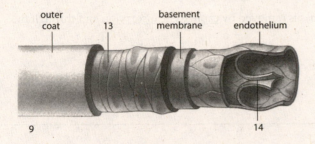

outer coat   13   basement membrane   endothelium

9   14

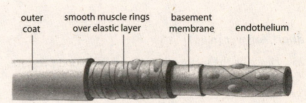

outer coat   smooth muscle rings over elastic layer   basement membrane   endothelium

.11

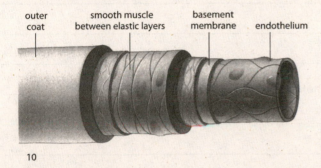

outer coat   smooth muscle between elastic layers   basement membrane   endothelium

10

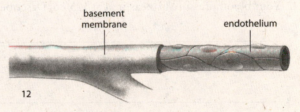

basement membrane   endothelium

12

True/False [pp.650-651]

If the statement is true, write a "T" in the blank. If the statement is false, make it correct by changing the underlined word(s) and writing the correct word(s) in the answer blank.

15. _____ Exchange of nutrients and wastes between cells and blood occurs in the <u>venules</u>.

16. _____ Since diffusion works <u>quickly</u>, cells do <u>not need</u> to be close to the blood vessel.

17. _____ Fluids with small solutes and ions leaks out of capillaries between <u>endothelial</u> cells.

18. _____ Fluid movement in the brain is prevented by <u>hyperendothelial</u> cells lining the capillaries.

19. _____ Because pressure in veins is so low, they have <u>one-way valves</u> to assist with venous return.

20. _____ During exercise there is less blood in the veins because the <u>valves open</u> to allow blood to move.

21. _____ Excess fluid accumulates in the tissues due to <u>vasodilatation</u>.

22. _____ Fluid that accumulates in the tissues is removed by <u>capillary reabsorption</u> and through the <u>lymphatic system</u>.

Short Answer [p.649]

23. Describe the factors that determine blood pressure. _____

_____

_____

_____

24. Your blood pressure is measured at 125/86. Describe what is happening in the heart at each number. _____

_____

_____

_____

Matching [pp.652-653]

Match each term to the most appropriate statement. (One letter is used twice.)

25. _____ atherosclerosis

26. _____ hemolytic anemias

27. _____ hemorrhagic anemias

28. _____ infectious mononucleosis

29. _____ leukemias

30. _____ polycythemias

31. _____ sickle-cell anemia

32. _____ thalassemias

33. _____ hypertension

34. _____ ventricular fibrillation

A. A category of cancers that suppress or impair white blood cell formation in bone marrow

B. When the muscles of the ventricles simply quiver and are unable to move blood out of the heart

C. When blood pressure is chronically high; can lead to an enlarged heart and kidney damage

D. Abnormal hemoglobin formed as a result of a gene mutation

E. Result from a sudden blood loss, as from a severe wound

F. Disorders caused by specific infectious bacteria and parasites as they replicate inside red blood cells and then lyse them

G. An Epstein–Barr virus causes this highly contagious disease, which results from too many monocytes and lymphocytes

H. Disorder involving sluggish blood flow caused by far too many red blood cells; occurs in "blood doping" and some bone marrow cancers

I. Occurs when material collects inside blood vessels and restricts blood flow

37.10. INTERACTIONS WITH THE LYMPHATIC SYSTEM [pp.654-655]

Selected Words: lymph vessels [p.654], lymph capillaries [p.655], tonsils [p.655], adenoids [p.655]

Boldfaced, Page-Referenced Terms

[p.654] lymph vascular system _____

_____

[p.654] lymph _____

_____

[p.655] lymph nodes _____

_____

[p.655] spleen _____

_____

[p.655] thymus gland _____

_____

Matching [pp.654-655]

Choose the correct term for each statement.

1. _____ a huge reservoir of red blood cells and a filter of pathogens and used-up blood cells from the blood

2. _____ delivers water and plasma proteins from capillary beds to the blood vascular system circulation; delivers fats from the small intestine to the blood; delivers pathogens, foreign cells, and material and cellular debris to the organized disposal centers

3. _____ immature T lymphocytes become mature here, and hormones are produced here

4. _____ Contain white blood cells that destroy invading bacteria and viruses gland as they are filtered from the lymph

A. Lymph nodes

B. Lymph

C. Spleen

D. Thymus

Identification/Fill-in-the-Blanks [p.654]

Refer to the illustration on the next page, then supply the missing terms indicated by each answer blank(s).

5. _____

6. _____ gland

7. _____ duct

8. _____

9. _____   _____

10. organized arrays of _____

11. _____   _____

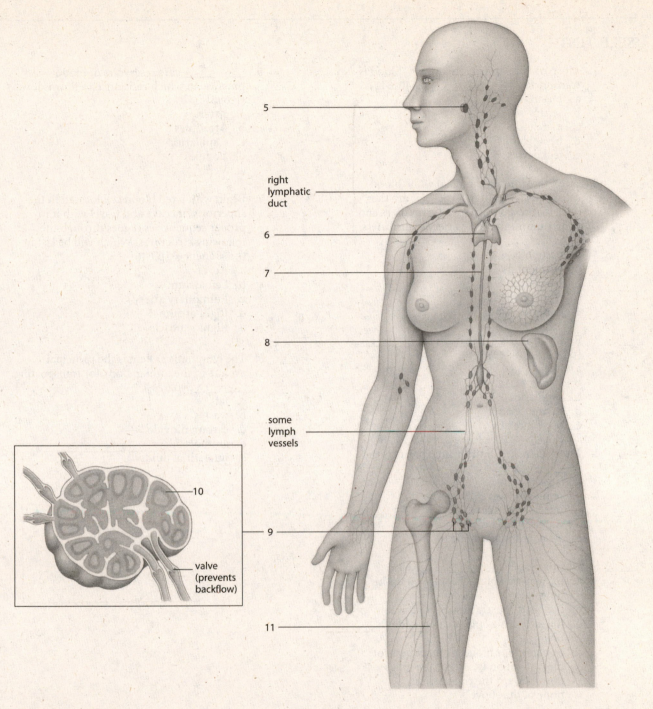

5

right
lymphatic
duct

6

7

8

some
lymph
vessels

10

valve
(prevents
backflow)

9

11

Short Answer [p.654]

12. List the three functions of the lymphatic system. _____

_____

_____

___ 1. Most of the oxygen in human blood is transported by _____. [pp.640-641]
   a. plasma
   b. serum
   c. platelets
   d. hemoglobin
   e. leukocytes

___ 2. Of all the different kinds of white blood cells, two classes of _____ are the ones that respond to specific invaders and confer immunity to a variety of disorders. [p.641]
   a. basophils
   b. eosinophils
   c. monocytes
   d. neutrophils
   e. lymphocytes

___ 3. Red blood cells originate in the _____. [p.641]
   a. liver
   b. spleen
   c. capillaries
   d. thymus gland
   e. bone marrow

___ 4. The pacemaker of the human heart is the _____. [p.647]
   a. sinoatrial node
   b. semilunar valve
   c. inferior vena cava
   d. superior vena cava
   e. atrioventricular node

___ 5. During systole, _____. [p.646]
   a. oxygen-rich blood is pumped to the lungs
   b. the heart muscle tissues contract
   c. the atrioventricular valves suddenly open
   d. oxygen-poor blood from all parts of the human body, except the lungs, flows toward the right atrium
   e. none of the above

___ 6. _____ are reservoirs of blood pressure in which resistance to flow is low. [p.648]
   a. Arteries
   b. Arterioles
   c. Capillaries
   d. Venules
   e. Veins

___ 7. Begin with a red blood cell located in the superior vena cava and travel with it in proper sequence as it goes through the following structures. Which will be last in the sequence? [p.646]
   a. Aorta
   b. Left atrium
   c. Pulmonary artery
   d. Right atrium
   e. Right ventricle

___ 8. The lymphatic system is the principal avenue in the human body for transporting _____. [p.674]
   a. fats
   b. wastes
   c. carbon dioxide
   d. amino acids
   e. interstitial fluids

Matching

9. _____ agglutination [p.642]
10. _____ atherosclerosis [p.652]
11. _____ carotid arteries [p.645]
12. _____ coronary arteries [p.645]
13. _____ edema [p.651]
14. _____ embolism, embolus [p.652]
15. _____ open circulatory system [p.638]
16. _____ hypertension [p.653]
17. _____ inferior vena cava [p.645]
18. _____ jugular veins [p.645]
19. _____ renal arteries [p.645]
20. _____ tachycardia [p.653]
21. _____ thrombosis, thrombus [p.652]

A. Receive(s) blood from the brain, tissues of the head, and neck

B. The heart's own blood supplier(s)

C. Much higher than normal rate of heartbeat; occurs during heavy exercising

D. High blood pressure

E. A blood clot that is on the move from one place to another

F. Deliver(s) blood to the head, neck, brain

G. The clumping of red blood cells, or of antibodies with antigens

H. Delivers blood to kidneys, where its composition and volume are adjusted

I. A blood clot that is lodged in a blood vessel and is blocking it

J. Receive(s) blood from all veins below the diaphragm

K. Progressive thickening of the arterial wall and narrowing of the arterial lumen (space)

L. When blood moves through vessels and mixes with the interstitial fluid

M. Accumulation of excess fluid in interstitial spaces; extreme in elephantiasis

---

CHAPTER OBJECTIVES/REVIEW QUESTIONS

1. Distinguish between open and closed circulatory systems. [p.638]
2. Describe the evolutionary changes that occurred in vertebrate circulatory systems as exemplified in fish, amphibians, and birds/mammals. [pp. 638-639]
3. Describe the composition of human blood, using percentages of volume. [p.640]
4. Distinguish the main types of leukocytes from each other in terms of structure and functions. [pp.640–641]
5. Describe the sequence of events that occurs when a blood vessel is damaged that leads to hemostasis. [p. 642]
6. Explain the ABO and Rh blood types and how they are inherited. [p642-643]
7. Trace the path of blood in the human body. Begin with the aorta and name all major components of the circulatory system through which the blood passes before it returns to the aorta. [pp.644-645]
8. Explain what causes a heart to beat. Then describe how the rate of heartbeat can be slowed down or speeded up. [p.647]
9. Describe how the structures of arteries, capillaries, and veins differ. [p.648]
10. Describe what blood pressure is and how it is measured. Correlate the numbers with physical movements in the heart. [pp. 648-650]
11. State the significance of high- and low-density lipoproteins to cardiovascular disorders. [p.652]
12. Describe how hypertension develops, how it is detected, and whether it can be corrected. [p.653]
13. Describe the composition and function of the lymphatic system. [pp.654–655]

INTEGRATING AND APPLYING KEY CONCEPTS

1. Suppose humans had two-chambered or three-chambered hearts. How would this affect metabolism and activity levels?
2. A person has elevated blood pressure. She takes two medications—one is a diuretic and the other inhibits calcium release in muscle fibers. Explain each of these can lower the blood pressure.
3. You observe some people who appear as though fluid had accumulated in their lower legs and feet. Their lower extremities resemble those of elephants. You inquire about what is wrong and are told that the condition is caused by the bite of a mosquito that is active at night. Construct a testable hypothesis that would explain (1) why the fluid was not being returned to the torso, as normal, and (2) what the mosquito did to its victims.

38
IMMUNITY

INTRODUCTION

Chapter 38 looks at how the immune system protects us through three layers of defense, all integrated with each other. The first layer is the external barriers that try to prevent entry of antigens. The second layer is an innate response. It has the same basic set of responses to draw from regardless of the antigen. Third, there is the adaptive response that custom builds a response to each individual antigen type. It also looks at problems that arise when the system is not functioning properly, either by not responding to antigens or misidentifying and responding incorrectly.

FOCAL POINTS

- Table 38.2 [p.660] lists some of the chemical weapons used by the immune system.
- Table 38.3 [p.662] describes the barriers to pathogen invasion found in vertebrates.
- Figure 38.7 [p.664] shows the results of complement action.
- Figure 39.8 [p.665] illustrates the inflammatory response.
- Figure 38.15 [p.671] diagrams the antibody-mediated immune response.
- Figure 38.17 [p.672] illustrates the cell-mediated immune response.

INTERACTIVE EXERCISES

Impacts, Issues: Frankie's Last Wish [p.658]

38.1. INTEGRATED RESPONSE TO THREATS [pp.660-661]

38.2. SURFACE BARRIERS [pp.662-663]

38.3. REMEMBER TO FLOSS [p.663]

38.4. INNATE IMMUNE RESPONSES [pp.664-665]

Selected Words: cervix [p 658], human papilloma virus [p 658], immunity [p 660], pathogen associated molecular pattern (PAMP) [p.660], antigen [p.661], mast cell [p.661], lmphocyte [p.661], Propionibacterium acnes [p.662], Clostridium tetni [p.662], Corynebacterium diptheriae [p.662], Staphylococcus aureus [p.662], meningitis [p.662], mucus [p.663], lysozyme [p.663], bile salts [p.663], diarrhea [p.663], glycoprotein's [p.663], periodontitis [p.663], inflammation [p.664], fever [p.665]

Boldfaced, Page-Referenced Terms

[p.660] complement _____

[p.660] innate immunity _____

_____

[p.660] adaptive immunity _____

_____

[p.660] antigen _____

_____

[p.661] cytokine _____

_____

[p.661] macrophage _____

_____

[p.661] dendritic cell _____

_____

[p.661] eosinophil _____

_____

[p.661] basophil _____

_____

[p.661] B lymphocyte _____

_____

[p.661] T lymphocyte _____

_____

[p.661] natural killer cell _____

_____

[p.665] inflamation _____

_____

[p.665] fever _____

_____

Fill-in-the-Blanks [pp.660-663]

Immune responses are provoked by the presence of non-self (1) _____ on the surface of invaders. There are approximately (2) _____ different molecules recognized as non-self. These are called (3) _____. An early chemical response is the release of (4) _____ that binds to invaders, killing them or tagging them for phagocytosis. This fast response is part of (5) _____ immunity. When lymphocytes evolved, they developed a customized response to specific antigens, a process called (6) _____ immunity.

The best way to deal with damaging invaders is to prevent their entry. Several barriers prevent pathogens from crossing the boundaries of your body. Intact (7) _____ and (8) _____ membranes are effective barriers.

(9) _____ is an enzyme that destroys the cell wall of many bacteria. (10) _____ fluid destroys many food-borne pathogens in the stomach. Normal (11) _____ residents of the skin, gut, and vagina outcompete pathogens for resources and help keep their numbers under control.

True/False [pp.660-663]

If the statement is true, write a "T" in the blank. If the statement is false, make it correct by changing the underlined word(s) and writing the correct word(s) in the answer blank.

12. _____ A PAMP is a molecular pattern unique to your immune system that(?) enables pathogens to be recognized.

13. _____ Since the body must have openings for nutrients to enter and waste to be removed, these portals are protected by complement molecules.

14. _____ Mucus membranes are sticky and trap many microorganisms; they are then swept away.

15. _____ Among the white blood cells, neutrophils make up the largest number.

16. _____ Immune system cells communicate with each other chemically by use of hormones.

Match each of the following cells with its correct function.

17. _____	Eosinophil	A.	Phagocytes that show antigens to naïve T cells
18. _____	Mast cell	B.	Coordinate immune response; recognize infected or abnormal cells and kill those cells by contact
19. _____	B Lymphocyte	C.	Contains enzymes that help remove parasitic worms; circulates
20. _____	Macrophage	D.	Contain histamine granules; contribute inflammation; fixed in tissues
21. _____	Basophil	E.	Produce antibodies against recognized antigen
22. _____	Dendritic cell	F.	Most abundant phagocyte; circulates
23. _____	Neutrophil	G.	Phagocyte that presents antigens; mature when in tissues
24. _____	T Lymphocyte	H.	Contains histamine granules; contribute inflammation; circulates

Sequence [pp.664-665]

Arrange the following steps in the inflammatory response in correct time sequence. Write the letter of the first step next to 25. The letter of the last step is written next to 30.

25. _____	A. Mast cells in tissue release histamine
26. _____	B. Fluid and proteins in the plasma leak into the tissues causing swelling and pain
27. _____	C. Neutrophil and macrophages are attracted to area and engulf invaders
28. _____	D. Capillaries dilate and become more permeable
29. _____	E. Tissue is damaged and antigens enter the tissue
30. _____	F. Complement proteins attach to bacteria; clotting factors build a barrier around inflamed area

Short Answer [p.665]

31. Explain how a fever helps fight an infection. _____

_____

_____

_____

38.5. OVERVIEW OF ADAPTIVE IMMUNITY [pp.666-667]

38.6. ANTIBODIES AND OTHER ANTIGEN RECEPTORS [pp.668-669]

38.7. THE ANTIBODY-MEDIATED IMMUNE RESPONSE [pp.670-671]

38.8. THE CELL-MEDIATED RESPONSE [pp.672-673]

Selected Words: self versus non-self recognition [p.666], specificity [p.666], diversity [p.666], memory [p.666], naïve B or T cell [p.666], MHC markers [p.666], antigen-MHC complex [p.666], antibody-mediated response [p.667], cell-mediated response [p.667], lymph node [p.667], variable region [p.668], constant region [p.668], IgA, IgG, IgD, IgE, IgM [p.668], dendritic cell [p.670], clonal selection [p.670], receptor-mediated endocytosis [p.670], helperT cell [p.671], effector cell [p.671], memory B and T cell [p.671], cytotoxic T cell [p.672], perforin [p.673], apoptosis [p.773]

Boldfaced, Page-Referenced Terms

[p.666] MHC markers _____

_____

[p.666] T cell receptor (TCR) _____

_____

[p.666] effector cells _____

_____

[p.667] antibody-mediated immune response _____

_____

[p.667] cell-mediated immune response _____

_____

[p.669] B cell receptors _____

_____

Fill-in-the-Blanks [pp.666-673]

If the (1) _____ immune response fails to repel the microbial invaders, then the body calls on its (2)

_____ immune response , which identifies *specific* targets to kill and *remembers* the identities of its targets.

Your own unique(3) _____ patterns identify your cells as "self" cells. Any other surface pattern is, by

definition, (4) _____, and doesn't belong in your body.

The principal actors of the adaptive immune system are (5) _____ descended from stem cells in the

bone marrow that have two different strategies of action to deal with their different kinds of enemies.(6) _____

mediate the antibody response and act principally against the extracellular enemies that are pathogens in blood or on

the cell surfaces of body tissues. (7) _____ defend principally against intracellular pathogens such as (8)

_____, and against any cells that are perceived as abnormal or foreign, such as (9) _____ cells and cells

of organ transplants.

Each kind of cell, virus, or substance bears particular molecular configurations (patterns) that give it a unique (10) _____. A(n) (11) _____ is any molecular configuration on the pathogen's surface that causes the formation of lymphocyte armies. Any cell that processes and displays a(n) (11) _____together with a suitable MHC molecule is known as a(n) (12) _____ cell and can activate lymphocytes to undergo rapid cell divisions. Lymphocyte subpopulations that fight and destroy enemies are known as (13) _____ cells. Long-lived (14) _____ cells are also produced and react to future encounters with the same antigen. The antibody-mediated response is primarily a function of (15) _____ cells, while the cell-mediated response relies on (16) _____ cells.

True/False [pp.666-667]

If the statement is true, write a "T" in the blank. If the statement is false, make it correct by changing the underlined word(s) and writing the correct word(s) in the answer blank.

17. _____ The first step in adaptive immunity is <u>production of antibodies</u>.

18. _____ Pieces of the antigen are displayed on the cell surface as an <u>antigen-MHC complex</u>.

19. _____ When a T cell binds to an antigen-MHC complex, it secretes <u>antibodies</u> to signal other B and T cells.

20. _____ Antibodies are <u>cells</u> that can inactivate antigens.

21. _____ <u>B cells</u> will attack intracellular pathogens.

22. _____ Lymph nodes swell due to an accumulation of <u>antigens</u> in the node.

Matching

Match the antibody type with its function. [pp.668-669]

23. _____ IgM

24. _____ IgG

25. _____ IgE

26. _____ IgA

27. _____ IgD

A. Produced in exocrine secretions; found on mucus-coated surfaces of the respiratory, digestive, and reproductive tracts where they neutralize pathogens

B. Triggers inflammation when pathogens attack the body; plays a role in allergic responses

C. Activate complement proteins; neutralize many toxins; long-lasting; can cross placenta and protect developing fetus; also present in colostrum (early milk) from mammary glands

D. First to be secreted during immune responses; after binding to antigen, trigger complement cascade; also tag invaders and bind them in clumps for later phagocytosis

E. Acts as the B cell receptor for antigens

Sequence [pp.670-671]

Arrange the following steps in the antibody response in correct time sequence. Write the letter of the first step next to 28. The letter of the last step is written next to 37.

28. _____

29. _____

30. _____

31. _____

32. _____

33. _____

34. _____

35. _____

36. _____

37. _____

A. Antibody molecules are secreted by effector cells

B. Antigen-presenting cell binds to naïve T cell

C. Naïve B cell covered with identical antibody molecules (receptors)

D. Effector T cell binds to antigen-MHC complex on B cell and secretes cytokines

E. Antigen binds to antibodies (receptors) on naïve B cell and is taken into cell

F. B cell becomes activated B cell with antigen-MHC complex on cell surface

G. Effector and memory B cells mature

H. Cytokine binds to B cells, triggering mitosis of B cell

I. Dendritic cell engulfs same antigen and becomes antigen-presenting cell

J. T cell divides to produce effector and memory T cells

Sequence [pp.672-673]

Arrange the following steps in the cell-mediated response in correct time sequence. Write the letter of the first step next to 38. The letter of the last step is written next to 44.

38. _____

39. _____

40. _____

41. _____

42. _____

43. _____

44. _____

A. Activated helper T cell secretes cytokines

B. Effector and memory cytotoxic T cells mature

C. Mature cytotoxic T cell binds to antigen-MHC complex of an infected cell

D. Cytotoxic T cells causes infected cell to die

E. Antigen-presenting cell binds to naïve helper T cell and naïve cytotoxic T cell, both become activated

F. Cytokines bind to activated cytotoxic T cell, causing mitosis

G. Dendritic cell engulfs a virus infected cell and becomes antigen-presenting cell

38.9. ALLERGIES [p.673]

38.10. VACCINES [p.674]

38.11. IMMUNITY GONE WRONG [p.675]

38.12. AIDS REVISITED—IMMUNITY LOST [pp.676-677]

Selected Words: antihistamine [p.673], anaphylactic shock [p.673], primary immune response [p.674], active immunization [p 674], secondary immune response [p.674], passive immunization [p.674], Edward Jenner [p.674], vaccination [p.674], autoimmune disorder [p.675], Grave's Disease [p.675], rheumatoid arthritis [p.675], multiple sclerosis [p.675], severe combined immunodeficiency(SCID) [p.675], retrovirus [p.676], reverse transcriptase [p.676], AZT [p.677], protease inhibitor [p.677]

Boldfaced, Page-Referenced Terms

[p.673] allergy _____

_____

[p.673] allergen _____

_____

[p.674] immunization _____

_____

[p.674] vaccine _____

_____

[p.675] autoimmune response _____

_____

[p.676] AIDS _____

_____

Matching [pp.673-677]

Match each of the following terms with its correct description.

1. _____ Immunization
2. _____ Allergy
3. _____ Anaphylactic shock
4. _____ Vaccine
5. _____ Autoimmune disorder
6. _____ SCID
7. _____ Rheumatoid arthritis
8. _____ Multiple sclerosis
9. _____ AIDS
10. _____ Passive immunity

A. A whole-body reaction where arterioles dilate and fluid leaks out causing a precipitous drop in blood pressure

B. A group of disorders that result when immune system is suppressed by a virus infection

C. An immune response against healthy body tissues

D. Antibodies produced in another person are administered

E. A process that induces immunity

F. Occurs when T cells damage myelin sheath of axons

G. A substance made to contain a weakened antigen that induces a primary immune response to the antigen without causing illness

H. An immune response to a harmless substance

I. An autoimmune disease that causes inflammation of soft tissue in joints

J. A primary immune deficiency

True/False [pp.673-677]

If the statement is true, write a "T" in the blank. If the statement is false, make it correct by changing the underlined word(s) and writing the correct word(s) in the answer blank.

11. _____ A vaccination works because it initiates a secondary immune response.

12. _____ In active immunization, antibodies are administered to produce immunity to an antigen.

13. _____ Failure to distinguish between self and non-self cells is the hallmark of autoimmunity.

14. _____ AIDS is a type of primary immune deficiency.

15. _____ HIV is a retrovirus; that means it requires protease to replicate its genome.

16. _____ Grave's disease has a constellation of symptoms caused by a deficiency of thyroid hormone.

17. _____ Allergens induce a variety of responses mediated by histamine release.

18. _____ The first recorded vaccine in Europe was made by Louis Pasteur.

SELF-TEST

___ 1. Pathogen-associated molecular patterns (PAMPs) include _____ . [p.660]
   a. bacterial flagellum proteins
   b. prokaryotic cell wall materials
   c. double stranded RNA
   d. bacterial pilus proteins
   e. all of the above

___ 2. The plasma proteins that are activated when they contact a bacterial cell are collectively known as the _____ system. [p.660]
   a. shield
   b. complement
   c. IgG
   d. MHC
   e. HIV

___ 3. _____ are divided into two groups: T cells and B cells. [p.661]
   a. Macrophages
   b. Lymphocytes
   c. Platelets
   d. Complement cells
   e. Cancer cells

___ 4. _____ produce and secrete antibodies that set up bacterial invaders for subsequent destruction by macrophages. [pp.661, 667]
   a. B cells
   b. Phagocytes
   c. T cells
   d. Bacteriophages
   e. Thymus cells

___ 5. Antibodies are shaped like the letter _____ . [p.668]
   a. Y
   b. W
   c. Z
   d. H
   e. E

___ 6. The markers for every cell in the human body are referred to by the letters _____ . [p.666]
   a. HIV
   b. MBC
   c. RNA
   d. DNA
   e. MHC

___ 7. Effector B cells _____ . [pp.671]
   a. fight against extracellular pathogens and toxins circulating in tissues
   b. develop from antigen-presenting cells
   c. manufacture and secrete antibodies
   d. do not divide and form clones
   e. all of the above

___ 8. The clonal selection hypothesis explains _____ . [p.670]
   a. how self cells are distinguished from non-self cells
   b. how B cells differ from T cells
   c. how so many different kinds of antigen-specific receptors can be produced by lymphocytes
   d. how memory cells are set aside from effector cells
   e. how antigens differ from antibodies

Matching [pp.661-674]

Choose the most appropriate description for each term.

9. _____ allergy

10. _____ antibody

11. _____ antigen

12. _____ macrophage

13. _____ clone

14. _____ complement

15. _____ histamine

16. _____ MHC marker

17. _____ effector B cell

18. _____ T cell

A. Begins its development in bone marrow, but matures in the thymus gland

B. Cells that have directly or indirectly descended from the same parent cell

C. A potent chemical that causes blood vessels to dilate and let protein pass through the vessel walls

D. Y-shaped immunoglobulin

E. A non-self marker

F. A progeny of a turned-on B cell

G. A group of about 30 proteins that participate in the inflammatory response

H. An altered secondary immune response to a substance that is normally harmless to other people

I. The basis for self-recognition at the cell surface

J. Principal perpetrator of phagocytosis

CHAPTER OBJECTIVES/REVIEW QUESTIONS

1. List the general types of cells that form the basis of the vertebrate immune system. [p.661]
2. List and discuss four nonspecific defense responses that serve to exclude microbes from the body. [pp.662-663]
3. Describe the sequence of events that occur during inflammatory responses. [p.665]
4. Explain why the immune system of mammals usually does not attack "self" tissues. Understand how vertebrates (especially mammals) recognize and discriminate between self and non-self tissues. [p.666]
5. Distinguish between the antibody-mediated response pattern and the cell-mediated response pattern. [pp.666-667]
6. Describe the clonal selection theory, and tell what it helps to explain. [pp.670-671]
7. Describe two ways that people can be immunized against specific diseases. [p.674]
8. Distinguish allergies from autoimmune disorders. [pp.673-675]
9. Describe some examples of immune failures, and identify as specifically as you can which weapons in the immunity arsenal failed in each case. [pp.675-677]
10. Describe how AIDS specifically interferes with the human immune system. [pp.676-677]

INTEGRATING AND APPLYING KEY CONCEPTS

1. Discuss the advantages of having resident populations of microbes on the skin.
2. What parts of the immune system might slow or stop AIDS if they were enhanced?

39
RESPIRATION

INTRODUCTION

This chapter covers the biological processes involved in the exchange of gases in animals. All animals require an input of oxygen to fuel aerobic respiration, and a release of the waste gas carbon dioxide. This chapter introduces the principles of a respiratory system, as well as the various forms of respiratory systems in animals. It also explores the basis of gas exchange and some of the more common diseases and disorders that may occur when normal gas exchange is interrupted.

FOCAL POINTS

- Figure 39.2 [p.682] illustrates how the respiratory system interacts with other organ systems of the human body.
- Figure 39.10 [p.686] shows a respiratory system adapted for under water use—gills in fish.
- Figure 39.13 [p.688] illustrates the major structures of the respiratory system in humans. You should note that some of these structures have more than one function in human physiology.
- Figure 39.15 [p.690] diagrams how air moves into and out of the lungs.

INTERACTIVE EXERCISES

Impacts, Issues: Up in Smoke [p.680]

39.1. THE NATURE OF RESPIRATION [pp.682-683]

39.2. GASPING FOR OXYGEN [p.683]

39.3. INVERTEBRATE RESPIRATION [pp.684-685]

39.4. VERTEBRATE RESPIRATION [pp.686-687]

Selected Words: hemerythrin [p.683], dissolved oxygen (DO) [p.683], gill filaments [p.684], spiracle [p.685], hemocyanin [p.685], pharynx [p.686], tetrapods [p.686], amniotes [p.687]

Boldfaced, Page-Referenced Terms

[p.682] respiration _____

_____

[p.682] partial pressure _____

_____

[p.682] respiratory surface _____

_____

[p.683] surface-to-volume ratio _____

_____

[p.683] ventilation _____

_____

[p.683] respiratory proteins _____

_____

[p.683] hemoglobin _____

_____

[p.683] myoglobin _____

_____

[p.684] integumentary exchange _____

_____

[p.684] gills _____

_____

[p.684] lung _____

_____

[p.685] tracheal system _____

_____

[p.685] book lungs _____

_____

[p.686] countercurrent exchange _____

_____

Labeling [p.682]

Identify the numbered parts of the accompanying diagram, which illustrates the interconnection of the respiratory system with other body systems.

1. _____
2. _____
3. _____
4. _____
5. _____
6. _____
7. _____

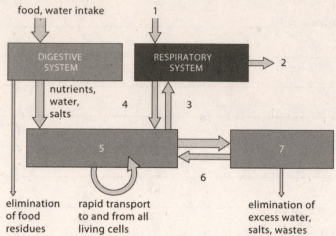

food, water intake

1

DIGESTIVE SYSTEM

RESPIRATORY SYSTEM

2

nutrients, water, salts

4   3

5

7

6

elimination of food residues

rapid transport to and from all living cells

elimination of excess water, salts, wastes

Matching [p.682-683]

Choose the most appropriate statement for each of the following terms.

8. _____ hemoglobin
9. _____ respiration
10. _____ surface-to-volume ratio
11. _____ myoglobin
12. _____ partial pressure
13. _____ ventilation
14. _____ respiratory surface

A. The name given to adaptations that increase gas exchange rates above the level of diffusion

B. The size of animal bodies is governed by this principle

C. The main respiratory pigment of humans

D. The sum of the physiological processes that move oxygen and carbon dioxide

E. The contribution of an individual gas to the total atmospheric pressure

F. Gases enter and leave an organism by first crossing this structure

G. The respiratory pigment found in most muscle cells

True/False [p.683]

If the statement is true, write a "T" in the blank. If the statement is false, make it correct by changing the underlined word(s) and writing the correct word(s) in the answer blank.

15. _____ Warm, slow flowing water contains larger amounts of dissolved oxygen.

16. _____ Over-enriched lake waters support rapid growth of algae; as nutrients are used, the algae die and the water is rapidly depleted of oxygen.

17. _____ When oxygen levels in a lake or stream are low, aquatic insect larvae are the last to disappear.

18. _____ Fish will survive in water until the dissolved oxygen level falls below two parts per million.

19. _____ Waters with the lowest oxygen concentrations often contain only sludge worms.

Choice [pp.684-685]

For items 16–20, choose from the following forms of invertebrate respiration.

    a. tracheal system   b. integumentary exchange   c. gills   d. book lungs

20. _____ The diffusion of a gas directly across the body covering

21. _____ A form of respiratory system commonly found in aquatic organisms

22. _____ An internal system of tubes that act as a respiratory surface

23. _____ Commonly found in insects, millipedes, and centipedes

24. _____ Thin-walled respiratory surfaces that exchange gases between a body fluid and its surroundings

25. _____ Found in some spiders in place of tracheal tubes

Choice [pp.686-687]

Match each respiratory system characteristic to the group of vertebrates that has it.

    a. fish   b. amphibians   c. birds

26. _____ Utilize a countercurrent flow to effectively exchange gases

27. _____ Utilize a flow-through respiratory system

28. _____ Have gills as larvae, then develop lungs

29. _____ Respiratory surfaces remove 80 to 90 percent of the oxygen from the environment

30. _____ These organisms use their skin to supplement gas exchange

39.5. HUMAN RESPIRATORY SYSTEM [pp.688-689]

39.6. CYCLIC REVERSALS IN AIR PRESSURE GRADIENTS [pp.690-691]

Selected Words: vocal cords [p.688], nostrils [p.689], "Adam's apple" [p.689], laryngitis [p.689], pleural membrane [p.689], inhalation [p.690], exhalation [p.690], medulla oblongata [p.691], chemoreceptors [p.691], partial pressure [p.691]

Boldfaced, Page-Referenced Terms

[p.689] pharynx _____

_____

[p.689] larynx _____

_____

[p.689] glottis _____

_____

[p.689] epiglottis _____

_____

[p.689] trachea _____

_____

[p.689] bronchus _____

_____

[p.689] bronchioles _____

_____

[p.689] alveoli _____

_____

[p.689] diaphragm _____

_____

[p.689] intercostal muscles _____

_____

[p.690] respiratory cycle _____

_____

[p.690] Heimlich maneuver _____

_____

[p.690] vital capacity _____

_____

[p.690] tidal volume _____

_____

Labeling and Matching [pp.688-689]

First, identify each structure indicated in the diagram on the next page. Then match the structure with its correct function and place the corresponding letter in the parentheses. [pp.688-689]

1. _____ (   )

2. _____ (   )

3. _____ (   )

4. _____ (   )

5. _____ (   )

6. _____ (   )

7. _____ (   )

8. _____ (   )

9. _____ (   )

10. _____ (   )

11. _____ (   )

12. _____ (   )

13. _____ (   )

A. A supplemental airway

B. The site of gas exchange

C. Smooth muscle that separates the thoracic and abdominal cavities

D. One of a pair of lobed, elastic organs involved in gas exchange

E. The site of sound production

F. Airway that connects the nasal cavity with the larynx

G. Warms and filters incoming air

H. Skeletal muscles with roles in breathing

I. Airway that connects the larynx to bronchi leading to lungs

J. The fine branches of the "bronchial tree"

K. A double membrane that contains a lubricating fluid

L. Groups of alveoli located at the end of a bronchiole

M. Separates the respiratory system from the digestive system

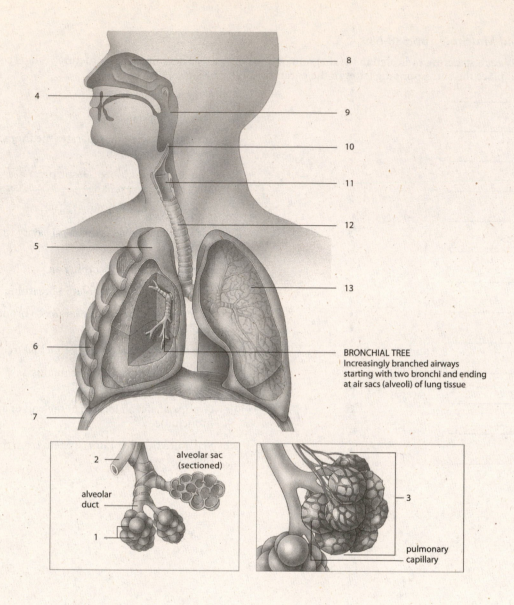

BRONCHIAL TREE
Increasingly branched airways starting with two bronchi and ending at air sacs (alveoli) of lung tissue

alveolar sac (sectioned)

alveolar duct

pulmonary capillary

Choice [pp.690-691]

Choose whether each of the following statements is associated with inhalation or exhalation.

  a. inhalation   b. exhalation

14. _____ Always requires muscle activity

15. _____ External intercostal muscles contract

16. _____ Occurs passively when one is breathing quietly

17. _____ Internal intercostal muscles contract

18. _____ Diaphragm flattens and moves down

19. _____ Requires muscle activity only during exercise

20. _____ Diaphragm relaxes and returns to dome shape

21. _____ Pressure in alveoli is below atmospheric pressure

22. _____ Pressure in alveoli is greater than atmospheric pressure

Matching [pp.690-691]

Match each of the following terms with its correct description.

23. _____ Tidal volume

24. _____ Medulla oblongata

25. _____ Heimlich maneuver

26. _____ vital capacity

27. _____ Carbonic acid

28. _____ Carotid arteries

29. _____ Sympathetic nerves

A. A process performed on someone to clear the airway of someone choking

B. Site of chemoreceptors to measure need for changes in breathing rate

C. Volume of air moved in and out lungs in a normal respiratory cycle

D. Produced when CO_2 combines with water in the blood

E. Nerves that increase breathing rate

F. Maximum volume of air that lungs can move in one breath

G. Site of neurons that are control center for breathing

39.7. GAS EXCHANGE AND TRANSPORT [pp.692-693]

39.8. RESPIRATORY DISEASES AND DISORDERS [pp.694-695]

39.9. HIGH CLIMBERS AND DEEP DIVERS [pp.696-697]

Selected Words: alveolus [p.692], pulmonary capillary [p.692], heme [p.692], carbaminohemoglobin [p.692], bicarbonate [p.692], apnea [p.694], sudden infant death syndrome (SIDS) [p.694], tuberculosis [p.694], pneumonia [p.694], bronchitis [p.694], emphysema [p.694], hypoxia [p.696], hyperventilate [p.696], "rapture of the deep" [p.696], "the bends" [p.696]

Boldfaced, Page-Referenced Terms

[p.692] respiratory membrane _____

_____

[p.692] oxyhemoglobin _____

_____

[p.693] carbonic anhydrase _____

_____

[p.693] carbon monoxide _____

_____

Choice [pp.692-693]

For each statement, choose the appropriate gas.

a. oxygen   b. carbon dioxide   c. carbon monoxide

1. _____ When bound to hemoglobin, forms carbaminohemoglobin
2. _____ Heme groups with iron bind with this gas
3. _____ This is released where the blood is warmer and pH is lower
4. _____ The enzyme carbonic anhydrase is involved in the transport of this gas
5. _____ When bound to hemoglobin it forms oxyhemoglobin
6. _____ In water, this gas forms carbonic acid
7. _____ Binds to hemoglobin the most tightly
8. _____ Binds with myoglobin in skeletal and cardiac muscles

Matching [pp.694-697]

Match each of the following disorders or conditions to its correct description.

9. _____ bronchitis
10. _____ tar
11. _____ altitude sickness
12. _____ emphysema
13. _____ nitrogen narcosis
14. _____ secondhand smoke
15. _____ hypoxia
16. _____ decompression sickness

A. The general condition when not enough oxygen reaches the cells

B. Ion balances in the cerebrospinal fluid are incorrect due to hyperventilation

C. Disruption of neural membranes due to incorrect nitrogen levels

D. Nitrogen bubbles form in the blood and tissues

E. Collection of toxic particles found in cigarette and marijuana smoke

F. Tobacco and marijuana smoke found in the air

G. Inflammation of the epithelium of the bronchioles

H. Destruction of the thin walls of the alveoli

SELF-TEST

Matching [pp.688-697]

1. _____ bronchioles
2. _____ bronchitis
3. _____ carbonic anhydrase
4. _____ emphysema
5. _____ apnea
6. _____ glottis
7. _____ hypoxia
8. _____ external intercostal muscles
9. _____ larynx
10. _____ oxyhemoglobin
11. _____ pharynx
12. _____ pleurisy
13. _____ tidal volume
14. _____ ventilation
15. _____ vital capacity

A. Membrane that encloses human lung becomes inflamed and swollen; painful breathing generally results

B. HbO_2

C. The amount of air inhaled and exhaled during normal breathing of a human at rest; generally about 500 ml

D. Throat passageway that connects to *both* the lower respiratory tract *and* the digestive tract

E. When breathing stops and then spontaneously starts repetitively

F. Inflammation of the two principal passageways that carry air into the human lungs

G. Contract when lungs are filling with air, relax when air is leaving the lungs

H. The opening into the "voicebox"

I. Finer and finer branchings that lead to alveoli

J. Maximum volume of air that can move out of your lungs after a single, maximal inhalation

K. An enzyme that increases the rate of production of H_2CO_3 from CO_2 and H_2O

L. Lungs have become distended and inelastic so that walking, running, and even exhaling are difficult

M. Where sound is produced by vocal cords

N. Movements that keep air or water moving across a respiratory surface

O. Too little oxygen is being distributed in the body's tissues

Multiple Choice

___ 16. The respiratory transport molecule that is abundant in muscle and skeletal cells of the human body is _____. [p.692]
a. myoglobin
b. hemoglobin
c. carbonic acid
d. carbonic anhydrase

___ 17. _____ is the most abundant gas in Earth's atmosphere. [p.682]
a. Water vapor
b. Oxygen
c. Carbon dioxide
d. Hydrogen
e. Nitrogen

___ 18. With respect to respiratory systems, countercurrent flow is a mechanism that explains how _____. [p.686]
a. oxygen uptake by blood capillaries in the lamellae of fish gills occurs
b. ventilation occurs
c. intrapleural pressure is established
d. sounds originating in the vocal cords of the larynx are formed
e. all of the above

___ 19. A flow-through respiratory system is found in _____. [p.687]
a. amphibians
b. reptiles
c. birds
d. mammals
e. humans

___ 20. During inhalation, _____. [p.690]
a. the pressure in the thoracic cavity (intrapleural pressure) is less than the pressure within the lungs (intrapulmonary pressure)
b. the pressure in the chest cavity (intrapleural pressure) is greater than the pressure within the lungs (intrapulmonary pressure)
c. the diaphragm moves upward and becomes more curved
d. the thoracic cavity volume decreases
e. all of the above

___ 21. Oxygen moves from alveoli to the bloodstream _____. [pp.692-693]
a. by diffusion when the concentration of oxygen is greater in alveoli than in the blood
b. by means of active transport
c. using the assistance of carbaminohemoglobin
d. principally due to the activity of carbonic anhydrase in the red blood cells
e. all of the above

___ 22. Immediately before reaching the alveoli, air passes through the _____. [p.688]
a. bronchioles
b. glottis
c. larynx
d. pharynx
e. trachea

___ 23. Oxyhemoglobin _____. [pp.692-693]
a. releases oxygen more readily in metabolically active tissues
b. tends to release oxygen in places where the temperature is lower
c. tends to hold onto oxygen when the pH of the blood drops
d. tends to give up oxygen in regions where partial pressure of oxygen exceeds that in the lungs
e. all of the above

___ 24. Which of the following is not involved in carbon dioxide transport? [pp. 692-693]
a. CO_2 is bound to hemoglobin to form carbaminohemoglobin.
b. CO_2 is dissolved directly in the blood.
c. CO_2 is transported as bicarbonate ions.
d. CO_2 binds to myoglobin for transport to alveoli.

___ 25. Which of the following respiratory ailments is common among smokers? [pp.694-695]
a. emphysema
b. bronchitis
c. cancer
d. impaired immune system
e. all of the above

CHAPTER OBJECTIVES/REVIEW QUESTIONS

1. Understand how the human respiratory system interacts with the digestive, circulatory, and urinary systems. [p.682]
2. Understand the physical properties of gases and the limitations of a respiratory surface. [p.682]
3. Recognize the difference between myoglobin and hemoglobin in human respiration. [p.683]
4. List the types of invertebrate respiratory surfaces that participate in gas exchange and give an example of each. [pp.684-685]
5. Define *countercurrent flow* and explain how it works. [p.686]
6. Recognize the difference in the respiratory system of birds compared to amphibians and mammals. [pp.686-687]
7. List all the principal parts of the human respiratory system, and explain how each structure contributes to transporting oxygen from the external world to the bloodstream. [pp.688-689]
8. Trace oxygen transport from the air to the tissues of the body. [pp.688-689]
9. Explain the factors that influence the release of oxygen to the tissues. [p.692]
10. Trace the transport of carbon dioxide from the tissues of the body to the lungs. [pp.688-689, 692]
11. Explain why carbon monoxide is a dangerous gas. [p.693]
12. Describe the respiratory cycle and the processes of inhalation and exhalation. [pp.690-691]
13. Explain the difference between vital capacity and tidal volume. [p.690-691]
14. Explain the causes of the major breathing disorders. [pp.694-695]
15. Describe the interaction of the nervous system with the respiratory system. [p.691]
16. Explain how bronchitis, emphysema, and lung cancer are all related to smoking. [pp.694-695]
17. List some of the challenges of respiration at high altitude and during diving. [pp.696-697]

INTEGRATING AND APPLYING KEY CONCEPTS

1. Consider the amphibians—animals that generally have aquatic larval forms (tadpoles) and terrestrial adults. Outline the respiratory changes that you think might occur as an aquatic tadpole metamorphoses into a land-going juvenile.
2. In the movie *Waterworld,* Kevin Costner's character develops gills to aid his underwater breathing. Explain what other changes would be necessary to his respiratory system in order to make this adaptation effective.
3. Sickle-cell anemia is often considered to be a disease of the circulatory system since it alters the shape of the red blood cells (RBCs) in the body. RBCs are the body's carriers of hemoglobin; thus the disease can also be considered a form of respiratory disease. Explain why.
4. Explain why the circulating fluid in insects does not have respiratory proteins.

40

DIGESTION AND HUMAN NUTRITION

INTRODUCTION

Chapter 40 describes the anatomy and physiology of the digestive system. Although it looks at systems in various animals, it concentrates on humans. It follows food through the system indicating where it is digested and where the resulting products are absorbed. The chapter also discusses what happens to each of the nutrients and human nutrition. Included in nutrition are what nutrients we need and what regulates appetite.

FOCAL POINTS

- Figure 40.2 [p.702] diagrams the relationship of the digestive system with other organ systems.
- Figure 40.5 [p.704] illustrates the human digestive system and its functions.
- Table 40.1 [p.706] summarizes the function and source of many digestive enzymes.
- Figures 40.8 and 40.9 [pp.707-708] detail the structure of the small intestine.
- Figure 40.10 [p.709] summarizes digestion and absorption in the small intestine.
- Figure 40.12 [p.711] summarizes major pathways in metabolism and liver functions. Figure 40.13 lists USDA nutritional guidelines.
- Tables 40.4 and 40.5 [pp.714-715] summarize information on major vitamins and minerals needed by humans.
- Section 40.10 [pp.716-717] illustrates how to calculate body mass index (BMI), desired caloric intake, and ideal weight.

INTERACTIVE EXERCISES

Impacts, Issues: Hormones and Hunger [p.700]

40.1. THE NATURE OF DIGESTIVE SYSTEMS [pp.702-703]

40.2. OVERVIEW OF THE HUMAN DIGESTIVE SYSTEM [pp.704-705]

40.3. FOOD IN THE MOUTH [p.705]

Selected Words: obesity [p.700], leptin [p.700], ghrelin [p.700], gastric bypass surgery [p.700], saclike gut [p.702], mechanical processing and motility [p.702], secretion [p.702], digestion [p.702], absorption [p.703], elimination [p.703], crop [p.703], gizzard [p.703], cloaca [p.703], molar [p.703], accessory organs [p.705], epiglottis [p.705], gastroesophageal reflux disease [p.705], mechanical digestion [p.705], incisors [p.705], canines [p.705], chemical digestion [p.705]

Boldfaced, Page-Referenced Terms

[p.702] digestive system _____

_____

[p.702] incomplete digestive system _____

_____

[p.702] complete digestive system _____

_____

[p.703] ruminants _____

_____

[p.705] gastrointestinal tract _____

_____

[p.705] pharynx _____

_____

[p.705] esophagus _____

_____

[p.705] peristalsis _____

_____

[p.705] stomach _____

_____

[p.705] sphincter _____

_____

[p.705] small intestine _____

_____

[p.705] large intestine _____

_____

[p.705] rectum _____

_____

[p.705] anus _____

_____

[p.705] dentin _____

_____

[p.705] enamel _____

_____

[p.705] salivary gland _____

_____

[p.705] salivary amylase _____

_____

Fill-in-the-Blanks [pp.700-705]

In humans, two hormones help control feeding behavior. (1) _____, secreted by adipose tissue, suppresses appetite and (2) _____, secreted by the stomach and brain, makes you feel hungry. In the U.S., (3) _____, an overabundance of fat in adipose tissue, is a major health problem. Some extremely obese people opt for (4) _____, in which the digestive tract is surgically shortened. Overall, (5) _____ encompasses the entire range of processes by which food is taken in, digested, absorbed and then used by the body.

A digestive system is able to break down food both (6) _____ and (7) _____ to molecules that are small enough for (8) _____. The (9) _____ system can then distribute nutrients to cells throughout the body. A(n) (10) _____ digestive system has only one opening and two-way traffic. Flatworms have a highly branched gut cavity that serves both digestive and (11) _____ functions. A(n) (12) _____ digestive system has a tube or cavity with regional specializations and a(n) (13) _____ at each end. (14) _____ that break up food involve the muscular contraction of the gut wall, and (15) _____ is the release into the lumen of enzyme fluids and other substances required to carry out digestive functions.

The pronghorn antelope feeds on plant material that breaks down slowly. Antelopes are (16) _____: hoofed mammals that have multiple stomach chambers; nutrients are released slowly when the animal rests. In comparison with human molars, the antelope molar has a much higher (17) _____; its teeth wear down rapidly because its plant diet is mixed with abrasive bits of dirt. The human digestive system is a tube, 21–30 feet long in an adult, that has regions specialized for different aspects of digestion and absorption; they are, in order, the mouth, pharynx, esophagus, (18) _____, (19) _____, large intestine, rectum, and (20) _____.

Various (21) _____ organs secrete enzymes and other substances that are also essential to the breakdown and absorption of nutrients; these include the salivary glands, liver, gallbladder, and (22) _____. Saliva contains water, mucus, and an enzyme called (23) _____ that breaks down starch. A cartilaginous flap called the (24) _____ closes off the trachea when food is swallowed. The (25) _____ is a muscular tube whose contractions propel food to the stomach.

Labeling [p.704]
Identify each numbered structure in the accompanying illustration.

26. _____

27. _____

28. _____

29. _____

30. _____

31. _____

32. _____

33. _____

34. _____

35. _____

36. _____

Matching [pp.702-705]

Match each of the following terms with its correct description.

37. _____ digestion

38. _____ secretion

39. _____ mechanical processing and motility

40. _____ elimination

41. _____ absorption

42. _____ enamel

43. _____ dentin

44. _____ molar

45. _____ canine

46. _____ salivary amylase

A. Bone-like substance that makes up teeth

B. An enzyme that breaks down starch

C. Movements that break up, mix and move food through digestive tract

D. Hard material that covers teeth

E. Breakdown of food into smaller particles and nutrient molecules

F. Tooth with flat, grinding surface

G. Release of digestive enzymes into digestive tract

H. Teeth that shear food

I. Take in digested molecules and water

J. Expulsion of undigested materials

40.4. FOOD BREAKDOWN IN THE STOMACH AND SMALL INTESTINE [pp.706-707]

40.5. ABSORPTION FROM THE SMALL INTESTINE [pp.708-709]

40.6. THE LARGE INTESTINE [p.710]

40.7. METABOLISM OF ABSORBED ORGANIC COMPOUNDS [p.711]

Selected Words: hydrochloric acid [p.706], pepsinogen [p.706], pepsin [p.706], gastrin [p.706], pyloric sphincter [p.706], ulcer [p.706], lipoproteins [p.709], micelle [p.709], cecum [p.710], colon [p.710], rectum [p.710], constipation [p.710], diarrhea [p.710], appendicitis [p.710], polyps [p.710], colonoscopy [p.710], hepatic portal vein [p.711]

Boldfaced, Page-Referenced Terms

[p.706] mucosa _____

_____

[p.706] gastric fluid _____

_____

[p.706] chyme _____

_____

[p.707] bile _____

_____

[p.707] gall bladder _____

_____

[p.707] emulsification _____

_____

[p.708] brush border cells _____

_____

[p.708] microvilli _____

_____

[p.710] feces _____

_____

[p.710] appendix _____

_____

[p.711] liver _____

_____

Matching [pp.706-710]

Match each organ with its main function.

1. _____ small intestine
2. _____ colon
3. _____ gallbladder
4. _____ mouth
5. _____ pancreas
6. _____ stomach
7. _____ rectum
8. _____ liver

A. Starts polysaccharide breakdown

B. Stores, mixes, dissolves food; kills many microorganisms; starts protein breakdown; empties in a controlled way

C. Digests and absorbs most nutrients

D. Produces enzymes that break down all major food molecules; produces buffers against hydrochloric acid from stomach

E. Secretes bile for fat emulsification

F. Stores, concentrates bile from liver

G. Stores, concentrates undigested matter by absorbing water and salts

H. Controls elimination of undigested and unabsorbed residues

Complete the Table [pp.706-710]

9. Complete the following table about the human digestive system.

Nutrient	Where Digested	Where Absorbed	Absorbed As
a. carbohydrate			
b. protein			
c. lipid			
d. nucleic acid			

True/False [pp.706-711]

If the statement is true, write a "T" in the blank. If the statement is false, make it correct by changing the underlined word(s) and writing the correct word(s) in the answer blank.

10. _____ Chyme is an <u>alkaline</u> mixture of food and chemicals secreted by the stomach.

11. _____ Gastrin is a hormone secreted by the lining of the <u>small intestine</u> in response to food entering it.

12. _____ Material entering the small intestine through the pyloric sphincter is immediately neutralized by the addition of <u>bicarbonate ions</u>.

13. _____ Cholecystokinin stimulates <u>stomach acid</u> secretion.

14. _____ The folding of the lining of the small intestine <u>increases</u> surface area for absorption.

15. _____ Water is absorbed across the wall of the intestine by <u>active transport</u>.

16. _____ Micelles are tiny droplets of <u>amino acids</u> that form and then can be absorbed.

17. _____ Movement of material from the colon to the rectum is stimulated by the hormone <u>secretin</u>.

18. _____ The large intestine compacts feces by absorbing <u>sodium ions</u>.

40.8. HUMAN NUTRITIONAL REQUIREMENTS [pp.712-713]

40.9. VITAMINS, MINERALS, AND PHYTOCHEMICALS [pp.714-715]

40.10. WEIGHTY QUESTIONS, TANTALIZING ANSWERS [pp.716-717]

Selected Words: saturated fats [p.712], trans-fatty acid [p.712], complex carbohydrates [p.712], soluble fiber [p.712], unsaturated fats [p.712], omega-3 fatty acid [p.712], olive oil [p.712], ketones [p.713], lutein [p.715], macular degeneration [p.715], body mass index [p.716], Calories (kilocalories) [p.716]

Boldfaced, Page-Referenced Terms

[p.712] essential fatty acids _____

_____

[p.713] essential amino acids _____

_____

[p.714] vitamins _____

_____

[p.714] minerals _____

_____

[p.715] phytochemicals _____

_____

Fill-in-the-Blanks [pp.712-715]

The newest USDA guidelines recommend eating more (1) _____ and _____, fat-free or low-fat (2) _____ products, and (3) _____. They also recommend lowering intake of (4) _____ grains, (5) _____ fats, (6) _____ fatty acids, and added (7) _____ and having no more than one teaspoon of (8) _____ each day. (7) sabotages a diet because it has a(n) (9) _____ index, which prevents cells from dipping into (10) _____ stores. (11) _____ is an especially bad sugar since, unlike most other sugars, it does not stimulate secretion of hormones that suppress appetite. A variety of diet plans exist. In the (12) _____ diet, (13) _____ oil represents most of the fat and 40% of the calories. (14) _____ diets use proteins and fats as the major calorie sources. In addition to energy sources, all diets need (15) _____, organic substances needed in small quantities for growth and survival, and (16) _____, essential inorganic compounds. (15) have many roles in cells but most of the B vitamins act as (17) _____ in metabolism.

Match each of the vitamins and minerals with their function.

18. _____ vitamin A		A.	Bone and tooth formation; blood clotting; nerve and muscle activity
19. _____ biotin		B.	Needed in synthesis of melanin and hemoglobin and in electron transfer chains
20. _____ folate			
21. _____ vitamin C		C.	Thyroid hormone formation
22. _____ vitamin D		D.	Found in hemoglobin and electron transfer proteins
23. _____ vitamin K		E.	Visual pigments; bone and teeth
24. _____ iodine		F.	Bone growth and mineralization; calcium absorption
25. _____ iron		G.	Blood clotting; ATP formation
26. _____ calcium		H.	Coenzyme in nucleic acid and amino acid metabolism
27. _____ copper		I.	Coenzyme in fat and glycogen metabolism
		J.	Collagen synthesis; carbohydrate metabolism

Complete the Table [p.734]

28. Complete the following table by determining how many kilocalories the people described should take in daily, given the stated exercise level, in order to maintain their weight.

Height	Age	Sex	Level of Physical Activity	Present Weight (lb)	Number of Kilocalories/Day
5' 6"	25	Female	Moderately active	138	a.
5' 10"	18	Male	Very active	145	b.
5' 8"	53	Female	Not very active	143	c.

Short Answer [p.734]

29. You are a 19-year-old male, very sedentary (TV, sleep, and computers), 6 feet tall, medium frame, and you weigh 195 pounds. Calculate the number of calories required to sustain your desired weight. Are you underweight, overweight, or just right? _____

_____

_____

_____

_____

___ 1. The process that moves nutrients into the blood or lymph is _____. [pp.702-703]
   a. ingestion
   b. absorption
   c. assimilation
   d. digestion
   e. none of the above

___ 2. The enzymatic digestion of proteins begins in the _____. [p.706]
   a. mouth
   b. stomach
   c. liver
   d. pancreas
   e. small intestine

___ 3. The enzymatic digestion of starches begins in the _____. [p.705]
   a. mouth
   b. stomach
   c. liver
   d. pancreas
   e. small intestine

___ 4. The greatest amount of absorption of digested nutrients occurs in the _____. [pp.708-709]
   a. stomach
   b. pancreas
   c. liver
   d. colon
   e. small intestine

___ 5. Glucose moves through the membranes of the small intestine mainly by _____. [p.709]
   a. peristalsis
   b. osmosis
   c. diffusion
   d. active transport
   e. bulk flow

___ 6. Which of the following is not found in bile? [p.707]
   a. lecithin
   b. salts
   c. digestive enzymes
   d. cholesterol
   e. pigments

___ 7. The element needed by humans for blood clotting, nerve impulse transmission, and bone and tooth formation is _____. [p.715]
   a. magnesium
   b. iron
   c. calcium
   d. iodine
   e. zinc

Matching [pp.712-715]

Match each numbered item with its description in the lettered list.

8. vitamins C and E

9. complex carbohydrates

10. essential amino acids

11. essential fatty acids

12. mineral

13. rickets

14. scurvy

15. vitamin

A. Linoleic acid is one example

B. Phenylalanine, lysine, and methionine are three of eight

C. Combine with free radicals; counteract their destructive effects on DNA and cell membranes

D. Vitamin C deficiency

E. Vitamin D deficiency in young children

F. Organic substance that helps enzymes to do their jobs; required in small amounts for good health

G. Inorganic substance required for good health

H. Long chains of simple sugars; in grains and white potatoes

CHAPTER OBJECTIVES/REVIEW QUESTIONS

1. Distinguish between incomplete and complete digestive systems and tell which is characterized by: (a) specialized regions, (b) two-way traffic, and (c) discontinuous feeding. [p.702]
2. Define and distinguish among *motility, secretion, digestion,* and *absorption.* [pp.702-703]
3. List all parts (in order) of the human digestive system through which food actually passes. Then list the accessory organs that contribute one or more substances to the digestive process. [pp.704-705]
4. Tell which foods undergo digestion in each of the following parts of the human digestive system and state what the food is broken into: mouth, stomach, small intestine, large intestine. [pp.706-707]
5. Describe the cross-sectional structure of the small intestine, and explain how its structure is related to its function. [pp.707-709]
6. List the items that leave the digestive system and enter the circulatory system during the process of absorption. [pp.708-709]
7. State which processes occur in the colon (large intestine). [p.710]
8. Construct an ideal diet for yourself for one 24-hour period. Calculate the number of calories necessary to maintain your weight [see pp.716-717] and then use the nutritional guidelines [p.712] to choose exactly what to eat and how much.
9. Name five vitamins and five minerals that are important in human nutrition, and state the specific role of each. [pp.714-715]

INTEGRATING AND APPLYING KEY CONCEPTS

1. Suppose you could not eat food for two weeks and you had only water to drink. List in correct sequential order the measures your body would take to try to preserve your life. Mention the command signals that are given as one after another critical point is reached, and tell which parts of the body are the first and the last to make up for the deficit.
2. What relationship do hormones have to human appetite? How might controlling some of these hormones be used to maintain a proper weight?
3. The bacterium *Vibrio cholorae* is a normal inhabitant of small crustaceans called copepods that live in salt water. It attaches to the intestinal wall and helps the organism regulate the osmotic movement of Na+ and water. In humans, it attaches to the intestinal wall and attempts to do the same thing. Predict the result of this activity in humans.

41

MAINTAINING THE INTERNAL ENVIRONMENT

INTRODUCTION

This chapter looks at kidney function and its relationship to homeostasis. It investigates what the extracellular fluid should contain. The structure of the human urinary system is described and the function of each portion is explained. It also briefly explores the process of temperature regulation.

FOCAL POINTS

- Figure 41.6 [p.724] describes how the urinary system functionally interacts with other human organ systems.
- Figures 41.9 and 41.10 [pp.726-727] diagram the human urinary system.
- Figure 41.11 [pp.728-729] describes the step-by-step process of urine formation.
- Figure 41.12 [p.730] illustrates urinary system regulation.
- Tables 41.2 and 41.3 [pp.734-735] list the mammalian responses to heat and cold.

INTERACTIVE EXERCISES

Impacts, Issues: Truth in a Test Tube [p.720]

41.1. MAINTENANCE OF EXTRACELLULAR FLUID [p.722]

41.2. HOW DO INVERTEBRATES MAINTAIN FLUID BALANCE [pp.722-723]

41.3. FLUID REGULATION IN VERTEBRATES [pp.724-725]

Selected Words: diabetes mellitus [p.720], extracellular fluid (ECF) [p.722], ammonia [p.722], interstitial fluid [p.722], planarian [p.722], coelomic fluid [p.723], isotonic [p.724], uric acid [p.724], "metabolic water" [p. 725]

Boldfaced, Page-Referenced Terms

[p.722] contractile vacuole _____

_____

[p.722] flame cell _____

_____

[p.723] nephridia _____

_____

[p.723] Malpighian tubules _____

_____

[p.724] urinary system _____

_____

[p.724] kidneys _____

_____

[p.724] urea _____

_____

Fill-in-the-Blanks [pp.720-724]

Urine has been an important diagnostic tool for centuries. (1) _____ was first diagnosed by the large amounts of sugar in the urine. Alkaline urine is often a sign of (2) _____. One sign of kidney damage is the excretion of (3) _____ in the urine. A rise in (4) _____ in a woman's urine signals ovulation, and a woman can also test her urine to see if she is (5) _____ or if she is entering (6) _____. If a person uses marijuana, testable levels may remain in the urine for up to (7) _____ days.

In most animals, the fluid that fills the spaces between cells and tissues is called (8) _____ fluid. The fluid called (9) _____ moves substances by way of the circulatory system. Combined, (8) and (9) are called (10) _____ fluids. In animals a well-developed (11) _____ system helps keep the (12) _____ and composition of this fluid within tolerable ranges. Other major organ systems interact with the urinary system to maintain this (13) _____ in the internal environment.

Matching [pp.722-723]

Match each of the following items with its correct description.

14. _____ extracellular fluid
15. _____ solute
16. _____ ammonia
17. _____ intracellular fluid
18. _____ contractile vacuole
19. _____ flame cell
20. _____ coelomic fluid
21. _____ nephridia
22. _____ Malpighian tubule
23. _____ uric acid

A. Cells in the excretory system of flatworms with a tuft of cilia

B. Fluid inside cells

C. Tubular excretory organs of earthworms and other annelids

D. Fluid between cells

E. Fluid inside the body cavity of an earthworm

F. Molecule formed when amino acids and nucleotides are broken down

G. Nitrogen waste product that does not need to be dissolved in large quantity of water to be excreted

H. Long, thin excretory tubes of terrestrial insects

I. Organelle that removes fluid from freshwater sponges

J. Molecules dissolved in water

True/False [pp.724-725]

If the statement is true, write a "T" in the blank. If the statement is false, make it correct by changing the underlined word(s) and writing the correct word(s) in the answer blank.

24. _____ Marine fish have body fluids that are _more_ salty than seawater.

25. _____ Marine bony fish are constantly _gaining_ water.

26. _____ Life on land requires highly efficient _kidneys_.

27. _____ Reptiles convert ammonia to _urea_ and mammals convert it to _uric acid_.

28. _____ Some desert animals are able to survive without taking in extra water; they utilize water produced by their _metabolism_.

29. _____ Marine mammals have _smaller_ kidneys because they live in sea water.

30. _____ Freshwater bony fish continually _lose_ water because of the osmotic imbalance between their body fluids and the environment.

31. _____ Amphibians lose water through the respiratory surface of their _lungs_.

41.4. THE HUMAN URINARY SYSTEM [pp.726-727]

41.5. HOW URINE FORMS [pp.728-729]

41.6. REGULATION OF WATER INTAKE AND URINE FORMATION [pp.730-731]

41.7. ACID-BASE BALANCE [p.731]

Selected Words: peritoneum [p.726], renal capsule [p.726], renal artery and vein [p.726], urinary bladder [p.726], renal medulla [p.727], efferent arteriole [p.727], afferent arteriole [p.727], glomerular capillaries [p.727], filtrate [p.728], filtration [p.728], reabsorption [p.728], thirst [p.730], aquaporins [p.730], rennin [p.731], angiotensinogen [p.731], angiotensin I and II [p.731], diabetes insipidus [p.731], buffer system [p.731], bicarbonate-carbonic acid buffer system [p.731], acidosis [p.731]

Boldfaced, Page-Referenced Terms

[p.726] renal _____

_____

[p.726] renal cortex _____

_____

[p.726] renal medulla _____

_____

[p.726] ureter _____

_____

[p.726] urethra _____

_____

[p.727] nephron _____

_____

[p.727] Bowman's capsule _____

_____

[p.727] proximal tubule _____

_____

[p.727] loop of Henle _____

_____

[p.727] distal tubule _____

_____

[p.727] collecting duct _____

_____

[p.727] glomerulus _____

_____

[p.727] peritubular capillaries _____

_____

[p.728] glomerular filtration _____

_____

[p.728] tubular reabsorption _____

_____

[p.729] tubular secretion _____

_____

[p.730] thirst center _____

_____

[p.730] antidiuretic hormone (ADH) _____

_____

[p.731] aldosterone _____

_____

[p.731] atrial natriuretic peptide _____

_____

[p.731] acid-base balance _____

_____

[p.731] buffer system _____

_____

[p.728] glomerular filtration _____

_____

Labeling [pp.726-727]

Identify each indicated part of the accompanying illustrations.

1. _____

2. _____

3. _____

4. _____

5. _____

6. _____

7. _____

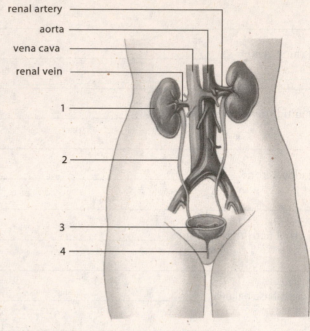

renal artery
aorta
vena cava
renal vein

1
2
3
4

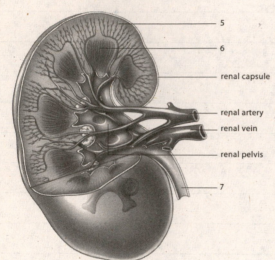

5
6
renal capsule
renal artery
renal vein
renal pelvis
7

Fill-in-the-Blanks [pp.728-729]

Mammalian (8) _____ filter water, mineral ions, organic wastes, and other substances from the blood. Each human kidney contains over a million (9) _____, where water and solutes are filtered from the blood. A nephron starts at the (10) _____, which together with the (11) _____ capillaries is called the (12) _____. Next is the tubular region closest to the capsule, called the (13) _____ tubule. Following this is a hairpin structure called the (14) _____ and a distal tubule. A(n) (15) _____ is a part of the duct system leading to the kidney's central cavity and into the (16) _____.

During this process, most of the filtered material is returned to the blood. The small portion of unclaimed water and solutes is called (17) _____. This fluid flows from each kidney into a(n) (18) _____, which empties into a muscular sac called the (19) _____. Urine is stored here before flowing into the (20)

_____, which opens at the body's surface. Flow from the bladder is a(n) (21) _____ action. A(n) (22) _____ muscle surrounds the urethra and is under (23) _____ control.

Labeling [pp.728-129]

Identify each indicated part of the illustration. Note that it depicts two nephrons, one of which extends deeply into the medulla, making it possible to greatly concentrate the urine it carries (see Section 41. 5).

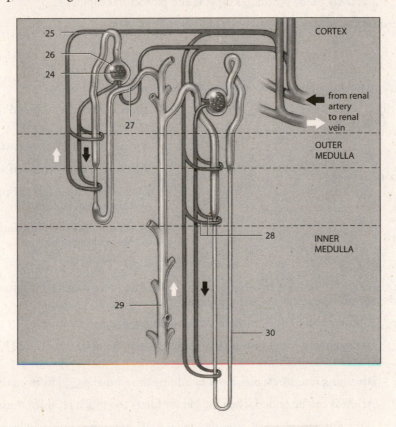

24. _____
25. _____
26. _____
27. _____
28. _____
29. _____
30. _____

Choice [pp.728-729]

Choose the process described in each of exercises 31-35.

    a. tubular secretion   b. tubular reabsorption   c. filtration

31. _____ Driven by blood pressure

32. _____ Excess ions are moved into the nephron

33. _____ Water and solutes are reclaimed

34. _____ Involves the activity of the Bowman's capsule and glomerular capillaries

35. _____ Releases drugs, toxicants, and other metabolites such as urea

Matching [pp.730-731]

Choose the most appropriate description for each of the following terms.

36. _____ aquaporins

37. _____ thirst center

38. _____ ADH

39. _____ aldosterone

40. _____ renin

A. Induces the adrenal gland to secrete aldosterone

B. Pressure-induced transporters of water

C. Promotes reabsorption of sodium

D. Hormone released by the pituitary to control water content of the urine

E. A section of the hypothalamus that induces water-seeking behavior

True/False [pp.730-731]

If the statement is true, write a "T" in the blank. If the statement is false, make it correct by changing the underlined word(s) and writing the correct word(s) in the answer blank.

41. _____ A buffer system helps maintain a constant <u>temperature</u>.

42. _____ In the bicarbonate-carbonic acid buffer system, the system will add <u>H$^+$</u> ions if the solution becomes too acid.

43. _____ Breathing rate affects blood pH by adding or removing <u>CO_2</u> to/from the blood.

44. _____ Acidosis can be caused when the kidney filters too <u>much</u> H$^+$ from the blood.

45. _____ If the pituitary gland secretes <u>too much</u> ADH, too much fluid will be retained, causing diabetes insipidus.

46. _____ Atrial natriuretic peptide is secreted by heart atrial muscle and when blood volume is <u>too low</u>.

41.8. WHEN KIDNEYS FAIL [p.732]

41.9. HEAT GAINS AND LOSSES [p.733]

41.10. TEMPERATURE REGULATION IN MAMMALS [pp.734-735]

Selected Words: kidney stones [p.732], hemodialysis [p.732], peritoneal dialysis [p.732], kidney transplant [p.732], core temperature [p.733], heat stress [p.734], sweat [p.734], panting [p.734], hyperthermia [p.734], fever [p.734], cold stress [p.734], hypothermia [p.735]

Boldfaced, Page-Referenced Terms

[p.733] thermal radiation _____

_____

[p.733] conduction _____

_____

[p.733] convection _____

_____

[p.733] evaporation _____

_____

[p.733] ectotherm _____

_____

[p.733] endotherm _____

_____

[p.735] shivering response _____

_____

[p.735] nonshivering heat production _____

_____

Matching [p.732]

Match each of the following terms or conditions with its correct description. One item is used twice.

1. _____ high protein diet
2. _____ diabetes mellitus
3. _____ hemodialysis
4. _____ kidney transplant
5. _____ peritoneal dialysis
6. _____ kidney function measure
7. _____ high blood pressure

A. Patients blood filtered in an artificial kidney

B. Permanent replacement of failed kidney

C. Kidney works extra hard to dispose of nitrogen-rich breakdown products

D. Wastes diffuse across membrane lining body cavity

E. One of the most common causes of kidney failure

F. Rate of glomerular filtration.

Matching [p.733]

Match the most appropriate description to each of the following terms.

8. _____ conduction

9. _____ convection

10. _____ ectotherm

11. _____ endotherm

12. _____ evaporation

13. _____ heterotherm

14. _____ thermal radiation

A. Body temperature determined more by heat exchange with the environment than by metabolic heat.

B. Heat transfer from or toward a body by currents in air or water.

C. Body temperature determined largely by metabolic activity and by precise controls over heat produced and heat lost.

D. Direct transfer of heat energy between two objects in direct contact with each other

E. The emission of energy in the form of infrared or other wavelengths that are converted to heat by the absorbing body

F. Body temperature fluctuating at some times and controlled at other times

G. In a change from the liquid state to the gaseous state, the energy required is supplied by the heat content of the liquid

Choice [pp.734-735]

For statements 15-23, choose from the following terms.

   a. fever   b. heat stress   c. cold stress   d. both b and c

15. _____ An inflammatory response to a tissue injury

16. _____ If a response is not successful, hypothermia may occur

17. _____ Response may involve the nonshivering response of brown adipose tissue

18. _____ Response involves peripheral vasodilation

19. _____ May be controlled by the use of anti-inflammatory drugs

20. _____ Response may include rhythmic tremors of skeletal muscles to generate heat

21. _____ Response involves the peripheral vasoconstriction of capillaries

22. _____ Response may involve behavioral responses such as panting

23. _____ If uncontrolled, can result in hyperthermia

24. _____ Response is partially controlled by the hypothalamus

SELF-TEST

1. An entire subunit of a kidney that purifies blood and restores solute and water balance is called a _____. [p.727]
   a. glomerulus
   b. loop of Henle
   c. nephron
   d. ureter
   e. none of the above

2. The last portion of the excretory system that urine passes through before being eliminated from the body is the _____. [p.726]
   a. renal pelvis
   b. bladder
   c. ureter
   d. collecting ducts
   e. urethra

3. Filtration of the blood in the kidney takes place in the _____. [p.728]
   a. loop of Henle
   b. proximal tubule
   c. distal tubule
   d. Bowman's capsule
   e. all of the above

4. _____ primarily controls the concentration of water in urine. [p.730]
   a. Aldosterone
   b. Antidiuretic hormone
   c. Angiotensin II
   d. Glucagon
   e. Renin

5. _____ primarily controls the concentration of sodium in urine. [p.731]
   a. Insulin
   b. Glucagon
   c. Antidiuretic hormone (ADH)
   d. Aldosterone
   e. Epinephrine

6. Hormonal control over the reabsorption or tubular secretion of sodium primarily affects _____. [pp.728-729]
   a. Bowman's capsules
   b. distal tubules and collecting ducts
   c. proximal tubules
   d. the urinary bladder
   e. loops of Henle

7. During reabsorption, sodium ions cross the proximal tubule walls into the interstitial fluid principally by means of _____. [pp.728-729]
   a. osmosis
   b. countercurrent multiplication
   c. bulk flow
   d. active transport
   e. all of the above

8. In humans, the thirst center is located in the _____. [p.730]
   a. adrenal cortex
   b. thymus
   c. heart
   d. adrenal medulla
   e. hypothalamus

9. Which of the following represents the process by which there is heat transfer between an animal and another object of differing temperature that is in direct contact with it? [p.733]
   a. evaporation
   b. conduction
   c. radiation
   d. convection

10. Which of the following terms represents a group of mammals whose high metabolic rates keep them active under a wide variety of temperature ranges? [p.733]
    a. endotherms
    b. ectotherms
    c. heterotherms
    d. none of the above

11. Which of the following is a response to heat stress? [p.734]
    a. panting
    b. peripheral vasodilation
    c. evaporative heat loss
    d. all of the above

CHAPTER OBJECTIVES/REVIEW QUESTIONS

1. List the factors that contribute to the gain and loss of water in mammals. [pp.722-723]
2. List the factors that contribute to the gain and loss of solutes in mammals. [pp.724-725]
3. Be able to identify the major components of the urinary system from a diagram. [pp.726-727]
4. Trace the flow of fluid through a human nephron, listing the major structures involved. [p.728]
5. Locate the processes of filtration, reabsorption, and tubular secretion along a nephron, and tell what makes each process happen. [pp.728-729]
6. Understand the interaction of the hypothalamus with the urinary system. [p.730]
7. Discuss kidney failure, its causes and treatments. [pp.732-733]
8. Distinguish between evaporation, radiation, conduction, and convection as methods of heat transfer. [p.733]
9. Explain how endotherms, ectotherms, and heterotherms regulate body temperature, and give an example of each. [p.733]
10. Explain the mechanisms by which a mammal may respond to heat stress. [p.734]
11. Explain the mechanisms by which a mammal may respond to cold stress. [pp.734-735]

INTEGRATING AND APPLYING KEY CONCEPTS

1. How might being ectothermic affect human activity? What about being heterothermic?
2. The hemodialysis machine used in hospitals is expensive and time consuming. So far, artificial kidneys capable of allowing people who have nonfunctional kidneys to purify their blood by themselves, without having to go to a hospital or clinic, have not been developed. Which aspects of the hemodialysis procedure do you think have presented the most problems in development of a method of home self-care? If you had an unlimited budget and were appointed head of a team to develop such a procedure and its instrumentation, what strategy would you pursue?
3. Explain why days with high humidity affect our ability to regulate body temperature properly.
4. Explain the symptoms expected in someone with hypothermis.

42

ANIMAL REPRODUCTIVE SYSTEMS

INTRODUCTION

This chapter investigates how animals reproduce. It begins with the distinction between asexual and sexual processes and the advantages of each. Reproductive adaptations required for land are described. The chapter focuses on human reproductive anatomy and the physiological processes that regulate reproductive cycles. Methods of preventing pregnancy are also examined.

FOCAL POINTS

- Figure 42.4 [p.742] illustrates the anatomy of the human male reproductive system.
- Figure 42.5 [p.744] diagrams gamete production in males.
- Figure 42.7 [p.745] illustrates hormone regulation of sperm production.
- Figure 42.9 [p.746] illustrates the anatomy of the human female reproductive system.
- Figure 42.10 [p.748] diagrams gamete production in females.
- Figure 42.11 [p.749] illustrates hormone regulation of menstrual cycle.
- Figure 42.14 and Table 42.3 [pp.752-753] list and compare the effectiveness of methods of contraception.

INTERACTIVE EXERCISES

Impacts, Issues: Male or Female? Body or Genes? [p.738]

42.1. MODES OF REPRODUCTION [pp.740-741]

42.2. REPRODUCTIVE SYSTEM OF HUMAN MALES [pp.742-743]

42.3. SPERM FORMATION [pp.744-745]

Selected Words: fragmentation [p.740], gamete [p.740], simultaneous hermaphrodite [p.740], sequential hermaphrodite [p.740], gonads [p.742], scrotum [p.742], secondary sexual traits [p.743], epididymis [p.743], ejaculatory duct [p.743], spongy tissue [p.743], seminal vesicle [p.743], prostate gland [p.743], bulbourethral gland [p.743], prostate cancer [p.743], prostate-specific antigen (PSA) [p.743], seminiferous tubules [p.744], primary spermatocyte [p.744], Sertoli cells [p.744], secondary spermatocyte [p.744], spermatid [p.744], testosterone [p.745], LH [p.745], FSH [p.745], GnRH [p.745], Leydig cells [p.745]

Boldfaced, Page-Referenced Terms

[p.740] asexual reproduction _____

_____

[p.740] sexual reproduction _____

_____

[p.740] hermaphrodites _____

_____

[p.741] external fertilization _____

_____

[p.741] internal fertilization _____

_____

[p.741] yolk _____

_____

[p.742] testosterone _____

_____

[p.743] vas deferens _____

_____

[p.743] penis _____

_____

[p.745] luteinizing hormone _____

_____

[p.745] follicle-stimulating hormone _____

_____

[p.745] gonadotropin-releasing hormone _____

_____

Matching [pp.740-741]

Match each of the following terms with its correct description.

1. _____ asexual reproduction
2. _____ sexual reproduction
3. _____ fragmentation
4. _____ sequential hermaphrodite
5. _____ simultaneous hermaphrodite
6. _____ external fertilization
7. _____ internal fertilization
8. _____ yolk
9. _____ internal development
10. _____ budding

A. Fluid rich in proteins and lipids
B. A piece of an individual organism forms an entire new organism
C. A single individual forms offspring genetically identical to itself
D. A new organism grows out of the original organism and then breaks free
E. The fertilized egg develops inside the female body, nourished directly from female's circulation
F. Produce offspring containing genetic material from two other organisms
G. Organism that produces both eggs and sperm at the same time
H. Egg fertilized outside the body, usually in water
I. Organisms switch from one sex to another during their life span
J. Egg fertilized inside the female body

Labeling [pp.742-743]

Identify each numbered part of the following illustration.

11. _____
12. _____
13. _____
14. _____
15. _____
16. _____
17. _____
18. _____
19. _____
20. _____
21. _____

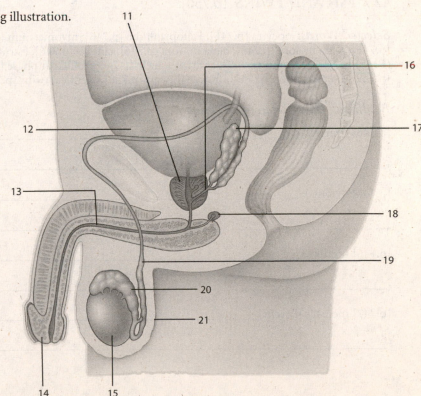

Matching [pp.743-745]

Match each of the following terms with its correct description.

22. _____ prostate-specific antigen
23. _____ Leydig's cells
24. _____ Sertoli cells
25. _____ secondary spermatocyte
26. _____ testosterone
27. _____ Luteinizing hormone (LH)
28. _____ Follicle stimulating hormone (FSH)
29. _____ prostate enlargement
30. _____ sperm head
31. _____ sperm midpiece

A. Contains DNA and "digestive" enzymes
B. Developing sperm cell at end of meiosis I
C. Hormone supports sperm development
D. Causes urethra to narrow; causes difficulty urinating
E. Cells in testes that secrete testosterone
F. Contains mitochondria for energy
G. Cells in testes that support developing sperm
H. Hormone that induces development of male secondary sex characteristics
I. Molecule measured to detect prostate cancer
J. Hormone that regulates testosterone production

42.4. REPRODUCTIVE SYSTEM OF HUMAN FEMALES [pp.746-747]

42.5. FEMALE TROUBLES [p.747]

42.6. PREPARATIONS FOR PREGNANCY [pp.748-749]

42.7 FSH AND TWINS [p.750]

Selected Words: oocytes [p.746], Fallopian tube [p.746], myometrium [p.746], endometrium [p.746], cervix [p.746], labia majora [p.746], labia minora [p.746], clitoris [p.746], estrous cycle [p.747], premenstrual syndrome (PMS) [p.747], fibroids [p.747], ovarian follicle [p.748], follicular phase [p.748], polar body [p.748], luteal phase [p.749], menstruation [p.749], fraternal twins [p.750], identical twins [p.750]

Boldfaced, Page-Referenced Terms

[p.746] ovaries _____

_____

[p.746] oviducts _____

_____

[p.746] uterus _____

_____

[p.747] menstrual cycle _____

_____

[p.748] zona pellucida _____

_____

[p.748] estrogens _____

_____

[p.748] secondary oocyte _____

_____

Labeling [pp.746-747]

Identify each numbered part of the following illustration.

1. _____

2. _____

3. _____

4. _____

5. _____

6. _____

7. _____

8. _____

9. _____

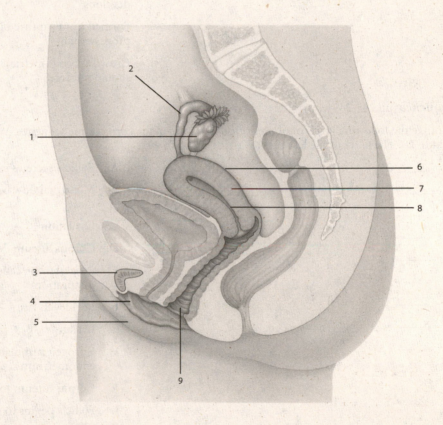

Match each of the following terms with its correct description.

10. _____	Premenstrual syndrome	A.	A primary oocyte and cells surrounding it
11. _____	Endometriosis	B.	Causes ovulation
12. _____	Fibroids	C.	Immature egg that has started meiosis but arrested in prophase I
13. _____	Primary oocyte	D.	Phase of ovarian cycle where follicle matures
14. _____	Ovarian follicle	E.	Body tissues swell and person may become irritable or depressed
15. _____	Surge of LH	F.	Loss of endometrial tissue and blood
16. _____	Corpus luteum	G.	Benign uterine tumors that may result in pain and long menstrual periods
17. _____	Menstrual flow	H.	Phase of ovarian cycle where corpus luteum develops
18. _____	Follicular phase	I.	Formed from ruptured follicle and produces progesterone
19. _____	Luteal phase	J.	Growth of endometrial tissues in areas where it should not form

Sequence [pp.748-749]

Arrange the following levels of organization in nature in the correct time sequence. Write the letter of the first event next to 20. The letter of the last event is written next to 29.

20. _____	A. Follicle secretes estrogen
21. _____	B. Menstrual flow begins as endometrium tissue dies
22. _____	C. Ovulation
23. _____	D. Corpus luteum degenerates
24. _____	E. Progesterone causes endometrium to complete maturation
25. _____	F. FSH levels rise
26. _____	G. LH surge
27. _____	H. Estrogen stimulates initial growth of endometrium
28. _____	I. Corpus luteum forms and secretes progesterone
29. _____	J. Follicle begins to mature

Short Answer [p.750]

30. Describe the relationship between FSH and twinning. _____

_____

_____

42.8. WHEN GAMETES MEET [pp.750-751]

42.9. PREVENTING OR SEEKING PREGNANCY [pp.752-753]

42.10. SEXUALLY TRANSMITTED DISEASES [pp.754-755]

Selected Words: intercourse [p.750], mechanoreceptors [p.750], oxytocin [p.750], endorphins [p.750], erectile disfunction [p.750], acrosomal enzymes [p.751], zygote [p.751], abstinence [p.752], vasectomy [p.752], tubal ligation [p.752], diaphragm [p.752], condom [p.752], intrauterine device [p.752], "The Pill" [p.752], "morning-after pill" [p.753], abortion [p.753], in vitro fertilization [p.753], pelvic inflammatory disease [p.754], human papilloma virus [p.754], trichomoniasis [p.754], Chlamydia [p.754], genital herpes [p.754], gonorrhea [p.754], syphilis [p.755], AIDS [p.755].

Boldfaced, Page-Referenced Terms

[p.754] sexually transmitted diseases _____

_____

True/False [pp.750-751]

If the statement is true, write a "T" in the blank. If the statement is false, make it correct by changing the underlined word(s) and writing the correct word(s) in the answer blank.

1. _____ During sexual arousal, <u>sympathetic</u> nervous system signals cause blood vessels to dilate and increase blood flow to penis

2. _____ <u>Seratonin</u> causes rhythmic contractions in male reproductive tract and female vagina

3. _____ <u>Erectile disfunction</u> is a condition where an erection cannot be attained or maintained

4. _____ When sperm reach the egg, <u>several</u> sperm may enter the egg

5. _____ Acrosomal enzymes break down the <u>*zona coporum*</u> to reach egg surface

6. _____ The initial cell after fertilization is a <u>haploid gamete</u>

7. _____ The second meiotic division of the egg occurs just <u>before ovulation</u>

8. _____ Ejaculation can provide 150 to 350 <u>billion</u> sperm

Matching [pp.752-753]

Match each of the following birth control methods with their correct description.

9. _____ rhythm method

10. _____ douching

11. _____ vasectomy

12. _____ condom

13. _____ spermicidal foam

14. _____ diaphragm

15. _____ intrauterine device

16. _____ tubal ligation

17. _____ birth control pill

18. _____ birth control patch

A. An object inserted into the uterus to prevent implantation

B. A mixture of synthetic estrogen and progesterone

C. A material placed in vagina to kill sperm

D. Fallopian tubes are severed

E. Rinsing of the vagina after intercourse

F. Timing intercourse when fertile egg not present

G. An adhesive material applied to skin that delivers estrogen and progesterone

H. Dome-shaped device that blocks the cervix

I. Vas deferens are severed

J. Sheath worn over the penis

Matching [pp.754-755]

Match each of the following sexually transmitted diseases with their correct description.

19. _____ genital herpes

20. _____ trichomoniasis

21. _____ pelvic inflammatory disease

22. _____ Syphillis

23. _____ Gonorrhea

24. _____ Human papilloma virus

25. _____ HIV

26. _____ AIDS

27. _____ Chlamydia

A. A virus that lowers immune system's ability to counteract infections

B. Vaginal discharge; painful urination in both sexes; may be contracted repeatedly

C. Most infected females undiagnosed; many will develop pelvic inflammatory disease

D. A complication of bacterial sexually transmitted diseases that scars the reproductive tract

E. Mucous membranes of genitals have small painful blisters; incurable

F. Compromised immune system leads to opportunistic infections; incurable

G. Causes genital warts and may lead to cervical cancer

H. Flagellated protist infects vagina and urethra

I. Multi stage disease process that starts with chancre and spreads to many other tissues

___ 1. Reproductive systems are essential for
_____. [p.740]
   a. maintaining homeostasis
   b. passing genes on to the next
      generation
   c. survival of the individual
   d. immediate responses to
      environmental change

___ 2. Sexual reproduction is advantageous
because it _____. [p.740]
   a. produces offspring genetically identical
      to parents
   b. requires finding a mate
   c. produces offspring genetically different
      than parents
   d. is optimal in a stable environment
   e. produces larger numbers of offspring
      in a shorter time

___ 3. To get from the testes to the outside, sperm
travel through _____ (in the correct
order). [pp.742-743]
   a. vas deferens, ureter, ejaculatory duct
   b. epididymis, urethra, vas deferens
   c. epididymis, vas deferens, ureter
   d. vas deferens, fallopian tubule, urethra
   e. epididymis, vas deferens, ejaculatory
      duct, urethra

___ 4. Sperm finishes maturing in the _____.
[p.744]
   a. epididymis
   b. seminiferous tubule
   c. seminal vesicle
   d. vas deferens
   e. bulbourethral gland

___ 5. Ovulation in humans is triggered by
_____. [p.748]
   a. high levels of luteinizing hormone
      (LH)
   b. high levels of progesterone
   c. the corpus luteum
   d. development of the endometrium
   e. decreasing levels of luteinizing
      hormone (LH)

___ 6. Menstruation is initiated when _____.
[p.749]
   a. luteinizing hormone (LH) levels
      increase
   b. progesterone levels decrease
   c. follicle stimulating hormone (FSH)
      levels increase
   d. the corpus luteum is established
   e. more than two weeks has passed
      since ovulation

___ 7. When a sperm penetrates the egg
membrane _____. [p.751]
   a. entry of other sperm is prevented
   b. meiosis II is completed
   c. nuclear membrane of sperm breaks
      down
   d. all of the above occur

___ 8. The most effective options for birth control
are _____. [p.752]
   a. condoms
   b. birth control pills
   c. vasectomy or tubal ligation
   d. rhythm method
   e. IUD with spermicide

___ 9. The estimated number of conceptions that
end early due a genetic problem is
_____. [p.753]
   a. less than 5%
   b. 10%
   c. 25%
   d. 50%
   e. 75%

___ 10. The most widespread and fastest growing
sexually transmitted disease in the United
States is _____. [p.754]
   a. AIDS
   b. Chlamydia
   c. gonorrhea
   d. genital herpes
   e. human papilloma virus (HPV)

CHAPTER OBJECTIVES/REVIEW QUESTIONS

1. Describe how asexual reproduction differs from sexual reproduction. Know the advantages and problems associated with having separate sexes. [pp. 740-741]
2. Explain why evolutionary trends in many groups of organisms tend toward developing more complex sexual strategies rather than retaining simpler asexual strategies. [pp. 740-741]
3. Describe accessory glands in the male reproductive system and explain the importance of their contributions to semen. [p.743]
4. Describe where immature sperm are found and the function of the cytoplasmic bridges between the cells. [p.744]
5. Explain how sperm production is controlled hormonally. [p.745]
6. Explain the sequence of hormonal changes through the human menstrual cycle and what event each hormone change causes. [pp.748-749]
7. Describe what causes erectile dysfunction in human males and the mechanisms of drugs that treat that disorder. [p.750]
8. Describe the sequence of events in fertilization from arrival of the sperm at the egg until the first mitotic division begins. [p.751]
9. Describe the various options for birth control and rate each in terms of its effectiveness. [pp.752-753]
10. List the most common agents of sexually transmitted diseases and the symptoms of each. [pp.754-755]

INTEGRATING AND APPLYING KEY CONCEPTS

1. What problems could you foresee for humans if they reproduced asexually instead of sexually?
2. How would evolution be affected if sexual reproduction did not exist?
3. There is increasing evidence that there are chemicals in the environment that are mimicking the affect of human hormones. Predict how estrogen mimics could effect human reproductive processes.
4. Discuss the social consequences of sexually transmitted diseases.

43

ANIMAL DEVELOPMENT

INTRODUCTION

Chapter 43 examines the intricate and fascinating process that leads from single cell zygote to a complex, mature animal composed of multiple tissues, organs, and organ systems. The events of each stage of development are described along with the controlling mechanisms of those events. The chapter then describes human development in more detail.

FOCAL POINTS

- Figure 43.2 [p.760] describes the stages of animal embryonic development.
- Figure 43.3 [pp.760-761] illustrates early embryonic development of vertebrates using frogs as an example.
- Figures 43.5 and 43.6 [p.763] examine cleavage patterns in several different organisms.
- Figures 43.8 and 43.9 [pp.765-765] describe some of the cell movements seen during gastrulation and formation of the neural tube.
- Table 43.1 [p.767] summarizes the stages of human development.
- Figure 43.13 [pp.768-769] describes and illustrates the first two weeks of human development.
- Figure 43.14 [p.770] shows human development for day 15 through day 25.
- Figure 43.15 [p.771] describes interaction between the circulatory systems of mother and fetus through the placenta.
- Figure 43.16 [pp.772-773] illustrates stages of human development through 16 weeks.
- Figure 43.17 [p.774] illustrates times during development when a fetus is sensitive to teratogenic agents.

INTERACTIVE EXERCISES

Impacts, Issues: Mind Boggling Births [p.758]

43.1. STAGES OF REPRODUCTION AND DEVELOPMENT [pp.760-761]

43.2. EARLY MARCHING ORDERS [pp.762-763]

Selected Words: multiple births [p.758], gamete formation [p.760], fertilization [p.760], zygote [p.760], blastomere [p.760], blastula [p.760], gastrula [p.760], organ formation [p.760], neural tube [p.761], metamorphosis [p.761], mRNA transcripts [p.762], vegetal pole [p.762], animal pole [p.762], gray crescent [p.762], spiral cleavage [p.763], radial cleavage [p.763], rotational cleavage [p.763], blastodisc [p.763]

Boldfaced, Page-Referenced Terms

[p.760] cleavage _____

[p.760] blastula _____

_____

[p.760] gastrulation _____

_____

[p.760] germ layers _____

_____

[p.762] cytoplasmic localization _____

_____

Fill-in-the-Blanks [pp.760-761]

(1) _____ is considered the first stage of animal development. When sperm and egg unite and their

DNA mingles and is reorganized, the process is referred to as(2) _____. At the end of (2), a(n) (3) _____

is formed. (4) _____ includes the repeated mitotic divisions of a zygote that segregate the egg cytoplasm into

cells known as (5) _____; the entire cluster of cells is known as a blastula. (6) _____ is the process that

arranges cells into three primary tissue layers. (7) _____ will eventually give rise to the nervous system and

integument; (8) _____ forms the inner lining of the gut and associated digestive glands;(9) _____ forms

many structures including the circulatory system, the muscles, and connective tissues.

Sequence [p.760]

Arrange the following events in correct chronological sequence. Write the letter of the first step next to 10. Write the
letter of the last step in the sequence next to 15.

10. _____   A. Gastrulation
11. _____   B. Fertilization
12. _____   C. Cleavage
13. _____   D. Growth, tissue specialization
14. _____   E. Organ formation
15. _____   F. Gamete formation

Matching [pp.762-763]

Match each of the following structures with its correct description.

16. _____ mRNA transcripts

17. _____ cytoplasmic localization

18. _____ vegetal pole

19. _____ animal pole

20. _____ gray crescent

21. _____ spiral cleavage

22. _____ radial cleavage

23. _____ blastomere

24. _____ blastula

25. _____ blastocyst

A. Portion of a yolk-rich egg where there is little yolk

B. Term for cells produced by cleavage process

C. Blastula of a mammal with outer cells and an inner cell mass

D. Messenger RNA made in egg prior to fertilization

E. Portion of a yolk-rich egg where most of the yolk is found

F. Collective name of cells produced by cleavage

G. Cleavage pattern common protostome animals (ie., arthropods)

H. Cytoplasmic components and molecules not evenly distributed in cytoplasm of oocyte

I. In amphibian zygote, region of cell cortex that is lightly pigmented and contributes to organization of development.

J. Cleavage pattern found in most chordates

Short Answer [p.762]

26. Explain how cytoplasmic localization can lead to blastomeres containing different information. _____

_____

_____

_____

43.3. FROM BLASTULA TO GASTRULA [pp.764-765]

43.4. SPECIALIZED TISSUES AND ORGANS FORM [p.765]

43.5. AN EVOLUTIONARY VIEW OF DEVELOPMENT [p.766]

Selected Words: primary tissue layers [p.764], dorsal lip cells [p.764], master genes [p.765], selective gene expression [p.765], localized gene expression [p.766], gradients [p.765], physical constraints [p.766], architectural constraints [p.766], phyletic constraints [p.766]

Boldfaced, Page-Referenced Terms

[p.764] ectoderm _____

_____

[p.764] mesoderm _____

_____

[p.764] endoderm _____

_____

[p.764] induction _____

_____

[p.765] cell differentiation _____

_____

[p.765] morphogen _____

_____

[p.765] morphogenesis _____

_____

[p.765] apoptosis _____

_____

[p.765] pattern formation _____

_____

[p.765] cell differentiation _____

_____

[p.766] homeotic gene _____

_____

[p.766] somite _____

_____

Fill-in-the-Blanks [pp.762-766]

By the process of (1) _____ fates of cells change when exposed to signaling molecules from adjacent tissues. Other signaling molecules called (2) _____ affect DNA in cells from a distance. They form a concentration (3) _____, and cells exposed to different levels of (2) develop differently. This pattern formation usually begins with (4) _____ during cleavage. The overall body plan is determined by specialized master genes called (5) _____ genes, which are activated by (6) _____. The actual patterns seen in living organisms are often limited by (7) _____ constraints like the (8) _____ ratio; (9) _____ constraints imposed by the body axes; and (10) _____ constraints imposed on each lineage by the interactions of genes.

True/False [pp.764-766]

If the statement is true, write a "T" in the blank. If the statement is false, make it correct by changing the underlined word(s) and writing the correct word(s) in the answer blank.

11. _____ Transplanting a small segment of the dorsal lip to a new location on the same organism <u>has no effect on development.</u>

12. _____ Gastrulation is characterized by many <u>movements of cells</u> from one place to another in the developing embryo.

13. _____ When one group of cells affects the development of another group of cells we call that <u>transduction</u>.

14. _____ When one group of cells expresses a different select group of genes compared to other cells <u>cell localization</u> is occurring.

15. _____ Master genes produce signaling molecules called <u>homeostats</u>.

16 _____ Morphogens have different effects on cells based on the <u>number of receptors</u>.

17. _____ Somites formed on either side of the neural tube differentiate into <u>spinal nerves</u>.

18. _____ Homeotic genes become active based on their <u>position</u> in the embryo.

19. _____ <u>Architectural</u> constraints on development relate to surface-to-volume ratios.

20. _____ Constraints based on the evolutionary history of an organism are <u>physical</u>.

Short Answer [pp.764-766]

21. Describe what happens in induction. _____

_____

_____

22. Describe the process of apoptosis and its role in development. _____

_____

_____

23. Give the three steps in the general model for development. _____

_____

_____

43.6. OVERVIEW OF HUMAN DEVELOPMENT [p.767]

43.7. EARLY HUMAN DEVELOPMENT [pp.768-769]

43.8. EMERGENCE OF THE VERTEBRATE BODY PLAN [p.770]

43.9. THE FUNCTION OF THE PLACENTA [p.771]

43.10. EMERGENCE OF DISTINCTLY HUMAN FEATURES [pp.772-773]

Selected Words: prenatal [p.767], postnatal [p.767], embryonic period [p.767], fetal period [p.767], trimester [p.767], premature [p.767], blastocoel [p.768], embryonic disk [p.768], yolk sac [p.768], primitive streak [p.770], notochord [p.770], spina bifida [p.770], pharyngeal arches [p.770], chorionic villi [p.771], umbilical cord [p.771], lanugo [p.773], vermix [p.773]

Boldfaced, Page-Referenced Terms

[p.767] fetus _____

_____

[p.768] morula _____

_____

[p.768] implantation _____

_____

[p.768] amnion _____

_____

[p.769] chorion _____

_____

[p.769] placenta _____

_____

[p.769] allantois _____

_____

[p.769] human chorionic gonadotropin _____

_____

Fill-in-the-Blanks [pp.767-769]

Six or seven days after conception, (1) _____ begins as the blastocyst sinks into the endometrium.

Extensions from the chorion, (2) _____, fuse with the endometrium of the uterus to form a(n) (3)

_____, the organ of interchange between mother and fetus. By the end of the (4) _____ week, the

embryonic stage is completed; the offspring is now referred to as a(n) (5) _____ .

Choice [p.767]

For each of the following stages in development, choose the correct period for that stage.

    a. prenatal   b. postnatal

6. _____ Pubescent

7. _____ Fetus

8. _____ Child

9. _____ Newborn

10. _____ Zygote

11. _____ Old age

12. _____ Embryo

13. _____ Adult

14. _____ Infant

15. _____ Adolescent

16. _____ Blastocyst

Sequence [pp.768-770]

Arrange the following events in the correct time sequence. Write the letter of the first event next to 14. The letter of the last event is written next to 25.

17. _____

18. _____

19. _____

20. _____

21. _____

22. _____

23. _____

24. _____

25. _____

26. _____

27. _____

28. _____

A. Morula forms

B. Blastocyst surface cells attach to endometrium

C. Blood-filled spaces form in maternal tissue

D. Somites become visible

E. Amniotic cavity and yolk sac form

F. A compact ball of cells form

G. Blastocoel forms

H. Pharyngeal arches forming

I. Eight loosely arranged cells form

J. Primitive streak forms

K. First cleavage furrow forms

L. Chorionic villi form

Matching [p.769]

Match each of the four extraembryonic membranes with its correct function.

29. _____ Allantois

30. _____ Amnion

31. _____ Chorion

32. _____ Yolk sac

A. Outermost membrane; will become part of the spongy, blood engorged placenta

B. Directly encloses the embryo and holds it a buoyant, protective fluid

C. In humans, some of this membrane becomes a site of blood cell formation; some will become the forerunner of gametes

D. In humans, this membrane serves in development of the umbilical circulation and formation of the urinary bladder

Choice [p.770]

Indicate from which primary tissue each of the following is derived.

   a. ectoderm   b. endoderm   c. mesoderm

33. _____ Muscle, circulatory organs

34. _____ Nervous tissues

35. _____ Inner lining of the gut

36. _____ Circulatory organs (blood vessels, heart)

37. _____ Outer layer of the integument

38. _____ Reproductive and excretory organs

39. _____ Organs derived from the gut

40. _____ Most of the skeleton

41. _____ Connective tissues of the gut and integument

Short Answer [p.771]

42. Explain the functions of the placenta. _____

_____

_____

43. Explain why the mother's immune system does not make antibodies against the fetal tissues. _____

_____

_____

Labeling [pp.772-773]

Identify each numbered part of the following illustration.

44. _____ _____

45. _____ _____

46. _____ _____

47. _____

48. _____ _____

49. _____

50. _____

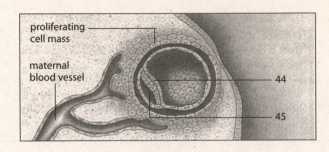

proliferating cell mass

maternal blood vessel

44

45

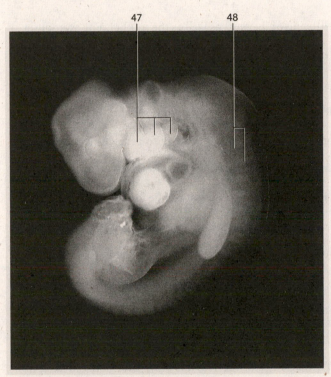

47   48

A human embryo at (46) weeks after conception.

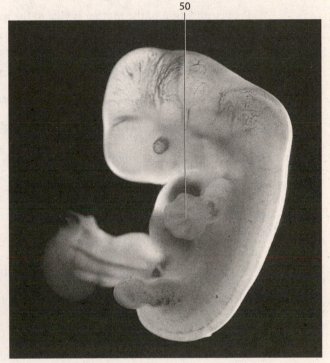

50

A human embryo at (49) weeks after conception.

43.11. MOTHER AS PROVIDER AND PROTECTOR [pp.774-775]

43.12. BIRTH AND LACTATION [p.776]

Selected Words: birth defect [p.774], teratogen [p.774], folate [p.774], cretinism [p.774], morning sickness [p.774], rubella (German measles) [p.775], toxoplasmosis [p.775], fetal alcohol syndrome [p.775], thalidomide [p.775], isotretinoin (Accutane) [p.775], paroxetine (Paxil) [p.775], amniotic fluid [p.776], positive feedback [p.776], "afterbirth" [p.776]

Boldfaced, Page-Referenced Terms

[p.776] labor _____

_____

[p.776] lactation _____

_____

[p.776] prolactin _____

_____

Fill-in-the-Blanks [pp.774-775]

Proper nutrition of the mother is vital to the well-being of the fetus as well as the mother. For example, a maternal diet deficient in the B vitamin (1) _____ often leads to neural tube defects in the fetus. A deficiency of (2) _____ can lead to the syndrome of defects in brain function in motor skills called cretinism. Many pregnant women also suffer from episodes of nausea called (3) _____ . This condition most often occurs during the time period when (4) _____ are developing in the fetus and is thought to protect the fetus from possible (5) _____ that might be found in certain foods. Maternal infections can also cause problems in the developmental process of the fetus. If a mother becomes infected with a protist that can be found in soil or cat feces she could have (6) _____ and the fetus could have (7) _____ or be (8) _____ . (9) _____ is viral disease that can be dangerous during the early weeks of development. Consumption of (10) _____ can result in a fetus with reduced brain size, facial deformities, mental impairment, and a variety of other problems. It has also become clear that consumption of (11) _____ in coffee or soft drinks leads to higher incidence of miscarriages. Smoking increases (12) _____ in the blood which outcompetes (13) _____ for binding sites on hemoglobin. The result is that the fetus receives less (13). Even some prescription drugs have caused birth defects. Probably the most infamous was (14) _____ (prescribed to relieve morning sickness) which caused severely deformed limbs when taken in the first trimester. Accutane, an effective treatment for (15) _____ , has caused heart problems and (16) _____ .

Sequence [p.776]

Arrange the following events in the correct time sequence. Write the letter of the first event next to 17. The letter of the last event is written next to 25.

17. _____   A. Oxytocin causes smooth muscles in uterus to contract

18. _____   B. Fetus passes through cervix and vagina

19. _____   C. Additional Oxytocin causes stronger contractions

20. _____   D. Strong muscle contractions cause placenta to detach from uterus

21. _____   E. Receptors in cervix sense pressure

22. _____   F. The umbilical cord is clamped and cut

23. _____   G. Cervix becomes flexible and stretches

24. _____   H. Added pressure from contractions signals more Oxytocin secretion

25. _____   I. "Afterbirth" delivered

26. _____   J. Hypothalamus signals posterior pituitary to release Oxytocin

Matching [p.776]

Match each of the following terms related to milk production with its correct role in nourishing a newborn.

27. _____ prolactin

28. _____ lactation

29. _____ oxytocin

30. _____ progesterone and estrogen

31. _____ suckling

A. Hormone(s) that causes smooth muscle contractions around milk glands

B. Physical stimulus that causes release of Oxytocin

C. Hormone(s) that causes milk synthesis

D. Milk production

E. Hormone(s) that decrease in amount causing milk production to increase

SELF-TEST

___ 1. The specialized blastula seen in humans is called a(n) _____. [p.763]
   a. blastocyst
   b. blastomere
   c. gastrula
   d. cleavage
   e. deuterostome

___ 2. The process of cleavage most commonly produces a(n) _____. [p.760]
   a. zygote
   b. blastula
   c. gastrula
   d. third germ layer
   e. organ

3. Incomplete cleavage of cells of very yolky eggs is characteristic of the cleavage pattern of _____. [p.763]
   a. frogs
   b. sea urchins
   c. chickens
   d. humans
   e. none of the preceding

4. The formation of three germ (embryonic) tissue layers occurs during _____. [p.764]
   a. gastrulation
   b. cleavage
   c. pattern formation
   d. morphogenesis
   e. neural plate formation

5. The differentiation of a body part in response to signals from an adjacent body part is _____. [p.764]
   a. contact inhibition
   b. ooplasmic localization
   c. embryonic induction
   d. pattern formation
   e. none of the above

6. A homeotic mutation _____. [p.766]
   a. may cause a leg to develop on the head where an antenna should grow
   b. may affect pattern formation
   c. affects morphogenesis
   d. may alter the path of development
   e. all of the above

7. Shortly after fertilization, the zygote is subdivided into a multicelled embryo during a process known as _____. [pp.760,762]
   a. meiosis
   b. parthenogenesis
   c. embryonic induction
   d. cleavage
   e. invagination

8. Muscles differentiate from_____ tissue. [p.770]
   a. ectoderm
   b. mesoderm
   c. endoderm
   d. parthenogenetic
   e. yolky

9. The gray crescent is _____ . [p.762]
   a. formed where the sperm penetrates the egg
   b. part of only one blastomere after the first cleavage
   c. the yolky region of the egg
   d. where the first mitotic division begins
   e. formed opposite from where the sperm enters the egg

10. The nervous system differentiates from _____ tissue. [p.770]
    a. ectoderm
    b. mesoderm
    c. endoderm
    d. yolky
    e. homeotic

11. Once implanted, the blastula releases _____ . [p.769]
    a. prolactin
    b. estrogen
    c. human chorionic gonadotropin
    d. amniotic fluid
    e. oxytocin

12. Which extraembryonic membrane lines the amniotic sac and the yolk sac, becoming part of the placenta? [p.769]
    a. amnion
    b. chorion
    c. allantois
    d. chorionic villi
    e. urinary bladder

13. In human development, at the end of the eighth week, all organs are formed and we define the individual as a(n) _____. [p.772]
    a. infant
    b. embryo
    c. zygote
    d. fetus
    e. blastocyst

14. Release of milk from the milk glands requires stimulus of the newborn suckling and _____. [p.776]
    a. prolactin
    b. progesterone
    c. estrogen
    d. human chorionic gonadotropin
    e. oxytocin

CHAPTER OBJECTIVES/REVIEW QUESTIONS

1. Describe the early stages of animal development and the events within each one. [pp.760-761]
2. Explain how cytoplasmic localization leads to blastomeres having different cytoplasmic contents and hence how that can lead to different developmental paths for the cells. [p.762]
3. Explain how the amount of yolk in an ovum can influence an animal's cleavage pattern. [pp.762-763]
4. Define the term *gastrulation* and state what process begins at this stage that did not happen during cleavage. [p.764]
5. Define the term *differentiation* and give two examples of cells in a multicellular organism that have undergone differentiation. [p.765]
6. Discuss the three constraints on development patterns and give one example of each. [p.766]
7. List the stages of human development and briefly describe how each is defined. [p.767]
8. Describe the role of the four extraembryonic membranes found in human development. [p.769]
9. Name each of the three embryonic tissue layers and the organs formed from each. [p.770]
10. Describe the events that occur during the first month of human development. State how much time cleavage and gastrulation require when organ formation begins, and what is involved in implantation and placenta formation. [pp.768-773]
11. Explain why the mother must be particularly careful of her diet, health habits, and lifestyle during the first trimester after fertilization (especially during the first six weeks). [pp.774-775]
12. Describe the series of events during labor that lead to childbirth. [p.776]
13. Describe the events required for successful production and release of milk. [p.776]

INTEGRATING AND APPLYING KEY CONCEPTS

1. If embryonic induction did not occur in a human embryo, how would the eye region appear? What would happen to the forebrain and epidermis? If controlled cell death did not happen in a human embryo, how would its hands appear? How would the face appear?
2. Discuss the implications of the fact that homeotic genes from one organism will also function in a completely different species. For example, human homeotic genes may be expressed in fruit flies.
3. With our increased knowledge of what events and conditions may adversely affect human development, how can a society encourage women to take proper care of themselves while pregnant.

44

ANIMAL BEHAVIOR

INTRODUCTION

This chapter of the text examines the basic forms of behavior, or responses to stimuli. After examining the forms of chemical behavior and signaling mechanisms, the chapter examines the costs and benefits of social behavior. The final sections of the chapter examine how human social behavior is based on the same fundamental principles of animal behaviors presented earlier in the chapter.

FOCAL POINTS

* Understand the differences between instinctive and learned behaviors.

INTERACTIVE EXERCISES

Impacts, Issues: My Pheromones Made Me Do It [p.780]

44.1. BEHAVIORAL GENETICS [pp.782-783]

44.2. INSTINCT AND LEARNING [p.784]

Selected Words: Marla Sokolowski [p.782], Stevan Arnold [p.782], tutor [p.784], species specific song [p.784]

Boldfaced, Page-Referenced Terms

[p.782] stimulus _____

_____

[p.784] instinctive behavior _____

_____

[p.784] fixed action pattern _____

_____

[p.784] learned behavior _____

_____

[p.784] imprinting _____

_____

[p.785] classical conditioning _____

_____

[p.785] operant conditioning _____

_____

[p.785] habituation _____

_____

[p.785] observational learning _____

_____

Matching [pp.782-785]

1. _____ observational learning
2. _____ stimulus
3. _____ instinctive behavior
4. _____ fixed action pattern
5. _____ learned behavior
6. _____ imprinting
7. _____ habituation

A. Behavior altered by experience

B. Environmental cue that is detected by a receptor

C. Imitating the behavior of another individual

D. Time-dependent form of learning; triggered by exposure to sign stimuli and usually occurring during sensitive periods of young animals

E. An innate response to a simple stimulus

F. Instinctive movements triggered by stimulus that will run to completion

G. Learning thru experience, not in response to stimulus

44.3. ADAPTIVE BEHAVIOR [p.786]

44.4. COMMUNICATION SIGNALS [pp.786-787]

44.5. MATES, OFFSPRING, AND REPRODUCTIVE SUCCESS [pp.788-789]

Selected Words: adaptive behavior [p.786], priming pheromones [p.786], tactile displays [p.787], "nuptial gifts" [p.788], lek [p.788]

Boldfaced, Page-Referenced Terms

[p.786] communication signals _____

_____

[p.786] pheromones _____

_____

[p.788] sexual selection _____

_____

1. _____ communication signals
2. _____ adaptive behavior
3. _____ reproductive success
4. _____ signal pheromones
5. _____ priming pheromones
6. _____ threat display
7. _____ courtship display
8. _____ tactile display
9. _____ acoustical signals
10. _____ alarm signals
11. _____ parental behavior

A. The signaler touches the receiver in a ritualized manner

B. Cues for social behavior between members of a species

C. Behavior used for attracting a mate or defining a territory

D. Behavior used to warn of a predator

E. Announces that the signaler is about to attack the receiver

F. Induces the receiver to respond quickly

G. Causes a physiological response

H. These rituals must be performed prior to forming a mating pair

I. Any behavior that helps perpetuate an individual's genes

J. Reproduction in which at least some of the offspring survive

K. Behavior that increases the survival of the young

Complete the Table [pp.788-789]

12. Complete the following table by supplying the common names of the animals discussed in the text that fit the examples of sexual selection.

Animals	Description of Sexual Selection
a.	Males hold strings of eggs around their legs, males without strings are rare later in the breeding season, females will fight for access to males without strings around their legs
b.	Females select the males that offer them superior material goods; females permit mating only after they have eaten the "nuptial gift" for about five minutes
c.	Females of a species cluster in defendable groups at a time they are sexually receptive; males compete for access to the clusters; combative males are favored
d.	Males congregate in a lek, or communal display ground; each male stakes out his territory; females are attracted to the lek to observe the male displays and usually select and mate with only one male

44.6. LIVING IN GROUPS [pp.790-791]

44.7. WHY SACRIFICE YOURSELF? [pp.792-793]

44.8. HUMAN BEHAVIOR [p.793]

Selected Words: alarm calls [p.790], cultural traits [p.790], "fishing sticks" [p.790], "self sacrifice" [p.792], eusocial [p.792]

Boldfaced, Page-Referenced Terms

[p.790] selfish herd _____

_____

[p.792] Theory of inclusive fitness _____

_____

[p.792] altruistic behavior _____

_____

Matching [pp.790-791]

Choose the most appropriate statement for each of the following items.

1. _____ disadvantages of sociality
2. _____ dominance hierarchy
3. _____ cooperative predator avoidance
4. _____ the selfish herd
5. _____ cooperative hunting

A. Competition for resources, rapid depletion of food resources, cannibalism, and greater vulnerability to disease

B. A simple society brought together by reproductive self-interest, in bluegill sunfish, the larger, more powerful males tend to claim the central nesting locations

C. Behavior used for attracting a mate or defining a territory

D. Some individuals of a wolf pack adopt a subordinate status with respect to the other members

E. Social groups of predatory animals

Fill-in-the-Blanks [pp.792-793]

A sexually reproducing (6) _____ parent caring for offspring is not helping exact (7) _____ copies of itself. Each of its gametes, and each of its offspring, inherits (8) _____ of its genes. Other individuals of the social group that have the same (9) _____ also share genes with their parents. Two siblings are as (10) _____ similar as a parent and its (11) _____. Nephews and nieces share about (12) _____ of their uncle's genes.

(13) _____ workers may be indirectly promoting genes for (14) _____ through altruistic behavior that will benefit their close (15) _____. All of the individuals in honeybees, termite and ant colonies are members of a great extended (16) _____. Non-breeding family members support siblings, a few of which are the future kings and (17) _____. Although a guard bee dies after she stings, her sacrifice preserves her (18) _____ within her hive mates.

__ 1. Information about the environment is detected by a sensory receptor is called a(n)_____? [p.782]
   a. instinct
   b. fixed action pattern
   c. stimulus
   d. learned behavior

__ 2. In _____, a particular behavior is performed without having been learned by actual experience in the environment. [p.784]
   a. natural selection
   b. altruistic behavior
   c. sexual selection
   d. instinctive behavior

__ 3. Newly hatched goslings follow any large moving object to which they are exposed shortly after hatching; this is an example of _____. [p.784]
   a. homing behavior
   b. imprinting
   c. piloting
   d. migration

__ 4. A young toad flips its sticky-tipped tongue and captures a bumble bee that stings its tongue; in the future, the toad leaves bumblebees alone. This is _____. [p.784]
   a. instinctive behavior
   b. a fixed action pattern
   c. altruistic behavior
   d. learned behavior

__ 5. Behavior that enhances another's reproductive success is called _____. [p.792]
   a. altruism
   b. instinctive
   c. selfish
   d. social

__ 6. Certain individuals receive a larger share of the resources than subordinates is known as _____. [p.791]
   a. cooperative predator avoidance
   b. the selfish herd
   c. dominance hierarchies
   d. self-sacrificing behavior

__ 7. A chemical odor in the urine of male mice triggers and enhances estrus in female mice. This chemical would be an example of a _____. [p.786]
   a. generic mouse pheromone
   b. signaling pheromone
   c. priming pheromone
   d. threat display

__ 8. When musk oxen form a "ring of horns" against predators, it is _____. [p.790]
   a. a selfish herd
   b. cooperative predator avoidance
   c. self-sacrificing behavior
   d. dominance hierarchy

__ 9. Competition among members of one sex for access to mates is called _____. [p.788]
   a. altruism
   b. social behavior
   c. inclusive fitness
   d. sexual selection

__ 10. A lek is a _____. [p.788]
   a. form of threat display
   b. type of pheromone
   c. communication signal
   d. communal display ground

CHAPTER OBJECTIVES / REVIEW QUESTIONS

1. What explains the fact that costal and inland garter snakes of the same species have different food preferences? [p.782]
2. Describe the intermediate response obtained in Arnold's experiment with costal and inland garter snakes. [p.782]
3. Explain instinctive behavior and give an example. [p.784]
4. Describe and cite an example of a fixed action pattern. [p.784]
5. Distinguish learned behavior from instinctive behavior. [p.784]
6. Explain imprinting. [p.784]

7. Define habituation. [p.785]
8. Explain the differences between classical and operant conditioning. [p.785]
9. Understand the various forms of communication signals and displays used by animals. [pp.786-787]
10. Distinguish between signaling and priming pheromones, and cite an example of each. [pp.786-787]
11. Explain the benefit of sexual selection. [p.788]
12. Define *lek*. [pp.788-789]
13. List the types of costs and benefits of social organisms. [pp.790-793]
14. What is the theory of inclusive fitness? [p.792]

INTEGRATING AND APPLYING KEY CONCEPTS

1. Think about communication signals that humans use, and list them. Do you believe a dominance hierarchy exists in human society? Think of examples of social situations in which dominance hierarchies may exist.
2. Apply the concept of self-sacrifice to parenting. Based on the information in this chapter, what are the benefits of a parent protecting its child?
3. Discuss the advantages and disadvantages of males using a *lek* to attract females. Which position in the *lek* is the most beneficial?

45

POPULATION ECOLOGY

INTRODUCTION

This chapter introduces the principles by which biologists study populations of a species. It is important as you progress through the chapter that you distinguish between the many terms associated with describing populations, such as size, density, and structure. The final sections of the chapter apply the knowledge of population structure to humans, and discuss some of the challenges facing human population growth.

FOCAL POINTS

* Figure 45.5 [p.800] and Figure 45.8 [p.802] together illustrate a key concept, the difference between exponential and logistic growth models.
* Figure 45.10 [p.805] explains the difference between the three forms of survivorship curves.
* Figure 45.16 [p.811] illustrates the different forms of age structure in populations based on their rate of growth.

INTERACTIVE EXERCISES

Impacts, Issues: The Numbers Game [p.796]

45.1. POPULATION DEMOGRAPHICS [p.798]

45.2. ELUSIVE HEADS TO COUNT [p.799]

45.3. POPULATION SIZE AND EXPONENTIAL GROWTH [pp.800-801]

Selected Words: pre-reproductive, reproductive, and post-reproductive ages [p.798], habitat [p.798], density [p.798], capita [p.800]

Boldfaced, Page-Referenced Terms

[p.798] demographics _____

_____

[p.798] population size _____

_____

[p.798] age structure _____

_____

[p.798] reproductive base _____

_____

[p.798] population density _____

_____

[p.798] population distribution _____

_____

[p.799] quadrats _____

_____

[p.799] capture-recapture methods _____

_____

[p.800] immigration _____

_____

[p.800] emigration _____

_____

[p.800] migration _____

_____

[p.800] zero population growth _____

_____

[p.800] per capita growth rate _____

_____

[p.800] exponential growth _____

_____

[p.801] doubling time _____

_____

[p.801] biotic potential _____

_____

Matching [p.798]

Choose the most appropriate statement for each term.

1. _____ demographics
2. _____ population size
3. _____ population density
4. _____ habitat
5. _____ population distribution
6. _____ age structure
7. _____ reproductive base
8. _____ density
9. _____ pre-reproductive, reproductive, and post-reproductive
10. _____ clumped distribution
11. _____ nearly uniform distribution
12. _____ random distribution

A. Includes pre-reproductive and reproductive age categories

B. The general pattern in which the individuals of the population are dispersed through a specified area

C. When individuals of a population are more evenly spaced than they would be by chance alone

D. The number of individuals in some specified area or volume of habitat

E. Occurs only when individuals of a population neither attract nor avoid one another, when conditions are fairly uniform through the habitat, and when resources are available all the time

F. The number of individuals in each of several to many age categories

G. The measured number of individuals in a specified area

H. The number of individuals within the population

I. The type of place where a species normally lives

J. Categories of a population's age structure

K. The vital statistics of a population

L. Individuals of a population form aggregations at specific habitat sites; most common dispersion pattern

Short Answer [p.799]

13. A zoologist wishes to estimate the population size of a species of salamander. Initially, 10 salamanders are caught and marked with an orange, waterproof dye. After six months the scientist returns and captures 5 marked salamanders out of a total catch of 50. What is the population size of the salamanders?_____

_____

_____

14. List three variables that may have caused error in the above estimate of population size.

_____

_____

_____

Matching [pp.800-801]

Match each of the following statements to the correct term.

15. _____ growth at a proportional rate

16. _____ the departure of individuals from a population

17. _____ net reproduction per individual per unit time

18. _____ a balanced number of births and deaths

19. _____ rates per individual

20. _____ the arrival of new individuals from other populations

21. _____ the maximum rate of increase per individual under ideal conditions

22. _____ the time it takes for a population to double its size

A. Per capita rate percentage of the population per year

B. *R*

C. Zero population growth

D. Biotic potential

E. Doubling time

F. Exponential growth

G. Emigration

H. Immigration

Problems [p.801]

23. Consider the equation $G = rN$, where G = the population growth rate per unit time, r = the net population growth rate per individual per unit time, and N = the number of individuals in the population. Assume that r remains constant at 0.2.

a. As the value of G increases, what happens to the value of N?

_____

b. If the value of G is negative, what happens to the value of N?

_____

c. If the net reproduction per individual stays the same and the population grows faster, then what must happen to the number of individuals in the population?

_____

24. Look at line (a) in the accompanying graph. After seven hours have elapsed, approximately how many individuals are in the population?

_____

25. Look at line (b) in the same graph.

   a. After 24 hours have elapsed, approximately how many individuals are in the population?

   _____

   b. After 28 hours have elapsed, approximately how many individuals are in the population? [p.801]

   _____

45.4. LIMITS ON POPULATION GROWTH [pp.802-803]

45.5. LIFE HISTORY PATTERNS [pp.804-805]

45.6. NATURAL SELECTION AND LIFE HISTORIES [pp.806-807]

Selected Words: *Type I* curves [p.804], *Type II* curves [p.805], *Type III* curves [p.805]

Boldfaced, Page-Referenced Terms

[p.802] limiting factor _____

_____

[p.802] carrying capacity _____

_____

[p.802] logistic growth _____

_____

[p.803] density-dependent factor _____

_____

[p.803] density-independent factor _____

_____

[p.804] life history pattern _____

_____

[p.804] cohort _____

_____

[p.804] survivorship curve _____

_____

[p.805] r-selection _____

_____

[p.805] K-selection _____

_____

Matching [pp.802-804]

Match each of the following terms to the most appropriate statement.

1. _____ limiting factor
2. _____ life history patterns
3. _____ carrying capacity
4. _____ survivorship curve
5. _____ logistic growth
6. _____ density-dependent factor
7. _____ density-independent factor

A. An essential resource that is in short supply

B. The maximum number of individuals of a population that the environment can sustain

C. A set of adaptations that influence survival, fertility, and age at first reproduction

D. A small population that initially grows slowly, then rapidly and then the numbers level off

E. Graph line of the age-specific survival of a cohort in a habitat

F. Biotic or abiotic factors that reduce the odds of an individual surviving or reproducing during overcrowding

G. Causes changes in population size regardless of density

Choice [pp.804-805]

Choose the most appropriate form of survivorship curve for each of the following descriptions.

   a. Type I   b. Type II   c. Type III

8. _____ Constant death rate at all ages
9. _____ Highest death rate at an early age
10. _____ High survivorship until late in life
11. _____ Most human populations follow this survivorship curve
12. _____ The survivorship of sea stars is an example
13. _____ Lizards and small mammals follow this pattern

45.7. HUMAN POPULATION GROWTH [pp.808-809]

45.8. FERTILITY RATES AND AGE STRUCTURE [pp.810-811]

45.9. POPULATION GROWTH AND ECONOMIC EFFECTS [pp.812-813]

45.10. RISE OF THE SENIORS [p.813]

Selected Words: agricultural productivity [p.808], preindustrial stage [p.812], industrial stage [p.812], postindustrial stage [p.812]

Boldfaced, Page-Referenced Terms

[p.810] total fertility rate (TFR) _____

_____

[p.812] demographic transition model _____

_____

Fill-in-the-Blanks [pp.808-809]

Early humans evolved in (1) _____, then in savannas. They were (2) _____, mostly, but also scavenged bits of meat. Bands of hunter-gatherers moved out of Africa about (3) _____ million years ago. By 44,000 years ago, their descendants were established in much of the world.

Starting about 11,000 years ago, many hunter-gatherer bands shifted to an (4) _____ lifestyle. Instead of following the (5) _____ herds, they settled in fertile valleys and other regions that favored seasonal harvesting of fruits and (6) _____. In this way, they developed a more dependable basis for life. A pivotal factor was the domestication of wild (7) _____, including the ancestral species to modern (8) _____ and (9) _____. People harvested, stored, and planted the seeds in the same location year after year. They also began domesticating (10) _____ for food and to help with labor. (11) _____ ditches were dug to help water their crops.

Until about 300 years ago, poor (12) _____ and infectious (13) _____ kept the death rate as high as the (14) _____ rate, keeping the population size nearly stable. Infectious diseases became density- (15) _____ controls. Epidemics swept through overcrowded settlements and outbreaks of (16) _____ and water-borne diseases resulted from poor sanitation. With the improvement of plumbing, access to (17) _____, and the development of vaccines and (18) _____, many diseases were no longer as deadly as before. The (19) _____ rates began to drop sharply. Birth rates began to exceed death rates and the rapid growth of the human population was underway.

Choice [p.811]

For the following choose the appropriate age structure diagram from the accompanying figure.

20. _____ United States
21. _____ China
22. _____ Canada
23. _____ Mexico
24. _____ Australia
25. _____ India

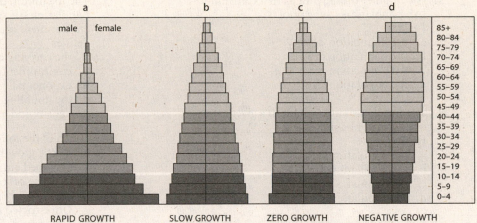

RAPID GROWTH   SLOW GROWTH   ZERO GROWTH   NEGATIVE GROWTH

Sequence [p.812]

Arrange the following stages of the demographic transition model in correct chronological sequence. Write the letter of the first step next to 26, the letter of the second step next to letter 27, and so on.

26. _____
27. _____
28. _____
29. _____

A. *Industrial stage*: population growth slows and industrialization is in full swing

B. *Preindustrial stage*: harsh living conditions, high birth and death rates, slow population growth

C. *Postindustrial stage*: zero population growth is reached, then birth rate falls below death rate, and population size slowly decreases

D. *Transitional stage*: industrialization begins, food production rises, and health care improves, death rates drop, birth rates remain high, resulting in rapid population growth

SELF-TEST

___ 1. The number of individuals that contribute to a population's gene pool is _____ [p.798]
   a. the population density
   b. the population growth
   c. the population birth rate
   d. the population size

___ 2. The number of individuals in a given area or volume of habitat is _____ [p.798]
   a. the population density
   b. the population growth
   c. the population birth rate
   d. the population size

___ 3. A population that is growing exponentially in the absence of limiting factors can be illustrated accurately by a(n) _____ [p.800]
   a. S-shaped curve
   b. J-shaped curve
   c. curve that terminates in a plateau phase
   d. tolerance curve

___ 4. Assuming the birth rate and death rate remain constant, both can be combined into a single variable, r, or _____. [p.800]
   a. the per capita rate
   b. the minus migration factor
   c. exponential growth
   d. the net reproduction per individual per unit time

___ 5. _____ is a way to express the growth rate of a given population. [p.801]
   a. Doubling time
   b. Population density
   c. Population size
   d. Carrying capacity

___ 6. The maximum rate of increase per individual under ideal conditions is called the _____. [p.801]
   a. biotic potential
   b. carrying capacity
   c. doubling time
   d. population size

___ 7. The maximum number of individuals of a population (or species) that a given environment can sustain indefinitely defines _____. [p.802]
   a. the carrying capacity of the environment
   b. exponential growth
   c. the doubling time of a population
   d. density-independent factors

___ 8. Which of the following is not characteristic of logistic growth? [p.802]
   a. S-shaped curve
   b. Leveling off of growth as carrying capacity is reached
   c. Unrestricted growth
   d. Slow growth of a low-density population followed by rapid growth

___ 9. The beginning of industrialization, a rise in food production, improvement of health care, rising birth rates, and declining death rates describes the _____ stage of the demographic transition model. [p.812]
   a. preindustrial
   b. transitional
   c. industrial
   d. postindustrial

___ 10. The survivorship curve typical of industrialized human populations is Type _____. [p.804]
   a. I
   b. II
   c. III
   d. none of the above

___ 11. Which does not fit a r-selection? [p.805]
   a. Reproduce when young
   b. Lots of offspring
   c. Significant parental care
   d. Maximize number of offspring

___ 12. Which factor had the greatest influence on increasing the human population? [p.808]
   a. Domestication of wild grasses
   b. Industrialization
   c. Decrease in the death rate
   d. Increase in the birth rates

___ 13. Which country will see the greatest increase in population in the near future? [p.810]
   a. India
   b. China
   c. Indonesia
   d. Japan

CHAPTER OBJECTIVES / REVIEW QUESTIONS

1. Define the term population. [p.798]
2. Define the following terms: *demographics, habitat, population size, population density, population distribution, age structure,* and *reproductive base.* [p.798]
3. List and describe the three patterns of distribution illustrated by populations in a habitat. [p.799]
4. Given data on a capture-recapture experiment, estimate the population size. [p.800]
5. Define *zero population growth* and describe how achieving it would affect population size. [p.800]
6. Distinguish immigration from emigration and define the term migration. [p.800]
7. Explain how $G = rN$ can be used to predict population growth. [p.801]
8. Explain the relationship between exponential growth and doubling time. [p.801]
9. Explain what is meant by *biotic potential.* [p.801]
10. List several examples of limiting factors, and explain how they influence population curves. [p.802]
11. Explain what is meant by the term *carrying capacity.* [p.802]
12. Explain the meaning of the logistic growth equation. [pp.802-803]
13. Compare logistic and exponential growth. [pp.800-803]
14. Define the term *density-dependent factors* on growth of populations; give an example. [p.803]
15. Define the term *density-independent factors* and list two examples and indicate how the factors would affect a population. [p.803]
16. Explain what is meant by a life history pattern. [p.804]
17. Explain the three survivorship curves. [pp.804-805]
18. Individuals in guppy populations targeted by killifish tend to be larger, less streamlined, and more brightly colored than individuals in populations targeted by pike-cichlids who tend to be smaller, more streamlined, and duller in color patterning. Other life history pattern differences exist between the two groups. After consideration of the research results obtained by Reznick and Endler, provide an explanation for these differences. [pp.806-807]
19. List three possible reasons why growth of the human population is out of control. [pp.808-809]
20. Define the term *total fertility rate.* [p.810]
21. Be able to analyze age structure diagrams to determine patterns of growth. [p.812]
22. List and describe the four stages of the demographic transition model. [p.812]

INTEGRATING AND APPLYING KEY CONCEPTS

1. The capture-recapture method of estimating population is widely recognized as being inaccurate. If given the resources, how would you go about estimating the size of a population without counting every individual in the population?
2. What measure could have been taken to prevent the reindeer population on St. Matthew's Island from going extinct?
3. If every less-developed country instituted laws / incentives to control population growth (like China), what impact would this have upon the world population?
4. How would changes in immigration laws in the United States influence the age structure diagrams? What are the long-term consequences of these decisions? Why would countries like Canada and Australia be actively increasing immigration?

46

COMMUNITY STRUCTURE AND BIODIVERSITY

INTRODUCTION

This chapter focuses on the variety of mechanisms by which populations of different species interact in a given habitat. You need to understand the principal categories of species interactions presented early in the chapter, which form the basis for the remainder of the chapter's material. The end of the chapter presents some of the consequences that can occur if these interactions are disturbed by human activity or the migration of species.

FOCAL POINTS

- Section 46.1 [p.818] introduces the categories of species interactions, which are discussed in more detail in subsequent sections.
- Section 46.8 [pp.828-829] examines the forms of ecological succession and the role of pioneer species.

INTERACTIVE EXERCISES

Impacts, Issues: Fire Ants in the Pants [p.816]

46.1. WHICH FACTORS SHAPE COMMUNITY STRUCTURE? [p.818]

46.2. MUTUALISM [p.819]

Selected Words: <u>Solenopsis invicta</u> [p.816], community structure [p.818], symbionts [p.818]

Boldfaced, Page-Referenced Terms

[p.818] habitat _____

_____

[p.818] community _____

_____

[p.818] niche _____

_____

[p.818] commensalisms _____

_____

[p.818] mutualism _____

_____

[p.818] interspecific competition _____

_____

[p.818] predation _____

_____

[p.818] parasitism _____

_____

[p.818] symbiosis _____

_____

[p.818] coevolution _____

_____

Fill-in-the-Blanks [p.818]

Many factors help give each (1) _____ an identifiable structure. First, (2) _____ and _____ influence a habitat's features including (3) _____, _____, and soil types. Second, the types, amounts, and seasonal availability of (4) _____ and other resources affect which (5) _____ live there. Third, species have (6) _____ that (7) _____ them to certain habitats. This is related to the fourth factor, species (8) _____ cause their population size to (9) _____. The fifth factor is (10) _____ and human-induced (11) _____ which alter the habitat.

Matching [p.818]

Match each of the following terms to the correct statement.

12. _____ coevolution

13. _____ community

14. _____ niche

15. _____ symbiosis

16. _____ habitat

17. _____ mutualism

18. _____ commensalism

19. _____ interspecific competition

A. An interaction between two species that is helpful, not essential

B. Close association between two species during part or all of their life cycle

C. Place where a species normally lives

D. The populations of all species in a given habitat

E. Relationship that hurts both species

F. An interaction that helps one species but does not affect the second species

G. Conditions required for the survival of a species

H. Two species that act as selective agents upon each other

46.3. COMPETITIVE INTERACTIONS [pp.820-821]

46.4. PREDATOR-PREY INTERACTIONS [pp.822-823]

46.5. AN EVOLUTIONARY ARMS RACE [pp.824-825]

Selected Words: coevolution [p.823], "play dead" [p.825]

Boldfaced, Page-Referenced Terms

[p.820] interference competition _____

_____

[p.820] exploitative competition _____

_____

[p.820] competitive exclusion _____

_____

[p.821] resource partitioning _____

_____

[p.821] character displacement _____

_____

[p.822] predators _____

_____

[p.822] prey _____

_____

[p.824] camouflage _____

_____

[p.824] mimicry _____

_____

[p.824] warning coloration _____

_____

Matching [pp.820-823]

Match each of the following terms to the appropriate statement.

1. _____ resource partitioning
2. _____ interference competition
3. _____ competitive exclusion
4. _____ coevolution
5. _____ exploitative competition
6. _____ predators

A. One species will reduce the amount of resources available for another species

B. A subdividing of resources that allows two species to coexist

C. Two species change over time due to close ecological interactions

D. Consumers that obtain energy and nutrients from living organisms

E. One species prevents or blocks another species' access to the same resources

F. Two species require the same limited resources; the stronger species will suppress the other species

Choice [pp.824-825]

Choose the most appropriate term for each evolutionary adaptation described in items 7-11.

a. mimicry   b. camouflage   c. chemical defense   d. last chance defense

7. _____ A blending of body form, color, or behavior with the environment
8. _____ Leaves that contain dangerous or hard-to-digest repellents
9. _____ Protection by pretending to be a dangerous organism
10. _____ Predators learn to avoid organisms that use this defense
11. _____ The use of one final trick to repel an attacker

46.6. PARASITE-HOST INTERACTIONS [pp.826-827]

46.7. STRANGERS IN THE NEST [p.827]

46.8. ECOLOGICAL SUCCESSION [pp.828-829]

46.9. SPECIES INTERACTIONS AND COMMUNITY INSTABILITY [pp.830-831]

Selected Words: novel host [p.826], pathogen [p.826], habitats [p.827], biological controls [p.827], climax community [p.828], equilibrium [p.830]

Boldfaced, Page-Referenced Terms

[p.826] parasites _____

_____

[p.826] parasitoids _____

_____

[p.826] social parasites _____

_____

[p.826] vectors _____

_____

[p.828] ecological succession _____

_____

[p.828] pioneer species _____

_____

[p.828] primary succession _____

_____

[p.828] secondary succession _____

_____

[p.829] intermediate disturbance hypothesis _____

_____

[p.830] keystone species _____

_____

[p.831] geographic dispersal _____

_____

[p.831] exotic species _____

_____

Fill-in-the-Blanks [p.826]

(1) _____ have a detrimental impact on populations. By draining (2) _____ from hosts, they alter the amount of energy and nutrients the host population demands from a(n) (3) _____. Also, weakened hosts are usually more vulnerable to (4) _____ and less attractive to potential (5) _____. Some parasitic infections cause (6) _____ while others shift the (7) _____ of the host species. In such ways, parasitic infections lower (8) _____ rates, raise (9) _____ rates, and affect intraspecies and interspecies (10) _____.

Sometimes the gradual drain of nutrients during a parasitic infection indirectly leads to the (11) _____ of the host. The host can become so weakened that they cannot fight off secondary (12) _____. Nevertheless, in evolutionary terms, killing a host too quickly is (13) _____ for a parasite's (14) _____ success. A parasitic infection must last long enough to give the parasite time to produce as many (15) _____ as possible. The longer it lives in the host, the more offspring it is likely to produce. We can thus expect selective agents to (16) _____ parasites that have a (17) _____-than-fatal effect on the host.

Choice [p.828]

For exercise 18-23 choose from the following terms.

a. primary succession   b. secondary succession

18. _____ Following a disturbance, a patch of habitat or a community recovers

19. _____ Successional changes begin when a pioneer population colonizes a barren habitat

20. _____ Many plants in this succession arise from seeds or seedlings that are already present when the process began

21. _____ A successional pattern that occurs in abandoned fields, burned forests, and tracts of land destroyed by volcanic eruptions

22. _____ May occur on a new volcanic island or on land exposed by the retreat of a glacier

23. _____ Once established, the pioneers improve conditions for other species and often set the stage for their own replacement

Matching [pp.828-831]

Choose the most appropriate statement for each term.

24. _____ pioneer species

25. _____ primary succession

26. _____ secondary succession

27. _____ climax community

28. _____ keystone species

29. _____ exotic species

30. _____ geographic dispersal

A. A species that has been dispersed from its home and become established elsewhere

B. Colonization of barren habitat that does not contain any soil

C. A community of species that will persist over time

D. Colonizers of newly vacated habitats

E. A successional mechanism in which a disturbed area recovers

F. A species that has a disproportionately large effect on a community relative to its abundance

G. The movement of a species from its home range

46.10. EXOTIC INVADERS [pp.832-833]

46.11. BIOGEOGRAPHIC PATTERNS IN COMMUNITY STRUCTURE [pp.834-835]

Selected Words: C. taxifolia [p.832], jump dispersal [p.832], Kudzu [p.832], *myxomatosis* [p.833], species richness [p.834], island biogeography [p.835]

Boldfaced, Page-Referenced Terms

[p.834] biogeography _____

_____

[p.835] area effect _____

_____

[p.835] equilibrium model of island biogeography _____

_____

[p.835] distance effect _____

_____

Short Answer [p.834]

1. List three factors responsible for creating the higher species-diversity values correlated with distance from the equator, both on land and in the seas.

_____

_____

_____

_____

_____

_____

2. After considering the distance effect, the area effect, and species-diversity patterns related to the equator answer the following question. Two islands (B and C) are the same size, topography and age and are equidistant from the African coast (A), as shown in the diagram. Which island will have the higher species-diversity values?

_____

_____

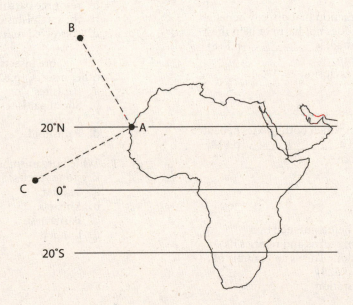

SELF-TEST

___ 1. All the populations of different species that occupy and are adapted to a given habitat are referred to as a(n) _____. [p.818]
   a. biosphere
   b. community
   c. ecosystem
   d. niche

___ 2. The range of all factors that influence whether a species can obtain resources essential for survival and reproduction is called the _____ of a species. [p.818]
   a. habitat
   b. niche
   c. carrying capacity
   d. ecosystem

___ 3. _____ is one of the one-way relationships in which one species benefits and the other is directly harmed. [p.818]
   a. Commensalism
   b. Competitive exclusion
   c. Parasitism
   d. Mutualism

___ 4. A lopsided interaction that directly benefits one species but does not harm or help the other much, if at all, is _____. [p.818]
   a. commensalism
   b. competitive exclusion
   c. predation
   d. mutualism

___ 5. An interaction in which both species benefit is best described as _____. [p.818]
   a. commensalism
   b. mutualism
   c. predation
   d. parasitism

___ 6. A striped skunk being pursued by a predator suddenly turns and releases its foul-smelling odor. This is an example of _____. [p.825]
   a. warning coloration
   b. mimicry
   c. camouflage
   d. last chance trick

___ 7. When an inexperienced predator attacks a yellow-banded wasp, the predator receives the pain of a stinger and will not attack again. This is an example of _____. [p.824]
   a. mimicry
   b. camouflage
   c. chemical defense
   d. warning coloration

___ 8. During the process of secondary succession, _____. [p. 828]
   a. pioneer populations grow in habitats that cannot support most species
   b. pioneers set the stage for their own replacement
   c. species will colonize bare soil
   d. a disturbed area recovers

___ 9. G. Gause used two species of Paramecium in a study that described _____. [p.820]
   a. interspecific competition and competitive exclusion
   b. resource partitioning
   c. the establishment of territories
   d. coevolved mutualism

___ 10. _____ are insects that lay their eggs in other insects. [p.826]
   a. Parasites
   b. Social parasites
   c. Vectors
   d. Parasitoids

___ 11. Which organism is the best example of a keystone species? [p.830]
   a. Sea star
   b. Mussels
   c. Barnacles
   d. Limpets

CHAPTER OBJECTIVES / REVIEW QUESTIONS

1. Define the terms *habitat* and *community*. [p.818]
2. Define the various "two species" interactions. [p.818]
3. Explain how mutualism helps support the theory of Endosymbiosis. [p.819]
4. Distinguish between interference competition and exploitative competition. [p.820]
5. Explain how G. Gause's work led to the formation of the competitive exclusion theory. [p.820]
6. Define *resource partitioning*. [p.821]
7. Explain why predator-prey cycles represent coevolution. [pp.822-823]
8. Understand the three models of predator responses to increases in prey density. [p.822]
9. Explain how camouflage, mimicry, chemical defenses, and last chance tricks all relate to the concept of an evolutionary arms race. Give examples of each. [pp.824-825]
10. Explain the influence of a parasite on a population. [p.826]
11. Explain the role of a pioneer species in ecological succession. [p.828]
12. Distinguish between primary and secondary succession. [p.828]
13. List the conditions that must be present in order to call a community a climax community. [p.828]
14. Explain the significance of the intermediate disturbance hypothesis. [p.829]
15. List and define the three variables that will influence succession. [pp.828-829]
16. Distinguish between a keystone species and an exotic species. [pp.830-831]
17. Explain the process and significance of geographic dispersal. [p.831]
18. List three reasons why distance form the equator corresponds to species richness. [p.834]
19. Distinguish between a distance effect and an area effect. [p.835]

INTEGRATING AND APPLYING KEY CONCEPTS

1. Describe the niche that you feel humans belong in? What do you think will happen to community stability as we expand our niche?
2. Give an example of a mutualistic relationship between humans and another species. What would happen to the human population if the other species were to go extinct?
3. What does the section on community instability tell you about the influence of humans on communities in equilibrium? What can be done to lessen this impact?

47

ECOSYSTEMS

INTRODUCTION

Ecosystems represent an interaction between the biotic and abiotic factors of a given region. In this chapter you will explore how the abiotic factors, namely energy and nutrients, influence the structure of an ecosystem. The first part of the chapter explains the flow of energy in an ecosystem; the second part takes a global perspective on the flow of nutrients in biogeochemical cycles.

FOCAL POINTS

- Figure 47.2 [p.840] demonstrates how energy flows through an ecosystem while nutrients are cycled within the ecosystem.
- Figure 47.3 [p.841] illustrates the interaction of organisms within a tall grass prairie food chain.
- Figures 47.11 [p.848] outlines the water cycle. It serves as the basis for later discussions of the carbon, nitrogen, and phosphorus cycles.

INTERACTIVE EXERCISES

Impacts, Issues: Bye-Bye, Blue Bayou [p.838]

47.1. THE NATURE OF ECOSYSTEMS [pp.840- 841]

47.2. THE NATURE OF FOOD WEBS [pp. 842- 843]

47.3. ENERGY FLOW THROUGH ECOSYSTEMS [pp. 844- 845]

Selected Words: herbivores [p.840], carnivores [p.840], parasites [p.840], omnivores [p.840], gross primary production [p.844], net primary production [p.844]

Boldfaced, Page-Referenced Terms

[p.840] ecosystems _____

_____

[p.840] primary producers _____

_____

[p.840] consumers _____

_____

[p.840] trophic levels _____

_____

[p.840] detritivores _____

_____

[p.840] decomposers _____

_____

[p.841] food chains _____

_____

[p.842] food webs _____

_____

[p.842] grazing food chain _____

_____

[p.843] detrital food chain _____

_____

[p.844] primary production _____

_____

[p.844] biomass pyramid _____

_____

[p.845] energy pyramid _____

_____

Matching [pp.840-845]

Choose the most appropriate statement for each term.

1. _____ primary producers
2. _____ consumers
3. _____ herbivores
4. _____ carnivores & parasites
5. _____ decomposers
6. _____ detritivores
7. _____ omnivores
8. _____ grazing food chain
9. _____ ecosystem
10. _____ detrital food chain
11. _____ primary production
12. _____ biomass pyramid
13. _____ energy pyramid
14. _____ trophic levels
15. _____ food chain
16. _____ food webs

A. Enter in producers flows to detritivores
B. Consumers that dine on animals and plants
C. Feed on the tissues of other organisms
D. Hierarchy of feeding relationships
E. A system of cross-connecting food chains
F. Feed on living animal tissues
G. Dry weight of all the organisms at each trophic level
H. The autotrophs
I. Energy stored in producer tissues that flows to the herbivores
J. Refers to a straight-line series of steps by which energy stored in autotroph tissues passes on through higher trophic levels
K. Consumers that eat only plants
L. Illustrates how the amount of useable energy decreases as it moves from one trophic level to the next
M. Break down of organic remains and wastes of producers and consumers
N. An array of organisms and their physical environment, interacting by a one-way flow of energy and a cycling of materials
O. Feed upon small particles of organic matter
P. The rate at which producers capture and store energy

Labeling [p.844]

Provide the name of the participant indicated by each number in the following diagram.

17. _____

18. _____

19. _____

20. _____

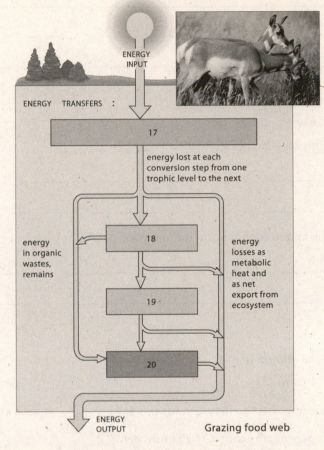

ENERGY
INPUT

ENERGY TRANSFERS :

17

energy lost at each
conversion step from one
trophic level to the next

energy
in organic
wastes,
remains

18

energy
losses as
metabolic
heat and
as net
export from
ecosystem

19

20

ENERGY
OUTPUT

Grazing food web

Short Answer [p.845]

21. Explain why energy transfers in aquatic systems are more efficient than terrestrial ones. _____

_____

_____

_____

47.4. BIOLOGICAL MAGNIFICATION [p.846]

47.5. BIOGEOCHEMICAL CYCLES [p.847]

Selected Words: DDT [p.846], Rachel Carson [p.846], mercury [p.846], water cycle [p.847], atmospheric cycle [p.847], sedimentary cycle [p.847], environmental reservoirs [p.847], nutrients [p.847]

Boldfaced, Page-Referenced Terms

[p.846] biological magnification _____

_____

[p.847] biogeochemical cycle _____

_____

Short Answer [p.846]

1. Define biological magnification. _____

    _____

    _____

2. Explain how the biological magnification of DDT is harmful. _____

    _____

    _____

Fill-in-the-Blanks [p.847]

In a(n) (3) _____ cycle, a(n) (4) _____ element moves from the environment , through an ecosystem, then back to the environment. (5) _____ such as carbon, oxygen, hydrogen, (6) _____, and phosphorus are essential to all forms of life.

In three types of biogeochemical cycles, parts of the environment act as (7) _____ for key elements. In the (8) _____ cycle, oxygen and hydrogen move, on a global scale, as molecules of (9) _____. In atmospheric cycles, some gaseous form of (10) _____ and nitrogen are available to the (11) _____. Phosphorus and other solid nutrients that have no gaseous form move in (12) _____ cycles. They accumulate on the (13) _____ and eventually return to land by way of geological uplifting, which often takes millions of years. The Earth's (14) _____ is the biggest reservoir for nutrients that have sedimentary cycles.

47.6. WATER CYCLE [pp.848-849]

47.7. CARBON CYCLE [pp.850-851]

47.8. GREENHOUSE GASES AND CLIMATE CHANGE [pp.852-853]

Selected Words: overdraft [p.849], carbon-oxygen cycle [p.851], "greenhouse gases" [p.851]

Boldfaced, Page-Referenced Terms

[p.848] water cycle _____

_____

[p.848] watershed _____

_____

[p.848] aquifers _____

_____

[p.848] groundwater _____

_____

[p.848] runoff _____

_____

[p.849] salinization _____

_____

[p.849] desalinization _____

_____

[p.850] carbon cycle _____

_____

[p.852] greenhouse effect _____

_____

[p.853] global warming _____

_____

Matching [pp.848-849]

Match each of the following terms to the appropriate statement.

1. _____ salinization
2. _____ groundwater
3. _____ desalinization
4. _____ water cycle
5. _____ transpiration
6. _____ watershed

A. A buildup of mineral salts
B. A region where precipitation flows into a specific waterway
C. The loss of water by plants into the atmosphere
D. The movement of water on a global scale
E. The removal of salt from seawater
F. The water that is found in soils and aquifers

Matching [pp.850-853]

Choose the most appropriate statement for each term.

7. _____ greenhouse gases
8. _____ carbon-oxygen cycle
9. _____ carbon cycle
10. _____ Earth's crust
11. _____ greenhouse effect
12. _____ carbon dioxide (CO_2)
13. _____ oceans & the atmosphere
14. _____ global warming

A. The primary reservoir of carbon
B. Where most annual carbon cycling occurs
C. The movement of carbon on a global scale
D. One of the main greenhouse gases
E. The fixation of carbon by autotrophs and the release of carbon by aerobic cells
F. The warming of the Earth's lower atmosphere as a result of greenhouse gases
G. A long-term increase in temperatures near Earth's surface
H. Prevents the escape of heat energy form the Earth

Short Answer [p.851]

15. Describe why the greenhouse effect is both necessary and harmful. _____

_____

_____

_____

16. What factor is the primary factor for the excess carbon dioxide in the atmosphere? _____

_____

_____

_____

Completion [pp.848-849]

For each of the following, place the correct numerical value in the blank.

17. _____ of the water humans use goes directly into agriculture.

18. About _____ of the United States' population relies on groundwater as a source of drinking water.

19. _____ of the Ogallala aquifer has been overdrafted.

47.9. NITROGEN CYCLE [pp.854-855]

47.10. PHOSPHORUS CYCLE [pp.856-857]

Selected Words: Rhizobium [p.854], ion exchange [p.855]

Boldfaced, Page-Referenced Terms

[p.854] nitrogen cycle _____

_____

[p.854] nitrogen fixation _____

_____

[p.855] ammonification _____

_____

[p.855] nitrification _____

_____

[p.855] denitrification _____

_____

[p.856] phosphorus cycle _____

_____

[p.857] eutrophication _____

_____

Matching [pp.854-857]

Choose the most appropriate statement for each term.

1. _____ nitrogen cycle

2. _____ nitrogen fixation

3. _____ ammonification

4. _____ ion exchange

5. _____ nitrification

6. _____ denitrification

7. _____ phosphorus cycle

8. _____ eutrophication

A. Ammonium is converted to nitrite by bacteria

B. Bacteria convert nitrate or nitrite to N_2 or nitrogen oxide

C. Microbes break down nitrogen-containing molecules and form ammonium

D. The sedimentary cycle of a nutrient that moves from land to oceans and slowly back again

E. A few kinds of bacteria convert N_2 to ammonia (NH_3) which is easily taken up by plants

F. The global movement of nitrogen in an ecosystem

G. The nutrient enrichment of an ecosystem

H. Nutrients that are bound to soil particles are replaced by hydrogen ions

Short Answer [p.854]

9. From the perspective of a plant, list the benefits and costs of an association with nitrogen-fixing bacteria.

_____

_____

_____

_____

SELF-TEST

___ 1. An array of organisms and their physical environment, all interacting by a one-way flow of energy and a cycling of materials, is a(n) _____. [p.840]
a. population
b. community
c. ecosystem
d. biosphere

___ 2. _____ ingest decomposing particles of organic matter. [p.840]
a. Herbivores
b. Parasites
c. Detritivores
d. Carnivores

___ 3. The members of feeding relationships are structured in a hierarchy, the steps of which are called _____. [p.840]
a. organism levels
b. energy source levels
c. eating levels
d. trophic levels

___ 4. In a grazing food chain, energy flows from _____. [pp.842-843]
a. producers to detritivores and decomposers
b. primary consumers to detritivores and decomposers
c. producers to herbivores
d. primary consumers to herbivores, then through carnivores

5. Which of the following lives in a host and feeds on its tissues? [p.840]
   a. Parasites
   b. Herbivores
   c. Producers
   d. Carnivores

6. In a natural community, the primary consumers are _____. [pp.840-841]
   a. herbivores
   b. carnivores
   c. scavengers
   d. decomposers

7. A straight-line series of steps of who eats whom in an ecosystem is sometimes called a(n) _____. [p.841]
   a. trophic level
   b. food chain
   c. ecological pyramid
   d. food web

8. Of the 1,700,000 kilocalories of solar energy that entered an aquatic ecosystem in Silver Springs, Florida, investigators determined that about _____ percent of incoming solar energy was trapped by photosynthetic autotrophs. [p.845]
   a. 1
   b. 10
   c. 25
   d. 74

9. A biogeochemical cycle that deals with phosphorus and other nutrients that do not have gaseous forms is the _____ type. [p.847]
   a. sedimentary
   b. hydrologic
   c. nutrient
   d. atmospheric

10. _____ is the nutrient enrichment of an ecosystem. [p.857]
    a. Eutrophication
    b. Salinization
    c. Nitrification
    d. Ammonification

11. In the carbon cycle, carbon enters the atmosphere through_____. [p.851]
    a. carbon dioxide fixation
    b. volcanic eruptions and fossil fuel burning
    c. oceans and accumulation of plant biomass
    d. release of greenhouse gases

12. _____ refers to an increase in concentration of a nondegradable (or slowly degradable) substance in organisms as it passes along food chains. [p.846]
    a. Ecosystem modeling
    b. Nutrient input
    c. Biogeochemical cycle
    d. Biological magnification

13. Which type of pollution concentrates itself in the bodies of fish? [p.846]
    a. Mercury
    b. Nitrogen
    c. Eutrophication
    d. Salinization

14. Which state in the U.S. has highly overdrawn its groundwater supplies? [p.849]
    a. Texas
    b. Nevada
    c. Washington
    d. Colorado

CHAPTER OBJECTIVES / REVIEW QUESTIONS

1. Distinguish among herbivores, carnivores, parasites, detritivores, and decomposers. [p.840]
2. Define the term ecosystem. [p.840]
3. Explain the difference between a food chain and a trophic level. [pp.840-841]
4. Understand the flow of energy and the connection between food chains and food webs. [pp.842-843]
5. Explain the process of biological magnification [p.846]
6. Describe how DDT damages ecosystems. [p.846]
7. Distinguish between primary productivity and net ecosystem production. [p.844]
8. Explain the differences between a biomass pyramid and an energy pyramid. [pp.844-845]
9. Using the Silver Springs ecosystem as an example, explain the path of energy flow. [p.845]
10. Explain the role of biogeochemical cycles in nutrient cycling. [p.847]
11. Distinguish between hydrologic, atmospheric, and sedimentary cycles with regards to the major types of reservoirs found in each. [p.847]
12. Explain the water cycle. [p.848]

13. Define watershed. [p.848]
14. Explain the threats to the global water supply. [pp.848-849]
15. Outline the carbon cycle. [pp.850-851]
16. Explain how greenhouse gases contribute to the greenhouse effect. [pp.852-853]
17. Outline the evidence for global warming. [p.853]
18. Explain the nitrogen cycle using the terms nitrogen fixation, ammonification, nitrification, and denitrification. [pp.854-855]
19. Explain how the use of fertilizers affects the process of ion exchange in soil. [p.855]
20. Outline the phosphorus cycle and give the biological importance of phosphorus. [pp.856-857]
21. Define eutrophication. [p.857]

INTEGRATING AND APPLYING KEY CONCEPTS

1. Pick an ecosystem and indicate 5 organisms that can be found in your ecosystem from each of the following categories: producers, herbivores, carnivores, omnivores, and detritivores. Draw food chains that show the feeding relationship between the various members of the community. Connect the food chains into a larger food web.
2. List everything that you have just eaten for dinner. Indicate the trophic level where each organism is found. How can you decrease your ecological impact by adjusting your meal?

48

THE BIOSPHERE

INTRODUCTION

This chapter takes a big-picture look at the biosphere, or Earth. Starting with an explanation of how climate is influenced by air and ocean currents, and land form formations such as mountains, the chapter progresses to a discussion of the primary surface biomes. The study of water-based biomes follows, from freshwater streams and lakes to the deep ocean. Throughout the chapter you should focus on the key characteristics of each of the biomes.

FOCAL POINTS

- Sections 48.1 and 48.3 explain how climate is influenced by three major factors: air currents [pp. 862-865], ocean currents, and landforms. [pp.866-867]
- Section 48.17 outlines the interrelatedness of many of the challenges facing the biosphere. [pp.886- 887]

INTERACTIVE EXERCISES

Impacts, Issues: Surfers, Seals, and the Sea [p.860]

48.1. GLOBAL AIR CIRCULATION PATTERNS [pp.862-863]

48.2. SOMETHING IN THE AIR [pp.864-865]

48.3. THE OCEANS, LANDFORMS AND CLIMATES [pp.866-867]

Selected Words: solar-hydrogen energy [p.863], "wind farms" [p.863], ozone hole [p.864], thermal inversion [p.864], industrial smog [p.864], photochemical smog [p.864], ocean currents [p.866], leeward [p.867]

Boldfaced, Page-Referenced Terms

[p.862] climate _____

_____

[p.864] pollutants _____

_____

[p.864] ozone layer _____

_____

[p.864] smog _____

_____

[p.864] acid rain _____

_____

[p.867] rain shadow _____

_____

[p.867] monsoon _____

_____

Fill-in-the-Blanks [p.862]

(1) _____ refers to the average weather conditions over time for a given area. (2) _____ & (3) _____ are two of the most influential factors in determining a region's climate. Due to the rotation of the Earth and its tilt, the intensity of (4) _____ will vary from location to location. As a result of these factors, the Earth will (5) _____ more at the (6) _____ than at the poles.

Regional differences in warming is the start of (7) _____ circulation patterns. In general, warm air holds more (8) _____ than cold air. This helps explain why the tropics are so humid.

Labeling [p.863]

Several terms in the accompanying diagram have been replaced by numbers. Identify each of the missing terms.

9. _____

10. _____

11. _____

12. _____

13. _____

14. _____

15. _____

16. _____

17. _____

18. _____

Air warms, picks up some 13, 14, cools, then gives up 15.

Cooled, dry air 16.

17 (winds from the east)

18 (winds from the west)

The cooled, drier air 12.

Air at the equator 9, picks up much moisture, and 10 to cool altitudes; then it gives up 11 (as precipitation).

60°N

30°N

northeast tradewinds

(doldrums)

equator

southeast tradewinds

30°S

The cooled, drier air 12.

60°S

18

17

Air warms, picks up some 13, 14, cools, and then gives up 15.

Cooled, dry air 16.

Deflections in the paths of air flow near Earth's surface

Initial pattern of air circulation

Matching [pp.864-865]

19. _____ seasonal declines in ozone concentration

20. _____ forms over cities that burn a lot of coal & fossil fuels in the winter

21. _____ this is caused by sulfur dioxides and nitrogen oxides in the atmosphere

22. _____ this is caused by the interaction of nitric oxide with sunlight in the atmosphere

23. _____ weather conditions that trap a layer of cool, dense air under a warm air layer

A. Industrial smog above the Earth's poles

B. Photochemical smog

C. Ozone thinning

D. Acid rain

E. Thermal inversion

Matching [pp.866-867]

24. _____ monsoons

25. _____ rain shadow

26. _____ leeward

27. _____ windward

28. _____ ocean current

A. Air circulation patterns caused by the interaction of the equatorial sun with trade winds

B. Mountain side facing away from the wind

C. Mountain side facing toward the wind

D. An arid region located leeward of a high mountain

E. Large volumes of water responding to trade winds and westerlies

48.4. BIOGEOGRAPHIC REALMS AND BIOMES [pp.868-869]

48.5. SOILS OF MAJOR BIOMES [p.870]

48.6. DESERTS [p.871]

Selected Words: climate [p.868], topsoil [p.870], CAM plants [p.871]

Boldfaced, Page-Referenced Terms

[p.868] biogeographic realms _____

_____

[p.868] biome _____

_____

[p.870] soils _____

_____

[p.870] soil profile _____

_____

[p.871] deserts _____

_____

Matching [pp.868-871]

1. _____ biogeographic realms
2. _____ biome
3. _____ soils
4. _____ deserts
5. _____ tropical forests
6. _____ CAM plants
7. _____ soil profile
8. _____ climate
9. _____ grasslands
10. _____ clay

A. Mixtures of mineral particles and variable amounts of decomposing organic material (humus)

B. Plants that have adapted to opening their stomata at night

C. Soil type that is richest in minerals but drains poorly

D. Land biome with a deep (up to 3 meters) layer of top soil

E. Temperature and rainfall patterns for a given area

F. Subdivision of a biogeographic realm

G. Areas where annual rainfall is less than 10 centimeters; evaporation rates are high

H. Six vast land areas on Earth, each with distinguishing plant and animals

I. Biome with a thin layer of topsoil and nutrient poor soils

J. The layered structure of soils

Matching [p.870]

11. _____ grasslands soil
12. _____ coniferous forest soil
13. _____ tropical rain forest soil
14. _____ deciduous forest soil
15. _____ desert soil

A. A horizon: alkaline, deep, rich in humus

B. O horizon: scattered litter; A horizon: rich in organic matter above humus layer unmixed with mineral

C. O horizon: sparse litter; A-E horizons: continually leached

D. O horizon: pebbles, little organic matter; A horizon: shallow, poor soil

E. O horizon: well-defined, compacted mat of organic deposits resulting mainly from activity of fungal decomposers

48.7. GRASSLANDS, SHRUBLANDS, AND WOODLANDS [pp.872-873]

48.8. MORE RAIN, BROADLEAF FORESTS [p.874]

48.9. YOU AND THE TROPICAL FORESTS [p.875]

48.10. CONIFEROUS FORESTS [p.876]

48.11. TUNDRA [p.877]

Selected Words: dust bowl [p.872], tallgrass prairie [p.872], taiga [p.876]

Boldfaced, Page-Referenced Terms

[p.872] grasslands _____

_____

[p.873] savannas _____

_____

[p.873] dry shrublands _____

_____

[p.873] dry woodlands _____

_____

[p.874] semi-evergreen forests _____

_____

[p.874] tropical deciduous forests _____

_____

[p.874] temperate deciduous forests _____

_____

[p.874] evergreen broadleaf forests _____

_____

[p.876] coniferous forests _____

_____

[p.877] arctic tundra _____

_____

[p.877] alpine tundra _____

_____

Matching [pp.872-877]

1. _____ alpine tundra
2. _____ artic tundra
3. _____ taigas
4. _____ coniferous forest
5. _____ evergreen broadleaf forests
6. _____ semi-evergreen forests
7. _____ temperate deciduous forests
8. _____ grasslands
9. _____ savannas
10. _____ dry shrublands
11. _____ dry woodlands

A. Characterized by an abundance of cone-bearing trees

B. Characterized by low rainfall (less than 25 centimeters per year), permafrost, and fast plant growth during a brief growing season

C. Also called "swamp forests," these are the boreal forests found in glaciated regions

D. This develops in the high mountain regions of the world; no permafrost

E. Characterized by high rainfall (more than 130 centimeters per year) and high humidity and temperature

F. Characterized by high rainfall, but longer dry seasons and slower decomposition rates than tropical rain forests

G. Characterized by a six-month growing season and approximately 50-150 centimeters of rainfall per year

H. Broad belt of grasslands with scattered shrubs and trees

I. Also known as chaparral, less than 25-60 centimeters of rain per year

J. Forms between deserts and temperate forests, 25-100 centimeters of rain per year

K. Drought tolerant trees that don't form a closed canopy, 40-100 centimeters of rain per year

48.12. FRESHWATER ECOSYSTEMS [pp.878-879]

48.13. "FRESH" WATER? [p.880]

48.14. COASTAL ZONES [p.880-881]

Selected Words: zones [p.878], oligotropic lakes [p.878], eutropic lakes [p.878], eutrophication [p. 878], pollutants [p.880], wastewater treatment [p.880], mangrove wetlands [p.880], intertidal zone [p.881]

Boldfaced, Page-Referenced Terms

[p.878] lake _____

_____

[p.879] spring overturn _____

_____

[p.879] thermocline _____

_____

[p.879] fall overturn _____

_____

[p.880] estuary _____

_____

Matching [pp.878-880]

Match each of the following terms to the appropriate statement.

1. _____ fall overturn
2. _____ eutrophic lakes
3. _____ lake
4. _____ spring overturn
5. _____ oligotrophic lake
6. _____ thermocline
7. _____ estuary
8. _____ mangrove wetlands

A. Newly formed lakes, deep, clear, and nutrient poor

B. Older, shallower, nutrient-rich lakes

C. Winds moving across a lake cause vertical movements of dissolved oxygen and nutrients

D. Cooling of the upper layer of a lake causes vertical movements of dissolved oxygen and nutrients

E. Coastal region where fresh and salt water mix

F. Salt-tolerant woody plants that live in sheltered areas along tropical coasts

G. Thermal stratification of a lake

H. Body of standing fresh water

Short Answer [pp.878-879]

9. List the key differences between an oligotrophic and eutrophic lake.

    _____

    _____

    _____

    _____

10. Explain what happens during the spring and fall overturns.

    _____

    _____

    _____

    _____

48.15. THE ONCE AND FUTURE REEFS [pp.882-883]

48.16. THE OPEN OCEAN [pp.884-885]

48.17. CLIMATE, COPEPODS, AND CHOLERA [pp.886-887]

Selected Words: bleaching [p.882], invasive species [p.883], "southern oscillation [p.886], *cholera* [p.886], *Vibrio cholerae* [p.887]

Boldfaced, Page-Referenced Terms

[p.882] coral reefs _____

_____

[p.884] Pelagic province _____

_____

[p.884] Benthic province _____

_____

[p.884] sea mounts _____

_____

[p.885] upwelling _____

_____

[p.884] hydrothermal vents _____

_____

[p.885] El Niño _____

_____

[p.885] ENSO _____

_____

[p.886] La Niña _____

_____

Matching [pp.882-887]

Match each of the following terms to the appropriate statement.

1. _____ cholera
2. _____ ENSO
3. _____ upwelling
4. _____ Pelagic Province
5. _____ Benthic Province
6. _____ seamounts
7. _____ hydrothermal vents
8. _____ coral reefs
9. _____ symbionts
10. _____ bleaching
11. _____ invasive species

A. Species that lives inside of another

B. Wave-resistant formation of the slowly accumulated remains of marine organisms

C. Ocean bottom

D. The location of superheated, mineral-rich water

E. The vertical movement of cold, deep, nutrient-rich water

F. The southern oscillation marked by increases in sea surface temperatures and changes in air circulation patterns

G. A disease caused by the bacteria Vibrio cholerae

H. When coral loses the dinoflagellates that live inside of its tissues

I. Undersea mountains over 1000 meters tall

J. Open waters of the ocean

K. Species that are not native to a region

Labeling [p.884]

Label each numbered item in the accompanying illustration.

12. _____ province

13. _____ province

14. _____ zone

15. _____ zone

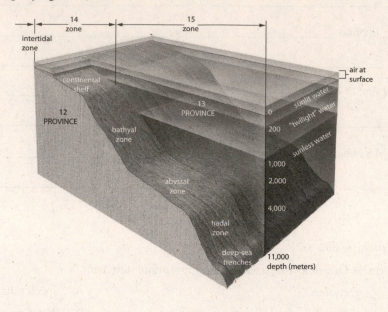

SELF-TEST

___ 1. The form of pollution that is the result of nitric oxide and sunlight is called _____. [p.864]
  a. photochemical smog
  b. industrial smog
  c. acid rain
  d. ozone thinning
  e. none of the above

___ 2. In a(n) _____, nutrient-rich fresh water draining from the land mixes with seawater carried in on tides. [p.880]
  a. pelagic province
  b. rift zone
  c. upwelling
  d. estuary

___ 3. A biome with broad belts of grasslands and scattered tress is known as a _____. [p.873]
  a. warm desert
  b. savanna
  c. tundra
  d. taiga

___ 4. The _____ biome is located about 300 north and south, has limited vegetation, and has rapid surface cooling at night. [p.871]
  a. shrublands
  b. savanna
  c. taiga
  d. desert

___ 5. In evergreen broadleaf forests, _____. [p.874]
  a. productivity is high
  b. litter does not accumulate
  c. soils are highly weathered and are poor nutrient reservoirs
  d. decomposition and mineral cycling are extremely rapid
  e. all of the above

___ 6. Premafrost would be found in which of the following? [p.877]
  a. Alpine tundra
  b. Artic tundra
  c. Savannas
  d. Grasslands
  e. Boreal forests

7. The lake's upper layer cools, the thermocline vanishes, lake water mixes vertically, and once again dissolved oxygen moves down and nutrients move up. This describes the _____. [p.879]
   a. spring overturn
   b. summer overturn
   c. fall overturn
   d. winter overturn

8. Rain shadows are located on the _____ side of high mountain ranges. [p.867]
   a. windward
   b. leeward
   c. both windward and leeward

9. _____ are air circulation patterns that influence the continents north pr south of warm oceans; low pressure causes moisture-laden air above the neighboring ocean to move inland, resulting in heavy rains. [p.867]
   a. Geothermal ecosystems
   b. Upwelling
   c. Taigas
   d. Monsoons

10. All of the water above the continental shelves is in the _____. [p. 884]
   a. neartic zone of the benthic province
   b. oceanic zone of the pelagic province
   c. neritic zone of the pelagic province
   d. oceanic zone of the benthic province

11. Forests that consist of a mixture of broadleaf trees that retain leaves year round and deciduous trees are called _____. [p.874]
   a. tropical deciduous forests
   b. temperate deciduous forests
   c. semi-evergreen forests
   d. evergreen broadleaf forests

12. Chemoautotrophic prokaryotes are the starting point for _____. [p.884]
   a. hydrothermal vent communities
   b. desert communities
   c. lake communities
   d. coniferous forest communities

13. Which soil type will have an alkaline, deep and rich humus layer? [p.870]
   a. grasslands
   b. desert
   c. coniferous forest
   d. deciduous forest

14. During which event does the water temperature of the Pacific Ocean rise, evaporation increases, and the air pressure drops? [p.886]
   a. ENSO
   b. La Nina
   c. Monsoons
   d. fall overturn

CHAPTER OBJECTIVES / REVIEW QUESTIONS

1. Explain the factors that contribute to climate. [p.862]
2. Describe the causes of ozone thinning, industrial smog, photochemical smog, and acid rain. [pp. 864-865]
3. Explain how ocean currents and mountain ranges influence climate. [pp.866-867]
4. Distinguish between a biogeographic realm and a biome. [p.868]
5. Understand the structure of soil and what is meant by a soil profile. [p.870]
6. Give the basic characteristics of a desert, dry shrubland, dry woodland, grassland, and savanna. [pp. 871-873]
7. Give the basic characteristics of the following forest types: evergreen broadleaf, semi-evergreen, temperate deciduous and tropical deciduous. [p.874]
8. Give the general characteristics of a coniferous forest. [p.876]
9. Distinguish between the arctic tundra and the alpine tundra. [p.877]
10. Explain why spring and fall overturns are important for lakes. [p.879]
11. Distinguish between an oligotrophic and eutrophic lake. [p.878]
12. Define eutrophication. [p.878]
13. Describe the structure of a stream. [p.879]
14. List the three stages of wastewater treatment. [p.880]
15. Define estuary and explain its importance. [p.880]
16. List the three zones of a rocky shore. [p.881]
17. Describe how coral reefs are formed and their importance. [p.882]
18. Report on the status of the coral reefs around the world. [pp.882-883]
19. Explain the difference between El Nino and La Nina. [pp.885-886]
20. Explain what is meant by a "southern oscillation." [p.886]

INTEGRATING AND APPLYING KEY CONCEPTS

1. Explain how altering the water quality (pH, heavy metal concentrations, pollutants, etc.) of Lake Michigan will have an impact upon citizens in China.
2. Explain how farming practices in the Midwest will impact the Gulf of Mexico. What can we do as individuals that will lessen the negative impact we are having on the Gulf of Mexico's estuary?
3. Compare the environmental conditions necessary to form a tropical rainforest versus a deciduous forest.

49

HUMAN IMPACTS ON THE BIOSPHERE

INTRODUCTION

Currently, humans are the dominant species on the planet. There is almost no habitat on the planet that we have not ventured into or impacted in some fashion. Many of our actions have led to damage, destruction or the alteration of many of the Earth's ecosystems. As a result of these changes, many species on Earth are facing extinction in the near future. This chapter focuses on some of the impacts that humans have had on the Earth and the many species that inhabit it. It also tries to address how we can alter our behaviors in order to sustain more of the planet's natural resources.

FOCAL POINT

- Section 49.2 [pp. 894 – 895] identifies the major reasons for species extinction.
- Section 49.8 [pp. 902 – 903] discusses several sustainable solutions to help maintain biodiversity.

INTERACTIVE EXERCISES

Impacts, Issues: A Long Reach [p.890]

49.1. THE EXTINCTION CRISIS [pp.892-893]

49.2. CURRENT THREATS TO SPECIES [pp.894-895]

49.3. THE UNKNOWN LOSSES [p.896]

Selected Words: Mass extinction [p.892], Dodo [p.893], habitat loss [p.894], habitat fragmentation [p.894], degrade habitat [p.894], over harvesting [p.894], poaching [p.894], exotic predators [p.895], exotic species [p.895]

Boldfaced, Page-Referenced Terms

[p.893] endangered species _____

_____

[p.893] threatened species _____

_____

[p.894] endemic species _____

_____

Fill-in-the-Blanks [pp.892-893]

(1) _____ is a natural process. When a species goes extinct it is (2) _____ by another similar species. It is estimated that (3) _____% of all species ever present on Earth have gone extinct. The Earth's (4) _____ is greater now than it has ever been in the past. (5) _____ mass extinctions have occurred and they mark the boundaries for (6) _____ time. It takes approximately (7) _____ million years for the diversity to return after a mass extinction. Scientists believe that we are in the midst of a (8) _____ mass extinction. The current extinctions are due primarily to (9) _____ and their impact upon the planet.

Matching [pp.894-895]

Match the current threat to species with the appropriate species. Some threats will have more than one species associated with it.

10. _____ habitat fragmentation

11. _____ habitat degradation

12. _____ overharvesting

13. _____ poaching

14. _____ exotic species

A. Kudzu
B. Decrease in the panda population
C. White abalone
D. Japanese honeysuckle
E. Decrease in the prairie fringed orchid population
F. Texas blind salamander
G. European brown trout
H. Rhino
I. Atlantic codfish

Short Answer [p.894]

15. List several steps that are being taken to try and help save the panda.

_____

_____

_____

_____

16. List several things humans are doing that degrade habitats.

_____

_____

_____

_____

49.4. ASSESSING BIODIVERSITY [pp.896-897]

49.5. EFFECTS OF DEVELOPMENT AND CONSUMPTION [pp.898-899]

Selected Words: urban and suburban development [p.898], hydroelectric power [p.899], surface mining [p.899]

Boldfaced, Page-Referenced Terms

[p.896] biodiversity _____

_____

[p.896] indicator species _____

_____

[p.896] hotspot _____

_____

[p.897] ecoregions _____

_____

Matching [pp.896-897]

Match the term with the correct description.

1. _____ bioindicator
2. _____ indicator species
3. _____ hot spots
4. _____ trout
5. _____ lichens
6. _____ ecoregions

A. Indicator of air pollution
B. Habitat that is rich in endemic species
C. Various types of organisms
D. Regions characterized by climate, geography, and species found within them
E. Indicator of stream quality
F. Species that alerts biologists to habitat degradation

Short Answer [p.896]

7. Indicate the 3 ways that Conservation Biology is trying to address the rate of decline in biodiversity.

_____

_____

_____

_____

Matching [p.897]

Match the correct ecoregion with the threat it is facing.

8. _____ northern prairie

9. _____ Pacific temperate rain forest

10. _____ Sierra Madre pine-oak forests

11. _____ Nevada coniferous forests

12. _____ California chaparral & woodlands

13. _____ southeastern coniferous & broadleaf

A. Logging, suppression of fire, & urban expansion

B. Logging, urban expansion

C. Conversion to pasture or farmland

D. Logging

E. Exotic species, overgrazing, & fire suppression

F. Overgrazing, logging, & overuse of forests for recreation

Fill-in-the-blank [p.899]

Fill in the blank with the appropriate percentage.

14. Petroleum supplies _____ % of our total energy.

15. Natural gas supplies _____ % of our total energy.

16. Renewable energy supplies _____ % of our total energy.

17. Nuclear energy supplies _____ % of our total energy.

18. Transportation consumes _____ % of our total energy.

19. Residential & Commercial use consumes _____ % of our total energy.

20. Electric Power generation consumes _____ % of our total energy.

21. Industry consumes _____ % of our total energy.

49.6. THE THREAT OF DESERTIFICATION [p.900]

49.7. THE TROUBLE WITH TRASH [p.901]

49.8. MAINTAINING BIODIVERSITY & HUMAN POPULATIONS [pp.902-903]

Selected Words: trash [p.901], forms of wealth [p.902], biological wealth [p.902], *Z. diploperennis* [p.902], Monteverde Cloud Forest [p.903]

Boldfaced, Page-Referenced Terms

[p.900] desertification _____

_____

[p.903] ecotourism _____

_____

[p.903] sustainable logging _____

_____

[p.903] responsible ranching _____

_____

[p.903] riparian zones _____

_____

Fill-in-the-Blanks [p.901]

The average U.S. citizen produces (1) _____ pounds of trash per day, only about (2) _____ of

this is recycled. Nonrecycled trash gets (3) _____ or disposed of in a (4) _____ that is lined to minimize

groundwater contamination. Disposable diapers will last over (5) _____ years, plastic bags will be around for

over (6) _____ years, and a cigarette filter will take over (7) _____ years to decompose. (8) _____

and proper (9) _____ of trash are two easy steps we all can take to help decrease our environmental impact.

Matching [pp.900-903]

Match the correct term with its description.

10. _____ desertification

11. _____ biological wealth

12. _____ ecotourism

13. _____ sustainable logging

14. _____ riparian zones

15. _____ responsible ranching

A. Food, medicine, and other products that have a value

B. Exclusion of cattle from riparian zones

C. Conversion of grasslands or woodlands to desert-like conditions

D. Setting aside species-rich habitat and encouraging tourism

E. Narrow corridors of vegetation along rivers or streams

F. Selectively cutting trees to reduce erosion and maximize biodiversity and profit

SELF-TEST

___ 1. How many major mass extinctions has the Earth experienced? [p.892]
   a. 5
   b. 7
   c. 2
   d. 8

___ 2. Which species status is of the greatest concern? [p.893]
   a. Threatened
   b. Vulnerable
   c. Endangered
   d. Stable

___ 3. Which is not a current threat to species? [p.893]
   a. Habitat loss
   b. Habitat fragmentation
   c. Overharvesting
   d. Ecotourism

___ 4. Which is an example of an exotic species? [p.895]
   a. Passenger pigeon
   b. Brown tree snake in Samoa
   c. Golden trout in California
   d. Kudzu in the American southeast

___ 5. An indicator species is a species that _____. [p.896]
   a. is endangered
   b. alerts biologists to habitat degradation
   c. has been around for a long time
   d. is evolving

___ 6. Renewable energy supplies _____ % of the U.S. total energy. [p.899]
   a. 6.8
   b. 10.3
   c. 21.6
   d. 8.2

___ 7. The average U.S. citizen produces _____ pounds of trash per day. [p.901]
   a. 2.1
   b. 4.6
   c. 11.6
   d. 8.2

___ 8. _____ is the conversion of grasslands or woodlands to desert-like conditions.
   a. Poaching
   b. Overharvesting
   c. Habitat destruction
   d. Desertification

___ 9. Which is not a sustainable use of biological wealth?
   a. Ecotourism
   b. Sustainable logging
   c. Responsible ranching
   d. Poaching

CHAPTER OBJECTIVES / REVIEW QUESTIONS

1. Understand how humans are accelerating the rate of extinction. [p.893]
2. Define mass extinction. [p.893]
3. Describe the criteria for a species to be listed as endangered or threatened. [p.893]
4. Explain how habitat loss, habitat fragmentation, habitat degradation, overharvesting, poaching, exotic predators, and exotic species are all contributing to species extinction. [pp.894-895]
5. Define endemic species. [p.894]
6. List the goals of conservation biologists. [p.896]
7. Explain the importance of an indicator species. [p.896]
8. Explain the importance of preserving hot spots. [p.896]
9. Identify the critical and endangered ecoregions around the world. [p.897]
10. Discuss the effects of urban and suburban development. [p.898]
11. Match the energy consumption to the source and sector of use. [p.899]
12. Describe the positive feedback cycle between drought and desertification. [p.900]
13. List the ecological effects of trash. [p.901]
14. Give examples of how countries can manage their biological wealth to produce economic opportunities. [p 902]
15. Explain how ecotourism, sustainable logging, and responsible ranching are sustainable uses of biological wealth. [p.903]

546   Chapter Forty-Nine

INTEGRATING AND APPLYING KEY CONCEPTS

1. Pick 5 endangered ecoregions and develop a plan to preserve them. Make sure you take into consideration the needs of the local populations.
2. How can we encourage more Americans to reduce, reuse, and recycle their resources? Devise a strategy to cut in half the amount of trash the average American produces each day.
3. Considering the rate of extinction on Earth today, how will humans be affected if this continues?

ANSWER KEY

Chapter 1   Invitation to Biology

Impacts, Issues: Lost Worlds and Other Wonders [p. 2]

1.1. LIFE'S LEVELS OF ORGAIZATION [pp.4-5]
1. E; 2. C; 3. F; 4. H; 5. D; 6. A; 7. I; 8. B; 9. J; 10. G;
11. K; 12. F; 13. H; 14. C; 15. B; 16. D; 17. E; 18. J;
19. I; 20. A; 21. K; 22. G

1.2. OVERVIEW OF LIFE'S UNITY [pp.6-7]
1. work; 2. nutrients; 3. producers; 4. photosynthesis;
5. consumers; 6. B; 7. D; 8. C; 9. A

1.3. OVERVIEW OF LIFE'S DIVERSITY [pp.8-9]
1. A; 2. C; 3. C; 4. B; 5. C; 6. B; 7. C

1.4. AN EVOLUTIONARY VIEW OF LIFE'S DIVERSITY [p.10]
1. natural selection; 2. artificial; 3. mutations in

1.5. CRITICAL THINKING IN SCIENCE [p.11]
1.6. HOW SCIENCE WORKS [pp.12-13]
1. G; 2. A; 3. D; 4. C; 5. F; 6. E; 7. B.; 8. A; 9. C; 10. D;
11. B; 12. H; 13. G; 14. F; 15. E

1.7. THE POWER OF EXPERIMENTAL TESTS [pp.14-15]
1.8. SAMPLING ERROR IN EXPERIMENTS [p.16]
1. possible biases might have included, gender, age,
weight, medications taken, or other plausible issues;
2. if variable are not isolated, sampling biases, like
those described in question 1, might be introduced;
3. T; 4. F, study the effects one variable at a time; 5. T

SELF-TEST
1. a; 2. c; 3. c; 4. a; 5. d; 6. a; 7. b; 8. b; 9. c; 10. b; 11. a

---

Chapter 2   Life's Chemical Basis

Impacts, Issues: What Are You Worth? [p. 20]

2.1. START WITH ATOMS [p.22]
2.2. PUTTING RADIOISOTOPES TO USE [p.23]
1. K; 2. C; 3. O; 4. D; 5. L; 6. B; 7. F; 8. I; 9. J; 10. E;
11. G; 12. A; 13. M; 14. H; 15. N; 16. False; 17. True;
18. False; 19. True; 20. False

2.3. WHY ELECTRONS MATTER [pp.24-25]
1. C; 2. D; 3. G; 4. A; 5. E; 6. B; 7. F; 8. I; 9. H

10.

Element	Atomic Number	Atomic Mass	Number of Protons	Number of Neutrons	Number of Electrons
Sodium	12	11	23	11	11
Fluorine	10	9	19	9	9
Carbon	6	6	12	6	6
Hydrogen	0	1	1	1	1
Oxygen	8	8	16	8	8
Helium	2	2	4	2	2
Chlorine	18 or 19	17	35.45	17	17

11. atom; 12. ion; 13. electrons; 14. bond;
15. compound; 16. mixture
17.

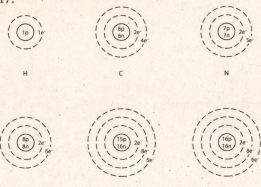

H   C   N

O   P   S

2.4. WHAT HAPPENS WHEN ATOMS INTERACT? [pp.26-27]
1. a. 2Na + S•Na2S; b. Mg + 2Cl•MgCl2; c. K + F•KF
2.

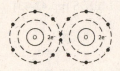

3. In a nonpolar covalent bond, the electrons are shared equally. Examples are hydrogen (H2) and nitrogen (N2). In a polar covalent bond, the atoms do not share the electrons equally, as in a water molecule (H2O).; 4. DNA contains many hydrogen bonds. Although on its own, each hydrogen bond is weak, together they support the structure well.

2.5. WATER'S LIFE-GIVING PROPERTIES [pp.28-29]
1. polarity; 2. hydrophilic; 3. hydrophobic; 4. Temperature; 5. hydrogen; 6. stabilize; 7. Evaporation; 8. ice; 9. solvent; 10. solute; 11. cohesion

2.6. ACIDS AND BASES [pp.30-31]
1. F; 2. C; 3. G; 4. E; 5. B; 6. A; 7. D; 8. H; 9. lemon juice (2), beer (5), black coffee (3), pure water (7), blood (8), seawater (8), ammonia (11), hair remover (12); 10. If the pH changes from 3 to 5, the hydrogen ion concentration will be less.

SELF-TEST
1. c; 2. c; 3. a; 4. d; 5. d; 6. c; 7. a; 8. e; 9. e; 10. c

Chapter 3   Molecules of Life

Impacts, Issues: Fear of Frying [p.34]

3.1. ORGANIC MOLECULES [pp.36-37]
3.2. FROM STRUCTURE TO FUNCTION [pp.38-40]
1. organic; 2. carbon; 3. hydrogen; 4. organisms;
5. inorganic; 6. laboratories; 7. bonding; 8. four;
9. hydrocarbon; 10. carbon; 11. methane;
12. functional; 13. a carbon atom of an organic molecule; 14. polar; 15. hydrogen; 16. True;

17. The hydroxyl and carbonyl functional groups are polar. 18. True; 19. True; 20. Carbon makes up more than half of the important elements in living organisms.; 21. A; 22. C; 23. D; 24. B; 25. E; 26. F;

27. methyl

28. carbonyl

29. carboxyl

30. carbonyl

31. hydroxyl

32. amino

33. phosphate

34. A; 35. G; 36. C; 37. B; 38. H; 39. F; 40. D; 41. I;
42. E

43.

amino acid   amino acid   dipeptide

3.3. CARBOHYDRATES [pp.40-41]

1.

glucose
(a monosaccharide)

glucose
(a monosaccharide)

enzyme
(synthesis)

(hydrolysis)
enzyme

maltose
(a disaccharide)

water

2. b; 3. b; 4. a; 5. c; 6. a; 7. c; 8. b; 9. b; 10. cellulose;
11. glucose; 12. hydrogen; 13. starch; 14. glycogen;
15. C; 16. E; 17. A; 18. G; 19. F; 20. B; 21. D

3.4. GREASY, OILY—MUST BE LIPIDS [pp.42-43]
1.

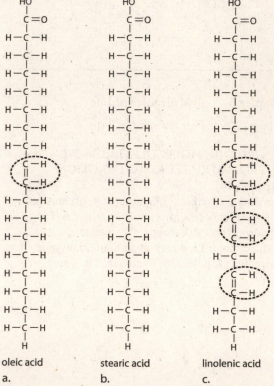

oleic acid
a.

stearic acid
b.

linolenic acid
c.

2. d; 3. a; 4. c; 5. d; 6. a; 7. b; 8. d; 9. b; 10. d; 11. c;
12. d; 13. a; 14. a; 15. a; 16. c; 17. a

3.5. PROTEINS—DIVERSITY IN STRUCTURE AND FUNCTION [pp.44-45]
3.6. WHY IS PROTEIN STRUCTURE SO IMPORTANT? [pp.46-47]
1. a. R group, b. amino group, c. carboxyl group; 2. H; 3. D; 4. B; 5. G; 6. A; 7. I; 8. K; 9. E; 10. C; 11. J; 12. F; 13. structure; 14. enzymes; 15. polypeptide; 16. membrane; 17. mutations;

18. variation; 19. traits; 20. evolution; 21. a; 22. c; 23. b; 24. b; 25. d; 26. c; 27. d
3.7. NUCLEIC ACIDS [pp.48-49]
1. a. nitrogenous base, b. five-carbon sugar, c. phosphate groups; 2. A; 3. C; 4. E; 5. B; 6. D

SELF-TEST
1. c; 2. a; 3. d; 4. b; 5. d; 6. a; 7. b; 8. a; 9. d; 10. a; 11. d; 12. e; 13. d; 14. c; 15. c; 16. b; 17. e; 18. c; 19. b

---

Chapter 4   Cell Structure and Function

Impacts, Issues: Food for Thought [p.52]

4.1. THE CELL THEORY [pp.54-55]
4.2. WHAT IS A CELL? [pp.56-57]
4.3. HOW DO WE SEE CELLS? [pp.58-59]
1. c; 2. g; 3. i; 4. f; 5. h; 6. j; 7. d; 8. b; 9. a; 10. e; 11. The four principles of the cell theory are: a) every organism consists of one or more cells, b) the cell is the smallest structural and functional unit of life, c) all cells come from pre-existing cells, d) cells pass hereditary material to their offspring when they reproduce.
12. d; 13. f; 14. b; 15. a; 16. c; 17. g; 18. e; 19. c; 20. f; 21. e; 22. b; 23. d; 24. a; 25 a. centimeter; 25b. 100; 26a. millimeter; 26b. 1,000; 27a. micrometer; 27b. 1,000,000; 28a. nanometer; 28b. 1,000,000,000

4.4. INTRODUCING PROKARYOTIC CELLS [pp.60-61]
4.5. MICROBIAL MOBS [p.61]
4.6. INTRODUCING EUKARYOTIC CELLS [p.62]
4.7. VISUAL SUMMARY OF EUKARYOTIC CELL COMPONENTS [p.63]
1. b; 2. f; 3. d; 4. e; 5. a; 6. h; 7. i; 8. c; 9. g; 10. d; 11. g; 12. a ; 13. h; 14. b; 15. c; 16. e; 17. i; 18. f; 19. cell wall; 20. chloroplast; 21. central water vacuole; 22. nucleus; 23. ribosomes; 24. rough endoplasmic reticulum (rough ER); 25. smooth endoplasmic reticulum (smooth ER); 26. Golgi body; 27. lysosome-like vesicle; 28. plasma membrane; 29. mitochondrion; 30. cytoskeleton; 31. nucleus; 32. ribosomes; 33. rough endoplasmic reticulum (rough ER); 34. smooth endoplasmic reticulum (smooth ER); 35. Golgi body; 36. lysosome; 37. plasma membrane; 38. centrioles; 39. mitochondrion; 40. cytoskeleton

4.8. THE NUCLEUS [pp.64-65]
4.9. THE ENDOMEMBRANE SYSTEM [pp 66-67]
4.10. LYSOSOME MALFUNCTION [p.68]
1. The DNA of a gene carries the code for correct assembly of a polypeptide like the one used to make the lysosomal enzyme to digest gangliosides. In Tay-Sachs disease, that enzyme misfolds and is destroyed. Gangliosides accumulate in nerve cells instead of being broken down. Within months the build-up begins to damage nerve cells in a baby born with Tay-Sachs disease. Death follows. 2. l; 3. a; 4. h; 5. g; 6. e; 7. m; 8. n; 9. d; 10. b; 11. j; 12. f; 13. k; 14. i; 15. c; 16. 4; 17. 2; 18. 6; 19. 3; 20. 1; 21 7; 22. 5

4.11. OTHER ORGANELLES [pp.68-69]
1. b; 2. a; 3. c; 4. a; 5. e; 6. d; 7. b; 8. c; 9. d; 10. a; 11. c; 12. b; 13. b; 14. a; 15. e; 16. Endosymbiosis is a theory that mitochondria evolved from aerobic bacteria that were eaten (or entered as parasites) but were not digested by a larger heterotrophic eukaryotic cell. Over time, the captive cell evolved inside the host cell into what are now known as mitochondria. The inner mitochondrial membrane is a remnant of the original prokaryotic cell membrane. Mitochondria have their own unique DNA and self-replicate.

4.12. CELL SURFACE SPECIALIZATIONS [pp.70-71]
4.13. THE DYNAMIC CYTOSKELETON [pp.72-73]
1. e; 2. d. 3. a; 4. k; 5. l; 6. i; 7. g; 8. h; 9. b; 10. c; 11. j; 12. f; 13. c; 14. g; 15. b; 16. i; 17. k; 18. l; 19. f; 20. j; 21. h; 22. a; 23. e; 24. d

SELF-TEST
1. d; 2. a; 3. d; 4. a; 5. c; 6. a; 7. b; 8. c; 9. b; 10. d; 11. c; 12. c; 13. d; 14. a

Chapter 5   A Closer Look at Cell Membranes

Impacts, Issues: One Bad Transporter and Cystic Fibrosis [p.76]

5.1. ORGANIZATION OF CELL MEMBRANES [pp.78-79]
5.2. MEMBRANE PROTEINS [pp.80-81]
1. D; 2. F; 3. H; 4. L; 5. A; 6. E; 7. M; 8. I; 9. G; 10. K; 11. B; 12. J; 13. C; 14. Cystic fibrosis is caused by a mutation that alters the structure of the protein CFTR. This causes a build-up of thick mucus in the respiratory tract.; 15. This is a key experiment that demonstrated that the all proteins in the cell membrane are not fixed in place, but are able to drift around quickly. This experiment supports the fluid mosaic model of the plasma membrane.; 16. The cell membrane is a mosaic of phospholipids, steroids, proteins and other molecules. The behavior of the phospholipids of the plasma membrane contributes to its fluidity.

5.3. DIFFUSION, MEMBRANES, AND METABOLISM [pp.82-83]
5.4. PASSIVE AND ACTIVE TRANSPORT [pp.84-85]
1. concentration gradient; 2. concentration gradient; 3. diffusion; 4. size; 5. temperature; 6. steepness of concentration gradient; 7. charge; 8. pressure; 9. permeability; 10. ions; 11. large, polar molecules; 12. gases; 13. nonpolar molecules; 14. passive; 15. active; 16. against; 17. endocytosis; 18. exocytosis

5.5. MEMBRANE TRAFFICKING [pp.86-87]
1. D; 2. E; 3. C; 4. A; 5. B

5.6. WHICH WAY WILL WATER MOVE? [pp.88-89]
1. F; 2. A; 3. H; 4. E; 5. B; 6. I; 7. D; 8. G; 9. Water moves across the membrane in both directions.; 10. Net movement is from side A to side B.; 11. Sucrose does not move across the membrane.; 12. The concentration of sucrose will increase because of net water loss on side A.; 13. Water level will get higher on side B.; 14. True; 15. False. The cell will shrivel up.; 16. False. Free – living organisms can counter shifts in tonicity by selectively transporting solutes across the plasma membrane.; 17. A plant wilts when the concentration of salt in the soil increases, causing the soil to become hypertonic.

SELF-TEST
1. d; 2. a; 3. d; 4. b; 5. d; 6. c; 7. d; 8. b; 9. c; 10. a

Chapter 6   Ground Rules of Metabolism

Impacts, Issues: A Toast to Alcohol Dehydrogenate [p.92]

6.1. ENERGY AND THE WORLD OF LIFE [pp.94-95]
1. alcohol; 2. alcohol dehydrogenase; 3. alcoholic hepatitis; 4. digests fats; 5. regulates the body's blood sugar level; 6. breaks down many toxic compounds; 7. binge drinking; 8. I; 9. I; 10. II; 11. I; 12. II; 13. I; 14. II; 15. II; 16. Energy transfers are never completely efficient. Some energy is converted to heat (thermal energy) which is not useful for doing work in biological systems. Therefore, the amount of useful energy is always decreasing.; 17. The measure of how much the energy of a system has become dispersed.; 18. False. The second law of thermodynamics relates to entropy.; 19. True; 20. True; 21. True; 22. False. In our world, energy can flow in only one direction, from the sun, to the producers, to the consumers.

6.2. THE ENERGY IN THE MOLECULES OF LIFE [pp.96-97]
6.3. HOW ENZYMES MAKE SUBSTANCES REACT [pp.98-99]
1. free energy; 2. endergonic; 3. exergonic; 4. activation energy; 5. catalysts; 6. exergonic; 7. exergonic; 8. exergonic; 9. endergonic; 10. exergonic; 11. triphosphate; 12. ribose; 13. adenine; 14. adenosine triphosphate; 15. C; 16. B; 17. D; 18. E; 19. A; 20. D; 21. B; 22. G; 23. C; 24. A; 25. E; 26. F; 27. activation energy without enzyme; 28. activation energy with enzyme; 29. energy released by the reaction; 30. catalyzed. 31. uncatalyzed; 32. substrate; 33. active site; 34. induced fit; 35. water; 36. transition state

6.4. METABOLISM—ORGANIZED, ENZYME-MEDIATED REACTIONS [pp.100-101]
6.5. NIGHT LIGHTS [p.102]

1. metabolism; 2. metabolic pathway; 3. anabolic (biosynthetic); 4. catabolic (degrative); 5. linear; 6. branched; 7. cyclic; 8. reverse; 9. reactants; 10. rate; 11. feedback mechanisms; 12. enzymes; 13. inhibit; 14. allosteric site; 15. feedback inhibition; 16. C;
17. D; 18. E; 19. B; 20. A; 21. electron; 22. oxidized; 23. reduced; 24. electron transfer chain; 25. most; 26. coenzymes; 27. ATP; 28. E; 29. B; 30. C; 31. A; 32. D; 33. luciferase; 34. bioluminescence; 35. metabolic activity; 36. living

SELF-TEST

1. d; 2. c; 3. c; 4. a; 5. a; 6. c; 7. d; 8. c; 9. d; 10. a

Chapter 7   Where it Starts – Photsynthesis

Impacts, Issues: Biofuels [p.106]

7.1. SUNLIGHT AS AN ENERGY SOURCE [pp.108-109]
7.2. EXPLORING THE RAINBOW [p.110]

1. highest; 2. shorter; 3. g; 4. k; 5. d; 6. e; 7. a; 8; b; 9. h; 10. j; 11. c; 12. i; 13. f; 14. l

7.3. OVERVIEW OF PHOTOSYNTHESIS [p.111]

1. CO_2; 2. O_2; 3. $C_6H_{12}O_6$; 4. 6; 5. carbon dioxide; 6. oxygen; 7. glucose; 8. light-dependent; 9. light-independent; 10. thylakoid; 11. pigments (photopigments); 12. photosystems; 13. ATP; 14. stroma

7.4. LIGHT-DEPENDENT REACTIONS [pp.112-113]
7.5. ENERGY FLOW IN PHOTOSYNTHESIS [p.114]

1. cyclic; 2. noncyclic; 3. ATP; 4. noncyclic; 5. NADPH; 6. oxygen; 7. Photosystem I; 8. Photosystem II; 9. Photosystem I; 10. 700; 11; 680; 12. electrons; 13. electron transfer chain; 14. hydrogen; 15. oxygen; 16. photolysis; 17. gradient; 18. stroma; 19. ATP synthases; 20. ATP; 21. Photosystem I; 22. H^+; 23. NADPH; 24. cyclic; 25. b; 26. c; 27. b; 28. c; 29. b; 30. a; 31. c; 32. b; 33. b; 34. c

7.6. LIGHT-INDEPENDENT REACTIONS: THE SUGAR FACTORY [p.115]
7.7. ADAPTATIONS: DIFFERENT CARBON-FIXING PATHWAYS [pp.116-117]

1. ribulose biphosphate (b); 2. phosphoglyceraldehyde (e); 3. phosphoglycerate (c); 4. plants with 3-carbon PGA as first stable intermediate (f); 5. plants with 4-carbon oxaloacetate as first stable intermediate (a); 6. crassulacean acid metabolism plants (d); 8. Calvin-Benson; 15. stroma; 9. ATP; 17. NADPH; 10. rubisco; 11. RuBP; 12. PGA; 13. ATP; 14. NADPH; 15. PGAL; 16. glucose; 17. RuBP; 18. sucrose; 19. starch; 20. C3 plants; 21. stomata; 22. CO_2; 23. O_2; 24. rubisco; 25. photorespiration; 26. carbon; 27. C4 plants; 28. mesophyll; 29. bundle-sheath cells; 30. Calvin-Benson; 31. CAM; 32. night; 33. C4; 34. CO_2

7.8. PHOTOSYNTHESIS AND THE ATMOSPHERE [p. 118]
7.9 A BURNING CONCERN [p. 119]

1. autotrophs; 2. photoautotroph; 3. photosynthesis; 4. sugar; 5. oxygen; 6. Heterotrophs; 7. chemoautotrophs; 8. methane (hydrogen sulfide); 9. hydrogen sulfide (methane); 10. billion; 11. cyclic photophosphorylation; 12. noncyclic photophosphorylation; 13. oxygen; 14. free radicals; 15. extinct; 16. aerobic respiration; 17. ozone; 18. ultraviolet; 19. ozone layer; 20. species; 21. carbon cycle; 22. industrial revolution; 23. carbon dioxide; 24. fossil fuels; 25. global warming

SELF-TEST

1. c, 2. b, 3. c, 4. a, 5. c, 6. d, 7. c, 8. b, 9. c, 10. d, 11. c, 12. b, 13. d, 14. d

Chapter 8 — How Cells Release Chemical Energy

Impacts, Issues: When Mitochondria Spin Their Wheels [p.122]

8.1. OVERVIEW OF CARBOHYDRATE BREAKDOWN PATHWAYS [pp.124-125]

1. Mitochondria are necessary because they produce cellular energy.; 2. <u>Similarities</u> – both aerobic respiration and anaerobic fermentation pathways begin with the same reactions, glycolysis, which breaks down glucose into two 3-carbon molecules called pyruvate. These reactions happen in the cytoplasm. <u>Differences</u> – Aerobic respiration ends in the mitochondria where oxygen serves as the final electron acceptor at the end of the pathway. Aerobic 20.

respiration is much more efficient, netting 36 molecules of ATP per molecule of glucose. Fermentation ends in the cytoplasm where a molecule other than oxygen serves as the final electron acceptor at the end of the pathway. Fermentation is much less efficient, netting 2 molecules of ATP per molecule of glucose.; 3. ATP; 4. metabolic pathways; 5. aerobic; 6. anaerobic; 7. glycolysis; 8. glucose; 9. pyruvate; 10. mitochondria; 11. cytoplasm; 12. 36; 13. 2; 14. Krebs cycle; 15. acetyl-CoA; 16. NAD^+; 17. FAD; 18. electron transport chain; 19. ATP

	In	*Out*
Glycolysis	glucose, 2 ATP	4 ATP (2 net), 2 NADH, 2 pyruvate
Krebs Cycle	2 pyruvate	6 CO_2, 2 ATP, 8 NADH, 2 $FADH_2$
Electron Transfer Phosphorylation	2 NADH, 8 NADH, 2 $FADH_2$	32 ATP

8.2. GLYCOLYSIS—GLUCOSE BREAKDOWN STARTS [pp.126-127]

1. autotrophic; 2. glucose; pyruvate; 4. ATP; 5. NADH; 6. glucose; 7. ATP; 8. PGAL; 9. phosphate; 10. hydrogen; 11. ATP; 12. NADH; 13. phosphate; 14. ATP; 15. substrate-level; 16. glucose; 17. pyruvate; 18. three; 19. D; 20. F; 21. B; 22. G; 23. H; 24. A; 25. E; 26. C

8.3. SECOND STAGE OF AEROBIC RESPIRATION [pp.128-129]
8.4. AEROBIC RESPIRATION'S BIG ENERGY PAYOFF [pp.130-131]

1. acetyl-CoA; 2. Krebs; 3. substrate-level; 4. 6 CO_2; 5. 2 ATP; 6. oxaloacetate; 7. electrons; 8. FAD; 9. NAD^+; 10. inner compartment; 11. inner membrane; 12. outer compartment; 13. outer membrane; 14. cytoplasm; 15. ATP; 16. oxygen (O_2); 17. $FADH_2$; 18. NADH; 19. electron transfer chain; 20. electron transfer phosphorylation; 21. active transport (pump); 22. inner; 23. inner; 24. outer; 25. ATP synthase; 26. 32; 27. oxygen; 28. water

8.5. ANAEROBIC ENERGY-RELEASING PATHWAYS [pp.132-133]
8.6. THE TWITCHERS [p.133]

1. fermenters; 2. botulism; 3. fermentation; 4. aerobic respiration; 5. glycolysis; 6. pyruvate; 7. CO2; 8. H2O; 9. yield of 2 ATP; 10. alcohol fermentation; 11. saccharomyces cerevisiae; 12. CO2; 13. alcoholic; 14. lactate fermentation; 15. lactobacillus acidophilus; 16. fast-twitch; 17. slow-twitch; 18. many; 19. aerobic respiration; 20. few; 21. lactate fermentation

8.7. ALTERNATIVE ENERGY SOURCES IN THE BODY [pp.134-135]
8.8. REFLECTIONS ON LIFE'S UNITY [p.136]

1. insulin; 2. glycogen; 3. glucagons; 4. liver; 5. brain; 6. fatty acids; 7. glycerol; 8. glycolysis; 9. amino acids; 10. Krebs cycle; 11. acetyl-CoA; 12. pyruvate; 13. e; 14. d; 15. d; 16. a; 17. c; 18. j; 19. e; 20. c; 21. h; 22. i; 23. e; 24. c; 25. h; 26. g

SELF-TEST

1. c; 2. c; 3. d; 4. b; 5. a; 6. c; 7. d; 8. a

Chapter 9   How Cells Reproduce

Impacts, Issues: Henrietta's Immortal Cells [p.140]

9.1. OVERVIEW OF CELL DIVISION MECHANISMS [pp.142-143]
1. D; 2. I; 3. A; 4. F; 5. E; 6. G; 7. B; 8. C; 9. H

9.2. INTRODUCING THE CELL CYCLE [pp. 144-145]
1. E; 2. D; 3. C; 4. F; 5. A; 6. B; 7. diploid;
8. descendant; 9. chromosomes; 10. two; 11. spindle;
12. microtubules; 13. chromosomes; 14. chromatids;
15. opposites; 16. G1 phase; 17. S phase; 18. G2 phase;
19. prophase; 20. metaphase; 21. anaphase;
22. telophase; 23. cytoplasmic division (cytokinesis);
24. mitosis; 25. daughter cells; 26. interphase;
27.

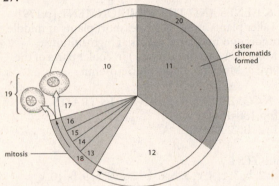

9.3. A CLOSER LOOK AT MITOSIS [pp.146-147]
1. interphase – daughter cells (F); 2. anaphase (A);
3. late prophase (G); 4. metaphase (D); 5. cell at
interphase (E); 6. early prophase (C); 7. transition to
metaphase (B); 8. telophase (H); 9. (5) interphase;
10. (6) early prophase; 11. (3) late prophase; 12. (7)
transition to metaphase; 13. (4) metaphase; 14. (2)
anaphase; 15. (8) telophase; 16. (1) interphase –
daughter cells

9.4. CYTOPLASMIC DIVISION MECHANISMS [pp.148-149]
1. D; 2. F; 3. E; 4. A; 5. C; 6. B; 7. A; 8. B; 9. A; 10. A;
11. B; 12. B; 13. C; 14. A; 15. A

9.5. WHEN CONTROL IS LOST [pp.150-151]
1. F; 2. A; 3. E; 4. C; 5. B; 6. D

SELF-TEST
1. a; 2. d; 3. a; 4. c; 5. c; 6. c; 7. e; 8. b; 9. c; 10. d

Chapter 10   Meiosis and Sexual Reproduction

Impacts, Issues: Why Sex? [p.154]

10.1. INTRODUCING ALLELES [p.156]
1. two; 2. two; 3. maternal; 4. paternal; 5. clones;
6. mutations; 7. combinations; 8. alleles

10.2. WHAT MEIOSIS DOES [pp.156-157]
1. H; 2. J; 3. G; 4. F; 5. B; 6. D; 7. E; 8. C; 9. I; 10. A;
11. true; 12. false; 13. false; 14. true; 15. true; 16. false;
17. false

10.3. VISUAL TOUR OF MEIOSIS [pp.158-159]
1. Prophase I – diploid; 2. Metaphase I – diploid;
3. Anaphase I – diploid; 4. Telophase I – diploid;
5. Prophase II – diploid; 6. Metaphase II – diploid;
7. Anaphase II – haploid; 8. Telophase II - haploid

10.4. HOW MEIOSIS INTRODUCES VARIATIONS IN TRAITS [pp.160-161]
1. crossing over; 2. Prophase I; 3. Daughter cells
produced are identical to the parent cell.; 4. H; 5. E;
6. D; 7. B; 8. G; 9. F; 10. A; 11. C

10.5. FROM GAMETES TO OFFSPRING [pp.162-163]
1. B; 2. C; 3. A; 4. C; 5. B; 6. A; 7. B; 8. A; 9. B; 10. A;
11. B; 12. A; 13. C; 14. A; 15. 1 (2n); 16. 4 (n); 17. 3
(n); 18. 2 (2n); 19. 4 (n); 20. 1 (2n); 21. 3 (n); 22. 2
(2n); 23. Both sperm and egg are haploid (n) and a
combination of both maternal and paternal DNA due
to crossing over. Four basically equal sperm are
formed in spermatogenesis where only one viable
ovum is formed with three polar bodies in oogenesis.

10.6. MITOSIS AND MEIOSIS—AN ANCESTRAL CONNECTION? [pp.164-165]
1. C; 2. F; 3. D; 4. A; 5. B; 6. E; 7. 4; 8. 8; 9. 4; 10. 8;
11. 2

SELF TEST
1. a; 2. a; 3. b; 4. c; 5. d; 6. c; 7. a; 8. c; 9. b; 10. d

Chapter 11 Observing Patterns in Inherited Traits

Impacts, Issues: The Color of Skin [p.168]

11.1. MENDEL, PEA PLANTS, AND INHERITANCE PATTERNS [pp.170-171]
1. M; 2. B; 3. D; 4. I; 5. H; 6. K; 7. E; 8. F; 9. A; 10. G; 11. N; 12. C; 13. J; 14. L; 15. The plant is self-fertilizing and a true-breeder; a true-breeder and has definite dominant and recessive traits.

11.2. MENDEL'S LAW OF SEGREGATION [pp.172-173]
1. Larger sample sizes reduce the effect of chance.; 2. true; 3. false; 4. true; 5. false; 6. true; 7. false; 8. true

11.3. MENDEL'S LAW OF INDEPENDENT ASSORTMENT [pp.174-175]
1. E; 2. B; 3. G; 4. A; 5. F; 6. C; 7. D; 8. I; 9. H; 10. a. 1:2 Tt, 1:2 tt b. 1:2 tall.1:2 dwarf; 11. a. homozygous dominant b. heterozygous; 12. heterozygous normal 1:2, Homozygous recessive (albino) 1:2; 13. a. F_1: black trotter 1:1, F_2 9 – black trotter/3 – black pacer/3 – chestnut trotter/1 – chestnut pacer, b. black trotter, c. BbTt, d. bbtt – chestnut pacer

11.4. BEYOND SIMPLE DOMINANCE [pp.176-177]
1. D; 2. G; 3. B; 4. C; 5. E; 6. F; 7. A; 8. Type A: AA or AO, Type B: BB or BO, Type AB: AB, Type O: OO; 9. Type O; 10. C; 11. D; 12. A; 13. B; 14. A; 15. a. 1:1 all pink, b. 1:2:1 red:pink:white; 16. a. A, B, AB/b. A, B, AB, O/c. A/d. O/e. AB, A, B; 17. a. 1:3 colored:white; b. 3:1 colored:white; 18. 3:1:4 black:brown:yellow

11.5. LINKAGE GROUPS [p.178]
1. genes; 2. chromosome; 3. linkage; 4. crossing over; 5. combinations; 6. proportions; 7. small; 8. CD – genes C & D are much closer together than genes A&B.

11.6. GENES AND THE ENVIRONMENT [p.179]
1. Taller plants at low altitude, short plants at middle altitude, and mid-sized plants at high altitude.; 2. More melanin (more color) in cooler parts of the body.; 3. Many predators.; 4. Serotonin

11.7. COMPLEX VARIATIONS IN TRAITS [pp.180-181]
1. B; 2. B; 3. A; 4. B; 5. A

SELF TEST
1. d; 2. b; 3. c; 4. c; 5. b; 6. b; 7. e; 8. a; 9. b; 10. c

Chapter 12 Chromosomes and Human Inheritance

Impacts, Issues: Strange Genes, Tortured Minds

12.1. HUMAN CHROMOSOMES [p.186-187]
1. diploid; 2. autosomes; 3. sex; 4. X; 5. Y; 6. XX; 7. XY; 8. male; 9. environmental; 10. frogs; 11. all; 12. one-half; 13. one-half; 14. SRY; 15. testosterone; 16. testes; 17. ovaries; 18. Colchicine interferes with assembly of mitotic spindles.; 19. Chromosomes are aligned according to centromere location, size, shape and length.; 20. female

12.2. EXAMPLES OF AUTOSOMAL INHERITANCE PATTERNS [pp.188-189]
1. E; 2. A; 3. F; 4. C; 5. D; 6. B; 7. Gg, Gg, gg; 8. 50%; 9. It is dominant since two individuals with the trait produced a normal (recessive) child. This male must be heterozygous because his mother was homozygous normal. You would expect 50% of the progeny would be heterozygous with the trait and 50% would be homozygous normal.; 10. A; 11. B; 12. A; 13. A; 14. B; 15. A; 16. B; 17. A; 18. B; 19. A; 20. B

12.3. TOO YOUNT TO BE OLD [p.189]
1. Cells age abnormally quickly. Affected individuals exhibit characteristics of extreme age very early in life: brittle bones, frailty, weakened muscles.; 2. lamin; 3. Most persons die in their teens from strokes or heart attacks.

12.4. EXAMPLES OF X-LINKED INHERITANCE PATTERNS [pp.190-191]
1. A; 2. E; 3. B; 4. C

12.5. HERITABLE CHANGES IN CHROMOSOME STRUCTURE [pp.192-193]
1. duplication (B); 2. inversion (C); 3. deletion (A); 4. translocation (D)

12.6. HERITABLE CHANGES IN THE CHROMOSOME NUMBER [pp.194-195]

1. 100%; 2. 50%; 3. polyploidy: cells have three or more of each type of chromosome, trisomic: cells have three chromosomes where normally there is a pair of chromosomes, monosomic: cells have one chromosome where normally there is a pair of chromosomes; 4. C; 5. B; 6. C; 7. D; 8. A; 9. B; 10. D; 11. C; 12. A; 13. D

12.7. HUMAN GENETIC ANALYSIS [pp.196-197]

1. E; 2. B; 3. G; 4. D; 5. A; 6. F; 7. C

12.8. PROSPECTS IN HUMAN GENETICS [pp.198-199]

1. phenotypic; 2. phenylketonuria (PKU); 3. prenatal; 4. amniocentesis; 5. chorionic villus sampling (CVS); 6. abortion

SELF-TEST

1. a; 2. d; 3. c; 4. c; 5. a; 6. a; 7. b; 8. b; 9. c; 10. e

Chapter 13   DNA Structure and Function

Impacts, Issues: Here Kitty, Kitty, Kitty, Kitty, Kitty [p.202]

13.1. THE HUNT FOR DNA [pp.204-205]

1. Miescher – found that nuclei contain an acidic substance composed mostly of nitrogen and phosphorus. Later than substance would be called deoxyribonucleic acid (DNA).; 2. Griffith – discovered that there is a substance in cells that encodes the information about traits that parents pass to offspring.; 3. Avery (*et. al)* – discovered that the "transforming principle" is DNA.; 4. Hershey and Chase – determined that DNA, not protein, is the material of heredity common to all life on Earth.; 5. virus; 6. bacterial; 7. hereditary; 8. genetic; 9. protein; 10. viral DNA; 11. ^{32}P; 12. bacteriophages; 13. ^{35}S; 14. ^{32}P; 15. DNA; 16. protein

13.2. THE DISCOVERY OF DNA'S STRUCTURE [pp.206-207]

1. deoxyribose sugar, phosphate group, nitrogen base.; 2. guanine (pu); 3. cytosine (py); 4. adenine (pu); 5. thymine (py); 6. deoxyribose (B); 7. phosphate group (G); 8. purine (C); 9. pyrimidine (A); 10. purine (E); 11. pyrimidine (D); 12. mucleotide (F)

13.3. DNA REPLICATION AND REPAIR [pp.208-209]

1. DNA unwinds in only one direction. Since the two DNA strands are antiparallel, only one side will add new nucleotides continuously. The other side must add segments of new DNA (Okazaki fragments) a little at a time and join them with DNA ligase.; 2. Most DNA polymerases immediately reverse synthesis and remove then replace mismatched nucleotides. Errors that are not corrected by the DNA polymerases and checked again by DNA repair mechanisms and most are corrected. Errors that remain even after these measures remain in the DNA as mutations.

3.
T-A	T-A
G-C	G-C
A-T	A-T
C-G	C-G
C-G	C-G
C-G	C-G

4. F, adenine bonds to thymine in DNA; 5. T; 6. T; 7. F, DNA polymerases govern the assembly of nucleotides; 8. T

13.4. USING DNA TO DUPLICATE EXISTING MAMMALS [pp.210-211]

1. 3; 2. 5; 3. 2; 4. 1; 5. 4; 6. F; 7. A; 8. C; 9. E; 10. H; 11. D; 12. B; 13. G; 14. Undifferentiated (stem) cells may be used to replace nerve cells in individuals with spinal cord damage. New organs might be "grown" to replace organs for transplant.

13.5. FAME AND GLORY [p.211]

1. Rosalind Franklin used x-ray crystallography to calculate DNA's diameter, the distance between its chains and between its bases, the pitch of the helix and the number of bases in each coil. This data confirmed the work of Watson and Crick, causing the recognition of the structure of DNA.; 2. The Nobel Prize is not given posthumously. Rosalind Franklin died of ovarian cancer before the prize was awarded.

SELF-TEST

1. d; 2. d; 3. a; 4. d; 5. c; 6. d; 7. b; 8. a; 9. d; 10. a

Chapter 14   From DNA to Protein

Impacts, Issues: Rincin and Your Ribosomes [p.214]

14.1. DNA, RNA, AND GENE EXPRESSION [pp.216-217]

1. sequence (linear order); 2. transcription;
3. translation; 4. transcription; 5. translation; 6. amino acids; 7. gene expression; 8. enzymes; 9. a. rRNA; RNA molecule that associates with certain proteins to form the ribosome, the "workbench" on which polypeptide chains are assembled; b. mRNA; RNA molecule that moves to the cytoplasm, complexes with the tRNA and ribosome, carries the code for the amino acid sequence of the protein; c. tRNA; RNA molecue that moves into the cytoplasm, picks up a specific amino acid, and moves it to the ribosome where tRNA pairs with a specific mRNA code word for that amino acid

14.2. TRANSCRIPTION: DNA TO RNA [pp.218-219]

1. RNA is a single helix where DNA is a double helix, RNA and DNA differ in the functional group on the 2' carbon of the sugar, uracil is used in RNA instead of thymine.; 2. Each nucleotide provides energy for its own attachment to the end of a growing strand. Transcription and DNA replication both use one strand of a nucleic acid as a template for synthesis of another.; 3. 1) In DNA replication only part of a DNA strand, not the whole molecule as in RNA, is used as a template, 2) DNA replication uses DNA polymerase where transcriptions uses RNA polymerase, 3) transcription results in a single strand of RNA; DNA replication results in two DNA double helices.

14.3. RNA AND THE GENETIC CODE [pp.220-221]

1. Introns are removed, remaining exons are spliced together, a guanine cap is attached to the 5' end of the mRNA, and a poly-A tail is added to the 3' end of the new mRNA.; 2. AUG-UUC-UAU-UGU-AAU-AAA-GGA-UGG-CAG-UAG; 3. met-phe-tyr-cys-asn-lys-gly-trp-gln-stop; 4. DNA (E); 5. intron (B); 6. cap (F); 7. exon (A); 8. tail (D); 9. [mature] mRNA (C)

14.4. TRANSLATION: RNA TO PROTEIN [pp.222-223]

1. translation; 2. initiation; 3. elongation;
4. termination; 5. messenger; 6. transfer; 7. start (AUG); 8. elongation; 9. polypeptide; 10. termination; 11. releasing; 12. enzyme; 13. polysomes; 14. phosphate-group; 15. GTP

14.5. MUTATED GENES AND THEIR PROTEIN PRODUCTS [pp.224-225]

1. Deletion is usually more damaging because it causes a frame shift in every codon past the mutation.; 2. Less likely because the third position is many times the "wobble" that still codes for the same amino acid.; 3. At the beginning because it would affect all of the codons from the mutation to the end of the mRNA strand.; 4. C; 5. D; 6. E; 7. A; 8. E; 9. D; 10. A; 11. DNA (H); 12. transcription (J); 13. intron (E); 14. exon (A); 15. mRNA (B); 16. rRNA (F); 17. tRNA (C); 18. ribosomal subunits (G); 19. amino acids (D); 20. anticodon (K); 21. translation (I); 22. polypeptide (L)

SELF-TEST
1. c; 2. b; 3. b; 4. c; 5. a; 6. b. 7. a; 8. c; 9. d; 10. d

Chapter 15   Controls Over Genes

Impacts, Issues: Between You and Eternity [p.228]

15.1. GENE EXPRESSION IN EUKARYOTIC CELLS [pp.230-231]
1. B; 2. E; 3. D; 4. A; 5. C; 6. Differentiation is the process by which cells become specialized.; 7. Many factors, including conditions in the cytoplasm and extracellular fluid, the type of cell, stage of development, and number of copies of the gene.; 8. A lot of protein is made quickly.; 9. Enzymes begin disassembling mRNA as soon as it arrives in the cytoplasm. The longer the poly-A tail, the longer the mRNA remains intact to be translated for production of protein.

15.2. A FEW OUTCOMES OF EUKARYOTIC GENE CONTROLS [pp.232-233]
1. X chromosome inactivation; 2. Barr body; 3. mosaic; 4. incontinentia pigmenti; 5. dosage compensation; 6. XIST; 7. RNA; 8. Each patch results from inactivation of one X chromosome. Since one patch results from inactivation of one chromosome where another patch results from inactivation of another. Orange or black fur results from expression of different alleles on the active X chromosome.; 9. Environmental cues such as daylight

15.3. THERE'S A FLY IN MY RESEARCH [pp.234-235]
1. E; 2. B; 3. G; 4. C; 5. H; 6. F; 7. D; 8. A; 9. Genes are inactivated by introducing a mutation. Individuals with the mutation are then compared to normal individuals.; 10. Maternal mRNAs are delivered to opposite ends of an unfertilized egg as it forms. mRNAs get translated right after the egg is fertilized and their protein products diffuse away on gradients that span the entire embryo.

15.4. PROKARYOTIC GENE CONTROL [pp.236-237]
1. False, do not; 2. True; 3. True; 4. True; 5. False, not absorbed directly by the intestine; 6. True; 7. False, repressors; 8. False, declines; 9. False, transcription; 10. False, not everyone is lactose intolerant; 11. True; 12. True; 13. regulatory gene (I); 14. operator (B); 15. lactose enzyme genes (F); 16. promoter (H); 17. lactose operon (A); 18. repressor protein (G); 19. repressor-operator complex (E); 20. lactose (C); 21. mRNA (J); 22. RNA polymerase (D)

SELF-TEST
1. a; 2. a; 3. d; 4. b; 5. c; 6. a; 7. a; 8. b; 9. d; 10. b

Chapter 16   Studying and Manipulation Genomes

Impacts, Issues: Golden Rice, or Frankenfood? [p.240]

16.1. CLONING DNA [pp.242-243]
1. bacteriophage; 2. restriction enzymes; 3. EcoRI; 4. recombinant; 5. restriction; 6. sticky ends; 7. DNA cloning; 8. plasmids; 9. cloning vectors; 10. clones; 11. mRNA; 12. introns; 13. DNA ligase; 14. reverse transcriptase; 15. cDNA; 16. B; 17. F; 18. E; 19. C; 20. A; 21. D

16.2. FROM HAYSTACKS TO NEEDLES [pp.244-245]
1. B; 2. E; 3. D; 4. A; 5. F; 6. C; 7. 6; 8. 3; 9. 4; 10. 2; 11. 5; 12. 1

16.3. DNA SEQUENCING [p.246]
1. Polymerase randomly adds either a regular nucleotide or a dideoxynucleotide to the end of a growing DNA strand. If a dideoxynuclotide is added, synthesis of that particular strand ends.; 2. An electric field is used to pull a mixture of different length DNA fragments through a semisolid gel matrix. Smaller fragments of DNA work their way through the gel faster than larger fragments. Over time this separates the fragments by size, revealing the DNA sequence.

16.4. DNA FINGERPRINTING [p.247]
1. PCR is used to copy a region of a chromosome known to have tandem repeats of 4 to 5 nucleotides. Electrophoresis is used to separate DNA fragments by size, revealing a unique banding pattern – the individual's DNA fingerpring.; 2. Just a small sample of DNA from blood, semen, hair follicle cells, etc. can be copied by PCR for DNA fingerprinting. Since DNA fingerprints are essentially unique, this evidence is compelling to convict criminals and also to determine if a suspect is innocent.

16.5. STUDYING GENOMES [pp.248-249]
1. F; 2. A; 3. C; 4. G; 5. B; 6. E; 7. D

16.6. GENETIC ENGINEERING [p.250]
1. production of human insulin, cheese production, improving taste and clarity of beer and juices, slows bread staling, modifies fats; 2. This research may help us to discover why some bacteria become resistant to antibiotics.

16.7. DESIGNER PLANTS [pp.250-251]
1. 2; 2. 5; 3. 1; 4. 4; 5. 3; 6. "Frankenfood" is used by groups opposed to genetic engineering.; 7. True; 8. True; 9. False – Ti plasmids; 10. False – Bt plasmids

16.8. BIOTECH BARNYARDS [pp.252-253]
1. Integrating the rat growth hormone gene into mice may allow us to incorporate beneficial genes.; 2. Goat milk producing lysozyme may protect infants and children from acute diarrheal disease.; 3. Studying knockout experiments in glucose metabolism of mice may help us to understand how diabetes works in humans.; 4. Xenotransplantation may provide a ready supply of transplantable organs of human use.; 5. Goat milk containing spider silk proteins may be valuable source for silk to produce fabrics, bulletproof vests, sports equipment and biodegradable medical supplies.

16.9. SAFETY ISSUES [p.253]
1. Host organisms are designed and used under a narrow range of conditions in the laboratory.; 2. DNA from pathogenic or toxic organisms for recombinant experiments are prohibited until proper containment facilities are developed.; 3. Release and import of genetically-modified organisms is carefully regulated.

16.10. MODIFIED HUMANS? [p.254]
1. Genes missing or malfunctioning may be introduced into persons suffering from these diseases, resulting in a cure or at least relief of symptoms.; 2. Cystic fibrosis, hemophilia A, several types of cancer, inherited retinal disease, inherited immune disorders.; 3. Unmutated copies of IL2RG gene were inserted into bone marrow for each of the boys. Those transgenic cells reproduced and restored a healthy immune system to 10 of the 11 boys.; 4. Jesse Gelsinger was allergic to the viral vector, an unexpected complication of the therapy. This is a warning because there is always risk to gene therapies and it should not be undertaken lightly.; 5. Benefits: potential cures for severe genetic disorders, Inappropriate uses: for monetary benefit and/or to choose preferences rather than legitimate needs

SELF-TEST
1. a; 2. b; 3. c; 4. b; 5. d; 6. b; 7. a; 8. d; 9. d; 10. a

Chapter 17   Evidence of Evolution

Impacts, Issues: Measuring Time [p.258]

17.1. EARLY BELIEFS, CONFOUNDING
      DISCOVERIES [pp.260-261]
17.2. A FLURRY OF NEW THEORIES
      [pp.262-263]
17.3. DARWIN, WALLACE, AND NATURAL
      SELECTION [pp.264-265]
1. D; 2. E; 3. F; 4. C; 5. B; 6. A; 7. b; 8. c; 9. b; 10. a; 11. c; 12. b; 13. b; 14. a; 15. b; 16. b; 17. a; 18. d; 19. b; 20. c; 21. a

17.4. GREAT MINDS THINK ALIKE [p.266]
17.5. ABOUT FOSSILS [pp.266-267]

17.6. DATING PIECES OF THE PUZZLE [p.268]
1. E; 2. D; 3. A; 4. B; 5. C; 6. G; 7. F; 8. 250,000; 9. rare; 10. decay; 11. eaten; 12. oxygen; 13. soft-bodied; 14. 0.5 grams, 16,110 years

17.7. A WHALE OF A STORY [p.269]
17.8. PUTTING TIME INTO PERSPECTIVE
      [p.270-271]
17.9. DRIFTING CONTINENTS, CHANGING SEAS
      [pp.272-273]
1. H; 2. I; 3. A; 4. E; 5. D; 6. F; 7. B; 8. G; 9. a; 10. b; 11. a; 12. b; 13. a; 14. a

SELF-TEST
1. d; 2. d; 3. d; 4. b; 5. a; 6. d; 7. b; 8. c; 9. a; 10. e

Chapter 18   Processes of Evolution

Impacts, Issues: Rise of the Super Rats [p.276]

18.1. INDIVIDUALS DON'T EVOLVE, POPULATIONS DO [pp.278-279]
18.2. A CLOSER LOOK AT GENETIC EQUILIBRIUM [pp.280-281]

1. g; 2. j; 3. h; 4. l; 5. a; 6. b; 7. c; 8. n; 9. i; 10. d; 11. e; 12. m; 13. o; 14. k; 15. f; 16. There is no mutation, the population is very large, the population is isolated from other populations of the same species, all members of the population survive and reproduce, mating is completely random; 17. a. 0.64 BB, 0.16 Bb, 0.16 Bb, 0.04 bb; b. genotypes: 0.64 BB, 0.32 Bb, and 0.04 bb; phenotypes: 965 black, 4% gray c.

Parents (F_1)	B sperm	b sperm
0.64 BB	0.64	0
0.32 Bb	0.16	0.16
0.04 bb	0	0.04
Totals =	0.80	0.20

18.a. 18% [$2pq = 2$ x (0.9) x (0.1) = 2 x (0.09) = 0.18 = 18% heterozygotes], b. 0.9 [$p^2 = 0.81$, $p = 0.9$ = the frequency of the dominant allele], c. 0.1 [$p + q = 1$, $q = 1.00 - 0.9 = 0.1$ = the frequency of the recessive allele]; 19.a. 128 [homozygous dominant = p^2 x 200 = $(0.8)^2$ x 200 = 0.64 x 200 = 128 individuals], b. 8 [$q = (1.00 - p) = 0.20$ = homozygous recessive, q^2 x 200 = $(0.2)^2$ x 200 = (0.04) x (200) = 8 individuals], c. 64 [heterozygotes = $2pq$ x 200 = 2 x 0.8 x 0.2 x 200 = 0.32 x 200 = 64 individuals] Check: 128 + 8 + 64 = 200; 20. 30% [If p=0.70, and p+q = 1, 0.70 + q = 1; then q = 0.30, or 30%]; 21. 48% [If p = 0.60, and p+q = 1, 0.60 +q=1; then q= 0.40; thus, 2pq = 0.48 or 48%]

18.3. NATURAL SELECTION REVISITED [p.281]
18.4. DIRECTIONAL SELECTION [pp.282-283]
18.5. SELECTION AGAINST OR IN FAVOR OF EXTREME PHENOTYPES [pp.284-285]

1. a. directional selection - the allele frequencies shift in a consistent direction, b. disruptive selection - the extreme ends of the variations are selected for, c. stabilizing selection - the average form of a trait is selected for 2. a; 3. b; 4. b; 5. c; 6. b; 7. a; 8. c; 9. b; 10. This is creating antibiotic resistant bacteria within a short period of time. We are selecting out the weak bacteria and leaving the "strong" ones. The "strong" ones have little competition for resources so they are able to reproduce easily creating new "antibiotic resistant" strains.

18.6. MAINTAINING VARIATION [pp.286-287]
18.7. GENETIC DRIFT – THE CHANCE CHANGES [pp.288-289]
18.8. GENE FLOW [p.289]
18.9. REPRODUCTIVE ISOLATION [pp.290-291]

1. b; 2. d; 3. c; 4. d; 5. d; 6. b; 7. a; 8. c; 9. b; 10. e; 11. c; 12. e; 13. frequency; 14. homozygous; 15. loss; 16. population; 17. small; 18. fixed; 19. frequencies; 20. gene; 21. a; 22. b; 23. a; 24. a; 25. b; 26. a

18.10. ALLOPATRIC SPECIATION [pp.292-293]
18.11. THER SPECIATION MODELS [pp.294-295]
18.12. MACROEVOLUTION [pp.296-297]

1. c; 2. g; 3. b; 4. f; 5. e; 6. a; 7. d; 8. h; 9. a; 10. c; 11. b; 12. a; 13. b; 14. b

SELF-TEST
1. a; 2. d; 3. a; 4. d; 5. e; 6. b; 7. c; 8. a; 9. a; 10. d; 11. a

---

Chapter 19   Organizing Information About Species

Impacts, Issues: Bye Bye Birdie [p.300]

19.1. TAXONOMY AND CLADISTICS [pp.302-303]
19.2. COMPARING BODY FORM AND FUNCTION [pp.304-305]

1. d; 2. g; 3. b; 4. e; 5. c; 6. f; 7. h; 8. a; 9. differences; 10. cladistics; 11. fewest; 12. parisomony; 13. matrix; 14. tree of life; 15. b; 16. d; 17. a; 18. c

19.3. COMPARING PATTERNS OF DEVELOPMENT [pp.306-307]

19.4. COMPARING DNA & PROTEINS [pp.308-309]
19.5. MAKING DATA INTO TREES [pp.310-311]
19.6. PREVIEW OF LIFE'S EVOLUTIONARY HISTORY [pp.312-313]

1. limb buds; 2. tail; 3. Dlx; 4. stick out; 5. appendage; 6. Hox; 7. Hox; 8. Dlx; 9. mutation; 10. neutral; 11. survival; 12. molecular clock; 13. A & C; 14. A & D; 15. A(oldest) evolved into C, C evolved into B, B evolved into D (youngest); 16. b; 17. c; 18. a; 19. a; 20. b; 21. c; 22. c

SELF-TEST
1. c; 2. a; 3. a; 4. b; 5. d; 6. c; 7. c; 8. d

Chapter 20   Life's Origin and Early Evolution

Impacts, Issues: Looking for Life in All the Odd Places [p.316]

20.1. IN THE BEGINNING….. [pp.318-319]
20.2. HOW DID CELLS EMERGE? [pp.320-321]
1. e; 2. f; 3. a; 4. b; 5. b; 6. d; 7. c; 8. d; 9. c; 10. a; 11. a; 12. e; 13. f

20.3. LIFE'S EARLY EVOLUTION [pp.322-323]

20.4. WHERE DID ORGANELLES COME FROM? [pp.324-325]
20.5. TIME LINE FOR LIFE'S ORIGIN AND EVOLUTION [pp.326-327]
20.6. ABOUT ASTROBIOLOGY [p.328]
1. b; 2. b; 3. a; 4. a; 5. b; 6. oxygen; 7. spontaneous; 8. aerobic; 9. releasing; 10. ozone; 11. UV radiation; 12. land; 13. c; 14. a; 15. a (c); 16. b; 17. b; 18. b; 19. D; 20. F; 21. E; 22. B; 23. C; 24. A; 25. I; 26.G; 27. H; 28. J

SELF-TEST
1. b; 2. b; 3. c; 4. d; 5. a; 6. d; 7. b; 8. a

Chapter 21   Viruses and Prokaryotes

Impacts, Issues: The Effects of AIDS [p.332]

21.1. VIRAL CHARACTERISTICS AND DIVERSITY [pp.334-335]
21.2. VIRAL REPLICATION [pp.336-337]
21.3. VIROIDS & PRIONS [p.338]
1. c; 2. d; 3. e; 4. b; 5. f; 6. a ; 7. The 3 hypotheses of viral origin are: a. viruses are descendants of cells that were parasites inside other cells, b. viruses are genetic elements that have escaped from cells, c. viruses represent a separate evolutionary branch and arose independently; 8. Steps of viral replication include: a. attachment, b. penetration, c. replication and synthesis, d. assembly, e. release; 9. The lytic cycle is when a virus attaches to a host cell and injects its DNA into the host. Viral genes direct the host cell to make viral DNA and proteins which assemble as viral particles. Eventually the host cell ruptures and the new viruses are released. The lysogenic cycle is when the virus attaches to the host cell but the virus enters a latent state that extends the multiplication cycle. Viral genes are integrated into the host DNA and are copied along with the host DNA. The viral DNA is passed to all of the descendent cells.; 10. The principal features of a virus are a. noncellular, no cytoplasm, ribosomes, or other typical cell components, b. genetic material may be DNA or RNA, c. can only replicate inside of a host cell, d. small (about 25 to 300 nanometers)

21.4. PROKARYOTES – ENDURING, ABUNDANT, & DIVERSE [p.339]
21.5. PROKARYOTIC STRUCTURE AND FUNCTION [pp.340-341]
1. d; 2. c; 3. a; 4. b; 5. The four main features of Prokaryotic cells are: (1) no nucleus; the chromosome is found in the nucleoid region, (2) generally they contain a single chromosome as well as circular loops of DNA called plasmids, (3) cell wall is present in most species, (4) ribosomes are distributed within the cytoplasm.; 6. Transduction is when a virus picks up DNA from one Prokaryotic cell that it is infecting and then transfers that DNA to another Prokaryotic cell. Transformation is when a Prokaryotic cell picks up DNA from the environment.; 7. c; 8. b; 9. d; 10. f; 11. e; 12. a; 13. twenty; 14. chromosome; 15. fission; 16. horizontal gene transfer; 17. conjugation

21.6. THE BACTERIA [pp.342-343]
21.7. THE ARCHAEANS [pp.344-345]
21.8. EVOLUTION AND INFECTIOUS DISEASE [pp.346-347]
1. nitrogen fixation; 2. heterocyst; 3. Rickettsias; 4. Gram-positive; 5. Gram-negative; 6. endospores; 7. vector; 8. b; 9. c; 10. d; 11. b; 12. a; 13. c; 14. b; 15. d; 16. c; 17. d; 18. a; 19. C; 20. E; 21. I; 22. B; 23. F; 24. A; 25. G; 26. H; 27. D

SELF-TEST
1. a; 2. b; 3. c; 4. d; 5. c; 6. a; 7. d; 8. a; 9. c; 10. a

Chapter 22 "Protists" – The Simplest Eukaryotes

Impacts, Issues: The Malaria Menace [p.350]

22.1. THE MANY PROTIST LINEAGES [p.352-353]
1. E; 2. P; 3. B; 4. E; 5. P; 6. E; 7. single;
8. heterotrophic; 9. water; 10. soil; 11. Autotrophic;
12. photosynthetic; 13. heterotrohic; 14. environment;
15. asexually; 16. sexually; 17. haploid; 18. zygote

22.2. FLAGELLATED PROTOZOANS [pp.354-355]
22.3. FORAMINIFERANS & RADIOLARIANS [p.356]
22.4. CILIATES [p.357]
22.5. DINOFLAGELLATES [p.358]
22.6. CELL-DWELLING APICOMPLEXANS [p.359]
22.7. THE STRAMENOPILES [pp.360-361]
1. Diplomonads; 2. parabasalids; 3. mitochondria;
4. Trypanosomes; 5. vector; 6. eyespot; 7. calcium
carbonate; 8. plankton; 9. alveolates; 10. Paramecium;
11. dinoflagellates; 12. apicomplexans; 13. anopheles;
14. plasmodium; 15. b, c; 16. h; 17. i, f; 18. e, g, j;
19. c, f; 20. c, k; 21. c, d, j; 22. a, c

22.8. THE PLANT DESTROYERS [p.361]
22.9. GREEN ALGAE [pp.362-363]
22.10. RED ALGAE DO IT DEEPER [p.364]
22.11. AMOEBOID CELLS AT THE CROSSROADS [p.365]
1. zygote, D; 2. resistant zygote, A; 3. meiosis and
germination, F; 4. asexual reproduction, B; 5. gamete
production, C; 6. gametes meet, E; 7. a; 8. b; 9. c;
10. a; 11. a; 12. c; 13. b; 14. b; 15. a; 16. b; 17. c; 18. a;
19. a; 20. b

SELF-TEST
1. b; 2. b; 3. b; 4. a; 5. c; 6. a; 7. d; 8. a; 9. d

Chapter 23 The Land Plants

Impacts, Issues: Beginnings, And Endings [p.368]

23.1. EVOLUTION ON A CHANGING WORLD STAGE [pp.370- 371]
23.2. EVOLUTIONARY TRENDS AMONG PLANTS [pp.372-373]
1. D; 2. F; 3. B; 4.G; 5. E; 6. A; 7. C; 8. C; 9. A; 10. F;
11. E; 12. G; 13. D; 14. B

23.3. THE BRYOPHYTES [pp.374-375]
23.4. SEEDLESS VASCULAR PLANTS [pp.376-377]
23.5. ANCIENT CARBON TREASURES [p.378]
1. mosses; 2. liverworts; 3. nonvascular; 4. lignin;
5. rhizoids; 6. water; 7. gametophytes; 8. sperm;
9. sporophytes; 10. gametophytes; 11. fertilization;
12. zygote; 13. D; 14. H; 15. H; 16. H; 17. D; 18. H;
19. c; 20. b; 21. e; 22. a; 23. d; 24. c; 25. e; 26. e; 27. b;
28. d; 29. e; 30. sporophyte (2n); 31. rhizome (2n);
32. sorus (2n); 33. spores (9n); 34. gametophyte (n);
35. sperm (n); 36. eggs (n)

23.6. SEED-BEARING PLANTS [p.379]
23.7. GYMNOSPERMS—PLANTS WITH NAKED SEEDS [pp.380-381]
1. C; 2. D; 3. A; 4. B; 5. E; 6. B; 7. D; 8. A; 9. B; 10. C;
11. B; 12. D; 13. D

23.8. ANGIOSPERMS—THE FLOWERING PLANTS [pp.382-383]
23.9. FOCUS ON A FLOWERING PLANT LIFE CYCLE [p.384]
23.10. THE WORLD'S MOST NUTRITIOUS PLANT [p.385]
1. F; 2. B; 3. C; 4. E; 5. D; 6. A; 7. seed coat; 8. embryo;
9. endosperm; 10. seed; 11. sporophyte; 12. pollen sac;
13. ovules within the ovary; 14. microspore; 15. egg;
16. female gametophyte; 17. male gametophyte;
18. eudicot; 19. seeds; 20. 16%; 21. amino acids;
22. drought; 23. frost; 24. salty soil

SELF-TEST
1. b; 2. e; 3. a; 4. d; 5. d; 6. e; 7. a; 8. d; 9. e; 10. d;
11. b

Chapter 24   Fungi

Impacts, Issues: High-Flying Fungi [p.388]

24.1. FUNGAL TRAITS AND CLASSIFICATION
[pp.390-391]
1. f; 2. g; 3. c; 4. e; 5. b; 6. d; 7. a; 8. heterotroph;
9. chitin; 10. yeast; 11. mycelium; 12. hyphae;
13. digestive enzymes; 14. saprobe; 15. cycled

24.2. THE FLAGELLATED FUNGI [p.391]
24.3. ZYGOSPORE FUNGI AND RELATIVES
[pp.392-393]
1. nuclear fusion; 2. zygospore (2n); 3. meiosis;
4. spores (n); 5. spore sac (n); 6. rhizoids (n);
7. asexual reproduction; 8. zygospore (n); 9. a; 10. c;
11. b; 12. c; 13. a; 14. b; 15. c; 16. a; 17. b

24.4. SAC FUNGI-ASCOMYCETES [pp.394-395]
24.5. CLUB FUNGI-BASIDIOMYCETES
[pp.396-397]
24.6. THE FUNGAL SYMBIONTS [pp.398-399]
24.7. AN UNLOVED FEW [p.399]
1. c; 2. e; 3. d; 4. a; 5. b; 6. nuclear fusion; 7. meiosis;
8. spore (n); 9. cytoplasmic fusion; 10. hyphae (n);
11. gill (n); 12. f; 13. d; 14. c; 15. a; 16. e; 17. b; 18. d;
19. b; 20. e; 21. c; 22. a; 23. f

SELF-TEST
1. b; 2. c; 3. a; 4. d; 5. c; 6. a; 7. c; 8. c; 9. d; 10. a;
11. b

Chapter 25   Animal Evolution – The Invertebrates

Impacts, Issues: Old Genes, New Drugs [p.402]

25.1. ANIMAL TRAITS & BODY PLANS
[pp.404-405]
25.2. ANIMAL ORIGINS & ADAPTIVE
RADIATION [pp.406-407]
1. conotoxins; 2. signaling molecules; 3. morphine;
4. gamma-glutmyl carboxylase; 5. repairing blood
vessels; 6. 500 million; 7. M; 8. B; 9. F; 10. D; 11. E;
12. I; 13. G; 14. C; 15. K; 16. J; 17. A; 18. H; 19. L;
20. N

25.3. THE SIMPLEST LIVING ANIMAL [p.408]
25.4. THE SPONGES [pp.408-409]
25.5. CNIDARIANS—TRUE TISSUES [pp.410-411]
1. D; 2. G; 3. B; 4. F; 5. C; 6. E; 7. A; 8. cnidarians;
9. Nematocysts; 10. medusa; 11. polyp;
12. gastrovascular cavity; 13. gastrodermis; 14. nerve
net; 15. mesoglea; 16. dinoflagellates; 17. coral
bleaching; 18. feeding polyp; 19. reproductive polyp;
20. female medusa; 21. ciliated bilateral larva

25.6. FLATWORMS--SIMPLE ORGAN SYSTEMS
[pp.412-413]
1. b, c; 2. c; 3. c; 4. a; 5. a; 6. b; 7. b; 8. a; 9. c; 10. a;
11. c; 12. c; 13. branching gut; 14. pharynx; 15. brain;
16. nerve cord; 17. ovary; 18. testis; 19. planaria;
20. no; 21. yes; 22. acoelomate; 23. bilateral;
24. Larvae; 25. Intermediate; 26. Humans; 27. Intestine;
28. Proglottids; 29. Organs; 30. Proglottids; 31. Feces;
32. Larval; 33. Intermediate

25.7. ANNELIDS—SEGMENTED WORMS
[pp.414-415]
1. f; 2. j; 3. b; 4. d; 5. g; 6. k; 7. c; 8. I; 9. a; 10. h; 11. e;
12. The earthworm uses its hydrostatic skeleton to
extend longitudinal muscles in the anterior end while
segments in the posterior end have their setae fixed
into the wall of the burrow. This pushes the anterior
end of the worm forward. Next the worm fixes the
anterior setae and releases the posterior setae while
shortening its segments. This pulls the posterior end
forward. The worm then fixes the posterior setae and
repeats the motion.;13. *Hirudo medicinalis* is used
post surgery to help remove blood clots from area
that have tiny capillary beds that need to heal.;
14. Clitellum; 15. Mouth; 16. Brain; 17. Crop; 18.
Intestine; 19. Ventral nerve cord; 20. Coelom;
21. Gut; 22. Pharynx; 23. Anus; 24. Longitudinal
muscle; 25. earthworm; 26; Annelida; 27. segmentation
and closed circulatory system; 28. yes; 29. bilateral;
30. yes

25.8. MOLLUSKS—ANIMALS WITH A MANTLE
[pp.416-417]
25.9. CEPHALOPODS FAST AND BRAINY [p.418]
1. mollusk; 2. shell; 3. mantle; 4. gills; 5. foot;
6. radula; 7. eyes; 8. tentacles; 9. nudibranchs;
10. nematocysts; 11. jet propulsion; 12. III; 13. I; 14. II;
15. Mouth; 16. Anus; 17. Gill; 18. Heart; 19. Radula;
20. Foot; 21. Shell; 22. Stomach; 23. Mouth; 24. Gill;
25. Mantle; 26. Retractor muscle; 27. Foot;
28. Radula; 29. Internal shell; 30. Mantle;
31. Reproductive organ; 32. Gill; 33. Ink sac; 34.
Tentacle; 35. d; 36. b; 37. a; 38. d; 39. d; 40. b; 41. c;
42. b; 43. d; 44. d; 45. b; 46. a

25.10. ROTIFERS & TARDIGRADES—TINY & TOUGH [p.419]

25.11. ROUNDWORMS—UNSEGMENTED WORMS THAT MOLT [p.420]

25.12. ARTHROPODS—ANIMALS WITH JOINTED LEGS [p.421]

1. b; 2. e; 3. f; 4. c; 5. g; 6. d; 7. a; 8. roundworm; 9. no;
10. pseudocoelom (false coelom); 11. d; 12. a; 13. d;
14. a; 15. d; 16. b; 17. a; 18. b; 19. c; 20. e; 21. e

25.13. CHELICERATES AND THEIR RELATIVES [p.422]

25.14. THE MOSTLY MARINE CRUSTACEANS [p.423]

25.15. MYRIAPODS—LOTS OF LEGS [p.424]

1. an exoskeleton; 2. spiders; 3. ticks; 4. lobsters and
crabs; 5. copepods; 6. barnacles; 7. barnacles; 8. molts;
9. millipedes; 10. centipedes; 11. centipedes;
12. Malphigian tubules; 13. horseshoe crabs
14. Poison gland; 15. Brain; 16. Heart; 17. Spinners;
18. Book lung; 19. chelicerates; 20. cephalothorax;
21. abdomen; 22. swimmerets; 23. walking legs; 24. 1st
legs; 25. antennae

25.16. A LOOK AT INSECT DIVERSITY [pp.424-425]

25.17. INSECT DIVERSITY & IMPORTANCE [pp.426-427]

1. f; 2. c; 3. b; 4. e; 5. d; 6. a; 7. e; 8. c; 9. b; 10. d; 11. a

25.18. THE SPINY-SKINNED ECHINODERMS [pp.428-429]

1. deuterostomes; 2. calcium carbonate; 3. radial;
4. bilateral; 5. brain; 6. nervous; 7. stomach; 8. enzymes;
9. tube; 10. water-vascular; 11. muscle; 12. ampulla;
13. anus; 14. sea urchin; 15. sea cucumber;
16. internal organs; 17. Lower stomach; 18. Upper
stomach; 19. Anus; 20. Gonad; 21. Coelom;
22. Digestive gland; 23. Eyespot; 24. starfish;
25. deuterostome

SELF-TEST

1. a; 2. d; 3. b; 4. a; 5. a; 6. c; 7. c; 8. a; 9. b; 10. a;
11. g, G; 12. b, DHJ; 13. f, BE; 14. h, F; 15. d, K; 16. I,
M; 17. e, CL; 18. a, I; 19. c, A; 20. Placozoans have
four different cell types that form two layers within
the body. Their cells have specialized functions
(secrete digestive enzymes, movement, etc.). They
also have genes that are similar to the ones that
encode for signaling molecules in human nerves.
These features suggest that placozoans represent an
early branch on the animal family tree.

Chapter 26   Animal Evolution – The Chordates

Impacts, Issues: Transitions Written in Stone [p.432]

26.1. THE CHORDATE HERITAGE [pp.434-435]

26.2. VERTEBRATE TRAITS AND TRENDS [pp.436-437]

26.3. JAWLESS LAMPREYS [p.438]

26.4. THE JAWED FISH [pp.438-439]

1. transitional forms; 2. Archaeopteryx;
3. Radiometric; 4. biological evolution; 5. nerve
cord; 5. gill slits; 7. notochord; 8. lancelets; 9. filter
feeding; 10. gut; 11. Tunicates; 12. tunic;
13. metamorphosis; 14. vertebral column; 15. fish;
16. jaws; 17. fins; 18. gills; 19. lungs; 20. circulatory;
21. lancelet; 22. invertebrate; 23. pharynx with gill
slits; 24. anus; 25. notochord; 26. tubular nerve
cord; 27. vertebrates; 28. jawed vertebrates;
29. tunicates; 30. lamprey; 31. ray-finned fish;
32. hagfish; 33. lamprey; 34. sharks; 35. gill slits;
36. Ray-finned; 37. Lungfish; 38. swim bladder;
39. Coelacanths; 40. amphibians

26.5. AMPHIBIANS—FIRST TETRAPODS ON LAND [pp.440-441]

26.6. VANISHING ACTS [p.441]

26.7. THE RISE OF AMNIOTES [pp.442-443]

26.8. SO LONG, DINOSAURS [p.443]

26.9. DIVERSITY OF MODERN REPTILES [pp.444-445]

1. land; 2. three chambered heart; 3. lungs; 4. inner
ear; 5. eyelids; 6. insects; 7. salamanders;
8. carnivores; 9. water; 10. reproduce; 11.
respiratory surface; 12. human; 13. Amniotes;
14. eggs; 15. water loss; 16. synapsids; 17. archosaurs;
18. birds; 19. Cretaceous; 20. dinosaurs; 21. E; 22. B;
23. D; 24. C; 25. A

26.10. BIRDS—THE FEATHERED ONES [pp.446-447]

26.11. THE RISE OF MAMMALS [pp.448-449]

26.12. MODERN MAMMALIAN DIVERSITY [pp.450-451]

1. embryo; 2. albumin; 3. yolk sac; 4. pectoral girdle;
5. pelvic girdle; 6. sternum; 7. humerus; 8. radius;
9. ulna; 10. dinosaurs; 11. Jurassic; 12. scales;
13. feathers; 14. sternum; 15. air cavities; 16. metabolic;
17. oxygen / blood; 18. four; 19. 1.6; 20. ostrich;
21. migration; 22. Antarctic; 23. hair; 24. mammary
glands; 25. teeth; 26. synapsids; 27. dinosaurs;
28. monotremes; 29. marsupials; 30. placental;
31. Morphological convergence; 32. c; 33. c; 34. b;
35. b; 36. a; 37. a; 38. b; 39. c

26.13. FROM EARLY PRIMATES TO HOMINIDS
  [pp.452-453]

26.14. EMERGENCE OF EARLY HUMANS
  [pp.454-455]

26.15. EMERGENCE OF MODERN HUMANS
  [pp.456-457]

1. primates; 2. prosimians; 3. Anthropoids;
4. monkeys; 5. humans; 6. daytime; 7. forward
facing; 8. smell; 9. prehensile; 10. opposable;
11. grip; 12. tools; 13. upright walking; 14. teeth;
15. brain; 16. behaviors; 17. culture; 18. Africa;
19. Australopiths; 20. Homo; 21. 2.5; 22. brain;
23. toolmaker; 24. Europe (Asia); 25. Asia (Europe);
26. 195,000; 27. Homo sapiens; 28. 100,000;
29. Neanderthals; 30. replacement; 31. Homo
erectus; 32. Homo sapiens; 33. multiregional;
34. Homo sapiens; 35. c; 36. a; 37. b; 38. b; 39. b;
40. a; 41. b

SELF-TEST

1. a; 2. a; 3. d; 4. b; 5. a; 6. d; 7. d; 8. c; 9. d; 10. a;
11. B; 12. H; 13. C; 14. D; 15. A; 16. E; 17. G;
18. F; 19. d, H; 20. b, B; 21. i, F; 22. g, D; 23. h, A;
24. c, E; 25. a, I; 26. f, C

Chapter 27   Plants and Animals – Common Challenges

Impacts, Issues: A Cautionary Tale [p.460]

27.1. LEVELS OF STRUCTURAL ORGANIZATION
  [pp.462-463]

27.2. RECURRING CHALLENGES TO SURVIVAL
  [pp.464-465]

1. e; 2. a; 3. d; 4. b; 5. c; 6. h; 7. g; 8. f
9. extracellular fluid; 10. volume; 11. composition;
12. homeostasis; 13. diffusion; 14. active transport ;
15. d; 16. a, d; 17. a; 18. b; 19. c; 20. b; 21. c; 22. d

27.3. HOMEOSTASIS IN ANIMALS [pp.466-467]
27.4. HEAT-RELATED ILLNESS [p.467]

27.5. DOES HOMEOSTASIS OCCUR IN PLANTS
  [pp.468-469]

**27.6. HOW CELLS RECEIVE AND RESPOND TO
SIGNALS [pp.470-471]**

1. a; 2. e; 3. c; 4. f; 5. d; 6. b; 7. b; 8. a; 9. a; 10. b;
11. a; 12. b; 13. a; 14. b; 15. b; 16. a; 17. a
18. multicelled; 19. external; 20. diffuse; 21. gap
junction; 22. plasmodesmata; 23. Molecular;
24. receptor; 25. transduced; 26. response; 27. proteins;
28. enzyme; 29. reactions; 30. amplify; 31. apoptosis

SELF-TEST

1. d; 2. b; 3. c; 4. d; 5. b; 6. c; 7. b; 8. e; 9. a; 10. c

Chapter 28   Plant Tissues

Impacts, Issues: Droughts Versus Civilization [p.474]

**28.1. COMPONENTS OF THE PLANT BODY
  [pp.476-477]**

1. ground tissues; 2. vascular tissues; 3. dermal tissues;
shoot system; 5. root system; 6. E; 7. H; 8. B; 9. F; 10.
A.; 11. D; 12. C; 13. G

**28.2. COMPONENTS OF PLANT TISSUES
  [pp.478-479]**

1. radial; 2. tangential; 3. transverse; 4. a; 5. c; 6. a;
7. a; 8. b; 9. b; 10. c; 11. a; 12. a; 13. c; 14. (a) tracheid,
(b) dead, (c) xylem; 15. (a) sieve tube member, (b)
alive, (c) phloem; 16. (a) companion cells, (b) alive,
(c) phloem; 17. (a) vessel member, (b) dead, (c)
xylem; 18. Epidermis is the first dermal tissue to
form. It secretes a waxy cuticle that conserves water.
It has specialized cells such as guard cells that control
the opening of stoma to allow gas passage. Periderm
replaces epidermis in mature woody stems and roots.
This dermal tissue is composed of parenchyma, cork,
and cork cambium.

**28.3. PRIMARY STRUCTURE OF SHOOTS
  [pp.480-481]**

1. epidermis; 2. vascular bundle; 3. pith;
4. collenchyma; 5. vessel; 6. sieve tube; 7. companion
cell; 8. epidermis; 9. cortex; 10. vascular bundle;
11. pith; 12. xylem; 13. vascular cambium; 14. phloem

28.4. A CLOSER LOOK AT LEAVES [pp.482-483]

1. blade; 2. petiole; 3. axillary bud; 4. node; 5. stem;
6. sheath; 7. eudicot; 8. monocot; 9. palisade
mesophyll; 10. spongy mesophyll; 11. lower
epidermis; 12. stoma; 13. leaf vein

**28.5. PRIMARY STRUCTURE OF ROOTS
  [pp.484-485]**

1. root hair; 2. endodermis; 3. pericycle; 4. cortex;
5. epidermis; 6. apical meristem; 7. root cap;
8 endodermis; 9. pericycle; 10. primary phloem;
11. primary root; 12. root cap; 13. root hairs;
14. surface area; 15. vascular cylinder; 16. pericycle;
17. lateral roots; 18. taproot; 19. fibrous

28.6. SECONDARY GROWTH [pp.486-487]
28.7. TREE RINGS AND OLD SECRETS [p.488]
1. C; 2. J; 3. K; 4. E; 5. L; 6. F; 7. B; 8. A; 9. H; 10. G; 11. I; 12. D; 13. Heartwood is the oldest xylem found toward the center of a stem. It is compressed, non-conducting, and is often dark due to deposits of resins, tannins, gums, and oils. Sapwood is younger and found closest to the vascular cambium. It is moist, functional, and usually lighter in color.; 14. The bark consists of all the living and dead tissues outside the vascular cambium. It is composed of secondary phloem and periderm (parenchyma, cork, and cork cambium).

28.8. MODIFIED STEMS [p.489]
1. e; 2. f; 3. b; 4. a; 5. d; 6. c

SELF-TEST
1. b; 2. a; 3. b; 4. c; 5. a; 6. d; 7. d; 8. c; 9. c; 10. b; 11. c; 12. b; 13. d

---

Chapter 29   Plant Nutrition and Transport

Impacts, Issues: Leafy Clean-Up Crews [p.492]

29.1. PLANT NUTRIENTS AND AVAILABILITY IN SOIL [pp.494-495]
1. f; 2. c; 3. i; 4. b; 5. h; 6. a; 7. j; 8. e; 9. g; 10. d; 11. The thin layers of water and negatively charged crystals that make up clay attract and bind the positively charged mineral ions (for example, iron, calcium, potassium, copper, and magnesium) found in soil water. The larger particles of sand and silt allow space for air and keep the tiny clay particles from packing too tightly together. 12. Soil erosion is the removal of soil due to wind, water, and glacial action. In areas of deforestation or poor plant cover, much of the organic topsoil can be lost impoverishing the soil for future plant growth. Leaching is the process by which water percolating through the soil removes nutrients to lower soil horizons where plants cannot access them. More leaching occurs in sandy, well-drained soils than in humus or clay soils that are less porous.

29.2. HOW DO ROOTS ABSORB WATER AND NUTRIENTS? [pp.496-497]
1. b; 2. a; 3. c; 4. d; 5. a; 6. d; 7. b; 8. d; 9. 2; 10. 4; 11. 1; 12. 3; 13. endodermis; 14. primary phloem; 15. primary xylem; 16. Casparian strip

29.3. HOW DOES WATER MOVE THROUGH PLANTS? [pp.498-499]
1. cohesion-tension; 2. xylem; 3. tension; 4. roots; 5. transpiration; 6. stomata; 7. negative pressure; 8. hydrogen bonds; 9. cohesion; 10. tension; 11. leaves; 12. stems; 13. roots; 14. soil

29.4. HOW DO STEMS AND LEAVES CONSERVE WATER? [pp.500-501]
1. g; 2. d; 3. f; 4. e; 5. a; 6. b; 7. h; 8. c

29.5. HOW DO ORGANIC COMPOUNDS MOVE THROUGH PLANTS? [pp.502-503]
1. source; 2. active transport; 3. water; 4. turgor; 5. solutes; 6. sink; 7. d; 8. f; 9. b; 10. a; 11. c; 12. e

SELF-TEST
1. c; 2. c; 3. d; 4. b; 5. c; 6. d; 7. b; 8. b; 9. a; 10. b; 11. c; 12. d

Chapter 30   Plant Reproduction

Impacts, Issues: Plight of the Honeybee [p.506]

30.1. REPRODUCTIVE STRUCTURES OF FLOWERING PLANTS [pp.508-509]

1.sporophyte; 2. meiosis; 3. male gametophyte; 4. female gametophyte; 5. sperm; 6. eggs; 7. fertilization; 8. seed; 9. Regular flowers are symmetrical around their central axis (daisy). Irregular flowers are not radially symmetrical around their central axis (orchid).; 10. A complete flower has all four sets of modified leaves: sepals, petals, stamens & carpals (pistils). An incomplete flower is missing one or more of the previously mentioned sets of modified leaves.; 11. A perfect flower has both male (stamens) and female (carpel or pistil) reproductive organs.; 12. filament; 13. anther; 14. stigma; 15. style; 16. ovary; 17. ovule; 18. receptacle; 19. sepal; 20. petal; 21. stamens; 22. carpel, pistil; 23. calyx; 24. corolla

30.2. FLOWERS AND THEIR POLLINATORS [p.510-511]

1. pollinators; 2. pollen; 3. Birds (insects, bats); 4. insects (birds, bats); 5. bats (birds, insects); 6. coevolved; 7. bees; 8. UV (ultraviolet); 9. bats (moths); 10. moths (bats); 11. beetles (flies); 12. flies (beetles); 13. nectar; 14. honey; 15. floral tube; 16. birds (butterflies, moths); 17. butterflies (birds, moths); 18. moths (butterflies, birds)

30.3. A NEW GENERATION BEGINS [pp.512-513]
30.4. FLOWER SEX [p.514]

1. 4; 2. 2; 3. 6; 4. 5; 5. 1; 6. 3; 7. haploid; 8. diploid; 9. haploid; 10. haploid; 11. haploid; 12. diploid; 13. 5; 14. 2; 15. 6; 16. 1; 17. 3; 18. 4; 19. sporopollenin; 20. recognition proteins; 21. adhesion proteins; 22. dormant; 23. pollen tube; 24. furrows (pores); 25. pores (furrows); 26. style; 27. chemical signals; 28. egg; 29. species-specific; 30. fertilize

30.5. SEED FORMATION [p.515]
30.6. FRUITS [pp.516-517]

1. shoot tip (shoot apical meristem); 2. cotyledons; 3. endosperm; 4. root tip (root apical meristem); 5. seed coat (integuments); 6. b; 7. a; 8. b; 9. a; 10. c; 11. A simple fruit develops from one ovary in one flower (pea pod). An aggregate fruit develops from several ovaries in one flower (strawberry). A multiple fruit forms from the fusion of ovaries from separate flowers (pineapple).; 12. A true fruit forms only from the ovary wall and its contents (cherry). Accessory fruits are composed of other flowers parts in addition to the ovary. An apple core is the ovary, the fleshy edible part is derived from the receptacle.; 13. c (peanut, peas); 14. e (acorn, corn); 15. a (cherry, peach); 16. d (grape, tomato); 17. b (apple, pear); 18. a; 19. c; 20. b; 21. a; 22. c; 23. b; 24. a; 25. b

30.7. ASEXUAL REPRODUCTION OF FLOWERING PLANTS [pp.518-519]

1. e; 2. d; 3. h; 4. g; 5. b; 6. i; 7. a; 8. f; 9. j; 10. c

SELF-TEST

1. b; 2. c; 3. d; 4. c; 5. c; 6. d; 7. b; 8. c; 9. c; 10. d; 11. d; 12. d

---

Chapter 31   Plant Growth and Development

Impacts, Issues: Foolish Seedlings, Gorgeous Grapes [p.522]

31.1. PATTERNS OF DEVELOPMENT IN PLANTS [pp.524-525]

1. shoot apical meristem; 2. root apical meristem; 3. dormancy; 4. Germination; 5. enzymes; 6. sugar monomers; 7. meristem; 8. radicle (root); 9. seed coat; 10. growth; 11. differentiation; 12. climate; 13. abrasion; 14. light; 15. smoke; 16. burning; 17. freezing; 18. seed coat; 19. radicle; 20. cotyledons; 21. hypocotyl; 22. primary root; 23. primary leaf; 24. branch root

31.2. PLANT HORMONES AND OTHER SIGNALING MOLECULES [pp.526-527]
31.3. EXAMPLES OF PLANT HORMONE EFFECTS [pp.528-529]

1. c; 2. a; 3. e; 4. b; 5. a; 6. d; 7. b; 8. a; 9. e; 10. b; 11. a; 12. a; 13. e; 14. a; 15. a; 16. hormones; 17. Environmental cues; 18. target cells; 19. gibberellins; 20. auxins; 21. abscisic acid (ABA); 22. cytokinins; 23. ethylene; 24. gibberellin; 25. amylase; 26. endosperm; 27. Indole-3-acetic acid (IAA); 28. proton pumps; 29. repressor proteins; 30. shoot tips; 31. young leaves; 32. phloem; 33. herbivores; 34. wasps; 35. jasmonates; 36. air; 37. egg; 38. larva

31.4. ADJUSTING THE DIRECTION AND RATES OF GROWTH [pp.530-531]

1. b; 2. c; 3. a; 4. a; 5. b; 6. c; 7. d; 8. c; 9. tropism; 10. stimulus; 11. hormones; 12. down; 13. up; 14. gravitropism; 15. statoliths; 16. auxin; 17. phototropism; 18. auxin; 19. phototropins; 20. blue light; 21. shaded; 22. elongate; 23. tendrils; 24. thigmotropism; 25. elongation; 26. Mechanical stress

31.5. SENSING RECURRING ENVIRONMENTAL CHANGES [pp.532-533]
31.6. SENESCENCE AND DORMANCY [p.534]

1. g; 2. b; 3. i; 4. d; 5. k; 6. c; 7. l; 8. f; 9. e; 10. j; 11, a; 12. h; 13. Sunlight; 14. phytochromes; 15. Red; 16. Far-red; 17. photosynthesis; 18. phototropism; 19. photoperiodism; 20. Short-day; 21. long-day; 22. day-neutral; 23. phytochrome; 24. abscission; 25. Ethylene

SELF-TEST

1. b; 2. d; 3. a; 4. c; 5. c; 6. a; 7. d; 8. c; 9. d; 10. d; 11. a; 12. c; 13. b; 14. d

---

Chapter 32   Animal Tissues and Organ Systems

Impacts, Issues: Why It Matters [p.538]

32.1. ORGANIZATION OF ANIMAL BODIES [p.540]

1. stem cells; 2. embryonic; 3. adult; 4. Anatomy; 5. physiology; 6. tight; 7. Adhering; 8. Gap; 9. Tight; 10. peptic

32.2. EPITHELIAL TISSUE [p.541]

1. Epithelial; 2. Simple epithelium; 3. Stratified epithelium; 4. Exocrine; 5. epithelial; 6. Endocrine; 7. hormones; 8. circulatory system; 9. F, cell layers; 10. T; 11. F, protection; 12. T; 13. F, the blood

32.3. CONNECTIVE TISSUES [pp.542-543]
32.4. MUSCLE TISSUES [pp.544-545]
32.5. NERVOUS TISSUE [p.545]

1. b; 2. c; 3. a; 4. c; 5. b; 6. a; 7. c; 8. c; 9. A; 10, C; 11. D; 12. B; 13. smooth; 14. smooth; 15. skeletal; 16. Skeletal; 17. skeletal; 18. striped; 19. cause movement in internal organs; 20. cardiac; 21. neurons; 22. Neuroglia; 23. neurons; 24. connective, D, h. j; 25. epithelial, G, a, e, k; 26. muscle, I, g, (j); 27. muscle, J, d, (j); 28. connective, E, f, j, (m); 29. connective, B, I; 30. epithelial, H, a, e, k; 31. connective, K, a, n; 32. nervous, L, b; 33. muscle, C, l; 34. epithelial, F, a, e, k; 35. connective, A, c, f, m

32.6. OVERVIEW OF MAJOR ORGAN SYSTEMS [pp.546-547]

1.C; 2. B; 3. A; 4. superior; 5. distal; 6. proximal; 7. posterior; 8. transverse; 9. inferior; 10. anterior; 11. frontal; 12. circulatory (D); 13. respiratory (E); 14. urinary (excretory)(A); 15. skeletal (C); 16. endocrine (J); 17. reproductive (G); 18. digestive (I); 19. muscular (H); 20. nervous (B); 21. integumentary (F); 22. lymphatic (K)

32.7. VERTEBRATE SKIN—EXAMPLE OF AN ORGAN SYSTEM [pp.548-549]
32.8. FARMING SKIN [p.549]

1. skin; 2. epidermis; 3. stratified squamous; 4. dermis; 5. connective tissue; 6. Keratinocytes; 7. keratin; 8. hair; 9. Melanocytes; 10. melanin; 11. ultraviolet; 12. melanocytes; 13. vitamin D; 14. elastic; 15. cultured; 16. burns; 17. Unbroken skin is an effective barrier against pathogens; 18. melanocytes, sweat glands, oil glands, any other differentiated structures.

SELF-TEST

1. d; 2. b; 3. c; 4. c; 5. d; 6. c; 7. c; 8. d; 9. a; 10. d; 11. A; 12. G; 13. E; 14. J; 15. H; 16. I; 17. F; 18. B; 19. D; 20. C

Chapter 33   Neural Control

Impacts, Issues: In Pursuit of Ecstasy [p.552]

33.1. EVOLUTION OF NERVOUS SYSTEMS [pp.554-555]
1. a; 2. f; 3. g; 4. e; 5. d; 6. d; 7. h; 8. c; 9. autonomic system; 10. somatic system; 11. peripheral nerves; 12. spinal cord; 13. brain

33.2. NEURONS—THE GREAT COMMUNICATORS [p.556]
1. dendrite (B); 2. input zone (A); 3. trigger zone (C); 4. conducting zone (E); 5. axon (D); 6. output zone (F); 7. Neurons have specialized extensions called axon and dendrites; they are modified for electrical and chemical processes for the transmission of signals; 8. Dendrites receive signals and axons transmit signals.

33.3. MEMBRANE POTENTIALS [p.557]
33.4 A CLOSER LOOK AT ACTION POTENTIALS [pp.558-559]
1. F, sodium; 2. T; 3. F, sodium; 4. F, threshold potential; 5. T; 6. -70mV; 7. T; 8. Na^+, K^+
33.5. HOW NEURONS SEND MESSAGES TO OTHER CELLS [pp.560-561]
33.6. A SMORGASBORD OF SIGNALS [p.562]
1. f; 2. h; 3. g; 4. c. 5. a; 6. e; 7. b; 8. d; 9. F, synaptic integration; 10. T; 11. F, synapse; 12. F, receptors; 13. T; 14. B; 15. A.; 16. G; 17. E; 18. C; 19. F; 20. D

33.7. DRUGS DISRUPT SIGNALING [p.583]
1. a. stimulant, block Ach receptors; b. stimulant, stop uptake of dopamine; c. stimulant, increase serotonin, norepinephrine, and dopamine; d. depressant, inhibit Ach output; e. depressant, inhibit Ach output; f. analgesic, slow clearing of synapse; g. analgesic, slow clearing of synapse; h. hallucinogen, binds to serotonin receptor—distort sensory data; i. hallucinogen, alters levels of dopamine, serotonin, norepinephrine, and GABA—relaxed and inattentive; 2. Drug takes on a vital biochemical function in the body's homeostasis; 3. Tolerance, habituation, inability to stop, concealing use, extreme or dangerous actions to get or use drug, deterioration of relationships, anger about use, prefer drug use over other activities.

33.8. THE PERIPHERAL NERVOUS SYSTEM [pp.564-565]
1. A; 2. E; 3. D; 4. C; 5. B; 6. Increase speed of nerve transmission; 7. They have opposite effects. Sympathetic alerts body to face stress while parasympathetic slows down activities for normal resting conditions; 8. optic nerve; 9. vagus nerve; 10. pelvic nerve; 11. midbrain; 12. medulla oblongata; 13. cervical nerves; 14. thoracic nerves; 15. lumbar nerves; 16. sacral nerves.

33.9. THE SPINAL CORD [pp.566-567]
1. spinal cord; 2. spinal nerve; 3. vertebra; 4. intervertebral disk; 5. meninges; 6. B; 7. C; 8. F; 9. A; 10. D; 11. E; 12. G; 13. c; 14. b; 15. d; 16. a; 17. c; 18. d; 19. d; 20. b; 21. d

33.10. THE VERTEBRATE BRAIN [pp.568-569]
1. D; 2. F; 3. G; 4. H; 5. C; 6. B; 7. A; 8. E; 9. a; 10. a; 11. a; 12. b; 13. b; 14. b; 15. c

33.11. THE HUMAN CEREBRUM [pp.570-571]
33.12. THE SPLIT BRAIN [p.572]
1. h; 2. i; 3. b; 4. e; 5. d; 6. h; 7. f; 8. a; 9. h; 10. c; 11. g; 12. Sperry demonstrated that signals that cross the corpus callosum coordinate the functions of the two cerebral hemispheres, each of which has responded to visual signals from the opposite side of the body; 13. Short; 14. long; 15. permanently; 16. skill; 17. prefrontal; 18. motor; 19. cerebellum; 20. basal ganglia; 21. Declarative; 22. sensory; 23. gatekeeper; 24. hippocampus; 25. association; 26. basal ganglia; 27. thalamus

33.13. NEUROGLIA—THE NEURON'S SUPPORT STAFF [p.573]
1. a. line the ventricles and produce cerebrospinal fluid; b. phagocytic cells within the central nervous system; c. form blood-brain barrier, take up neurotransmitters, form lactate, and make nerve growth factor; d. produce myelin sheath on axons in central nervous system.

SELF-TEST
1. a; 2. a; 3. b; 4. d; 5. a; 6. b; 7. c; 8. d; 9. b; 10. b; 11. d; 12. d; 13. c; 14. e; 15. b

Chapter 34 Sensory Perception

Impacts, Issues: A Whale of a Dilemma [p.576]

34.1. OVERVIEW OF SENSORY PATHWAYS [pp.578-579]
1. T; 2. T; 3. chemoreceptors; 4. chemoreceptors; 5. T; 6. fewer; 7. sensory adaptation; 8. somatic; 9. d; 10. c; 11. a; 12. b; 13. a; 14. e; 15. e; 16. b; 17. d; 18. b; 19. b; 20. a; 21. b

34.2. SOMATIC AND VISCERAL SENSATIONS [pp.580-581]
34.3. SAMPLING THE CHEMICAL WORLD [p.582]
1. somatosensory cortex; 2. lips/mouth (fingers/hand); 3. fingers/hand (lips/mouth); 4. Free nerve endings; 5. pain; 6. Meissner's corpuscles; 7. pain; 8. endorphins (enkephalins); 9. enkephalins (endorphins); 10. sensory; 11. Olfactory (chemo-); 12. nose; 13. five; 14. olfactory; 15. Pheromones; 16. olfactory (chemo-); 17. vomeronasal; 18. chemoreceptors

34.4. SENSE OF BALANCE [p.583]
34.5. SENSE OF HEARING [pp.584-585]
34.6. NOISE POLLUTION [p.586]
1. fluid in the canals; 2. T; 3. static equilibrium; 4. T; 5. T; 6. basilar membrane; 7. cochlea; 8. T; 9. middle ear bones (hammer, anvil, stirrup); 10. cochlea; 11. auditory (vestibulocochlear) nerve; 12. eardrum; 13. oval window; 14. basilar membrane; 15. tectorial membrane

34.7. SENSE OF VISION [pp.586-587]
34.8. A CLOSER LOOK AT THE HUMAN EYE [pp.588-589]
34.9. FROM RETINA TO THE VISUAL CORTEX [pp.590-591]
34.10. VISUAL DISORDERS [p. 592]
1. lens; 2. cornea; 3. iris; 4. sclera; 5. lens; 6. visual accommodation; 7. rods; 8. cones; 9. fovea; 10. rhodopsin; 11. blue-green; 12. optic nerve; 13. blind spot (optic disk); 14. vitreous body; 15. cornea; 16. iris; 17. lens; 18. aqueous humor; 19. ciliary muscles; 20. retina; 21. fovea; 22. optic nerve; 13. blind spot/optic disk; 24. sclera; 25. Farsightedness; 26. macula; 27. T; 28. lens; 29. cornea; 30. T; 31. red; 32. roundworms

SELF-TEST
1. c ; 2. c ; 3. d ; 4. e ; 5. d ; 6. a ; 7. c ; 8. b; 9. b; 10. a ; 11. d; 12. c

---

Chapter 35 Endocrine Control

Impacts, Issues: Why It Matters [p. 596]

35.1. INTRODUCING THE VERTEBRATE ENDOCRINE SYSTEM [pp.598-599]
1. Atrazine; 2. endocrine disruptor; 3. hermaphrodites; 4. feminized (abnormal); 5. kepone (DDT); 6. DDT (kepone); 7. E; 8. A; 9. B; 10. F; 11. C; 12. D; 13.a. 1, several releasing and inhibiting hormones, synthesizes ADH and oxytocin; b. 1, ACTH, TSH, LH, somatotropin (STH), prolactin (PRL); c. 1, stores and secrete two hypothalamic hormones, ADH and oxytocin; d. 2, cortisol, aldosterone, (sex hormones of opposite sex); e. 2, epinephrine, norepinephrine; f. 2, estrogens, progesterone; g. 2, testosterone; h. 1, melatonin; i. 1, thyroxine and triiodothyronine; j. 4, parathyroid hormone (PTH); k. 1, thymosins; l. 1, insulin, glucagon

35.2. THE NATURE OF HORMONE ACTION [pp.600-601]
1. a; 2. b; 3. a; 4. b; 5. b; 6. a; 7. b; 8. b; 9. b; 10. b; 11. A second messenger carries the signal from the cell membrane to the inner parts of the cell when the hormone attaches to a receptor on the cell membrane surface. Only steroid (lipid) hormones can pass directly through the membrane and into the nucleus of the cell. All others must attach to receptors on cell surface and then need a means to transfer the signal to inner parts of the cell—second messenger.; 12. If a cell that is normally a target cell for a particular hormone lacks the receptor for that hormone, then the cell will be unable to respond to the signal. It will act as though no hormone was secreted.

35.3. THE HYPOTHALAMUS AND PITUITARY GLAND [pp.602-603]
1. A (H); 2. P (G); 3. A (A); 4. A (I); 5. A (D); 6. A (B); 7. P (E); 8. A (C); 9. A (F); 10. T; 11. hypothalamus; 12. hypothalamus; 13. anterior pituitary; 14. T; 15. anterior pituitary; 16. glandular, nerve; 17. prolactin and oxytocin

35.4. GROWTH HORMONE FUNCTION AND DISORDERS [p.604]
35.5. SOURCES AND EFFECTS OF OTHER VERTEBRATE HORMONES [p.605]
1. F; 2. A; 3. E; 4. C; 5. B; 6. D

35.6. THYROID AND PARATHYROID GLANDS [pp.606-607]
35.7. TWISTED TADPOLES [p.607]
1. negative feedback; 2. hypothalamus; 3. TSH (thyroid stimulating hormone); 4. thyroid; 5. iodine; 6. simple goiter; 7. hyperthyroidism; 8. parathyroid; 9. calcium; 10. rickets; 11. metamorphosis; 12. thyroid disruptors

35.8. PANCREATIC HORMONES [p.608]
35.9. BLOOD SUGAR DISORDERS [p.609]
35.10. THE ADRENAL GLANDS [p.610]

35.11. TOO MUCH OR TOO LITTLE CORTISOL [p.611]
35.12. OTHER ENDOCRINE GLANDS [p.612]
1. c, B; 2. e, G; 3. b, D; 4. d, A; 5. a, G; 6. g, C; 7. f, E; 8. a, B, 9. d, A; 10. c, B; 11. e, C; 12. b, D; 13. juvenile (child); 14. low; 15. T; 16. elevated; 17. T; 18. long-term

35.13. A COMPARATIVE LOOK AT A FEW INVERTEBRATES [p.613]
1. molt; 2. ecdysone; 3. steroid; 4. X-; 5. eye stalk; 6. insects; 7. molt-inhibiting hormone

SELF-TEST
1. a; 2. d; 3. b; 4. e; 5. d; 6. b; 7. a; 8. c; 9. L; 10. E; 11. N; 12. D; 13, F; 14. A; 15. I; 16. B; 17. O; 18. G; 19. J; 20. C; 21. M; 22. K; 23. H

Chapter 36   Structural Support and Movement

Impacts, Issues: Pumping up Muscles [p.616]

36.1. INVERTEBRATE SKELETONS [pp.618-619]
36.2. THE VERTEBRATE ENDOSKELETON [pp.620-621]
1. c; 2. a; 3. b; 4. b; 5. a; 6. c; 7. B; 8. D; 9. E; 10. A; 11. C; 12. cranial bones; 13. vertebrae; 14. intervertebral disks; 15. clavicle; 16. scapula; 17. humerus; 18. radius; 19. ulna; 20. carpals; 21. metacarpals; 22. phalanges; 23. pelvic girdle; 24. femur; 25. patella; 26. tibia; 27. fibula; 28. tarsals; 29. metatarsals; 30. phalanges

36.3. BONE STRUCTURE AND FUNCTION [pp. 622-623]
36.4. SKELETAL JOINTS—WHERE BONES MEET [p.624]
36.5. THOSE ACHING JOINTS [p.625]
1. F; 2. C; 3. E; 4. D; 5. H; 6. G; 7. A; 8. B; 9. c; 10. b; 11. c; 12. a; 13. c; 14. b; 15. c; 16. a; 17. Ligaments holding bones together in a joint stretch too far or tears; 18. Crucuate ligaments are in the knee joint. They stabilize the knee. When damaged, the bones shift and may give out when trying to stand or walk.; 18. Arthritis is inflammation of a joint. Bursitis is when fluid filled sacscalled bursae that cushion joints become inflamed.

36.6. SKELETAL-MUSCULAR SYSTEMS [pp.626-627]
36.7. HOW DOES SKELETAL MUSCLE CONTRACT? [pp.628-629]
1. muscle fiber; 2. tendon; 3. bone; 4. lever; 5. fixed; 6. joint; 7. force; 8. skeletal; 9. opposition; 10; resists

(reverses); 11. Smooth; 12. stomach; 13. heart; 14. tendon; 15. triceps brachii (C); 16. pectoralis major (G); 17. serrattus anterior (I); 18. external oblique (M); 19. rectus abdominus (K); 20. adductor longus (D); 21. Sartorius (L); 22. quadriceps femoris (F); 23. tibialis anterior (A); 24. biceps brachii (O); 25. deltoid (N); 26. trapezius (P); 27. latissimus dorsi (J); 28. gluteus maximus (H); 29. biceps femoris (E); 30. gastrocnemius (B); 31. F; 32. D; 33. G; 34. E; 35. A; 36. C; 37. B; 38. actin; 39. T; 40. closer together; 41. stays the same; 42. smaller; 43. T; 44. actin; 45. T

36.8. FROM SIGNAL TO RESPONSE: A CLOSER LOOK AT CONTRACTION [pp.630-631]
36.9. ENERGY FOR CONTRACTION [p.631]
36.10. PROPERTIES OF WHOLE MUSCLES [p.632]
36.11. DISRUPTION OF MUSCLE CONTRACTION [p.633]
1. E; 2. D; 3. F; 4. A; 5. B; 6. C; 7. H; 8. G; 9. I; 10. L; 11. J; 12. K; 13: Troponin and tropomyosin cover up the binding sites on actin. When calcium is released from the sarcoplasmic reticulum, troponin and tropomyosin move out of the way to expose the myosin-binding sites.; 14. An isotonic contraction is one where you would observe motion. In an isometric contraction, the muscle is flexed and released, but no actual movement is observed.

SELF-TEST
1. b; 2. d; 3. c; 4. d; 5. b; 6. c; 7. b; 8. a

Chapter 37   Circulation

Impacts, Issues: And Then My Heart Stood Still [p.636]

37.1 THE NATURE OF BLOOD CIRCULATION [pp.638-639]
1. Sudden cardiac arrest; 2. Cardiopulmonary resuscitation (CPR); 3. defibrillator; 4. closed; 5. Blood; 6. heart; 7. interstitial fluid; 8. heart; 9. rapidly; 10. capillary; 11. slows; 12. single; 13. pulmonary; 14. systemic; 15. pressure; 16. heart; 17. heart(s); 18. Open circulatory system: Blood is pumped into short tubes that open into spaces in the body's tissues, mixes with tissue fluids, then reenters the heart at openings in the heart wall.; 19. Closed circulatory system: Blood flow is confined within blood vessels that have continuously connected walls. Blood does not mix with interstitial fluid. The blood is moved by the action of five pairs of "hearts."

37.2. CHARACTERISTICS OF BLOOD [pp.640-641]
37.3. HEMOSTASIS [p.642]
37.4. BLOOD TYPING [pp.642-643]
1. a. Solvent; b. Plasma proteins; c. Red blood cells; d. Neutrophils; e. Lymphocytes; f. Defense against parasitic worms; g. Platelets; 2. Blood carries oxygen, nutrients, and other solutes to cells and transports the cells' metabolic wastes and secretions. It transports hormones to all parts of the body. It helps stabilize pH and regulate body temperature. It is the transport system for cells and proteins that help protect and repair tissues.; 3. T; 4. 6 to 8 percent; 5. plasma; 6. bone marrow; 7. T; 8. short lived; 9. neutrophils (macrophages, monocytes); 10. platelets; 11. C; 12. B; 13. F; 14. D; 15. A; 16. E; 17. stem (F); 18. red blood (E); 19. platelets (C); 20. neutrophils (A, B); 21. B (D); 22. T (D); 23. monocytes (A); 24. macrophages (A, B); 25. a, b, e, g; 26. a, b, c, d, e, f, g, h; 27. c, g; 28. g; 29. f, g; 30. Anti Rh antibodies can pass across the placenta and damage red blood cells in the fetal circulation resulting in anemia or death of the fetus.

37.5. HUMAN CARDIOVASCULAR SYSTEM [pp.644-645]
37.6. THE HUMAN HEART [pp.646-647]
1. body systems; 2. portal; 3. T; 4. oxygenated or deoxygenated (pulmonary arteries); 5. T; 6. D; 7. E; 8. C; 9. A; 10. F; 11. B; 12. jugular; 13. superior vena cava; 14. pulmonary; 15. renal; 16. inferior vena cava; 17. iliac; 18. femoral; 19. femoral; 20. iliac; 21. abdominal; 22. renal; 23. brachial; 24. coronary;

25. pulmonary; 26. carotid; COLORING: all arteries are red except for the pulmonary trunk and arteries, which are blue. All veins are blue except for the pulmonary veins which are red. 27. aorta; 28. left pulmonary veins; 29. left semilunar valve; 30. left ventricle; 31. inferior vena cava; 32. right atrioventricular valve; 33. superior vena cava; COLORING: Both vevna cavae, the right atrium, right ventricle, and the pulmonary trunk and arteries are blue. The left atrium, left ventricle, the aorta, and the pulmonary veins are red.

37.7. PRESSURE, TRANSPORT, AND FLOW DISTRIBUTION [pp.648-649]
37.8. DIFFUSION AT CAPILLARIES, THEN BACK TO THE HEART [pp.650-651]
37.9. BLOOD AND CARDIOVASCULAR DISORDERS [pp.652-653]
1. aorta (arteries); 2. slows; 3. friction; 4. Arterioles; 5. vasodilation; 6. vasoconstriction; 7. Receptors; 8. kidneys; 9. vein; 10. artery; 11. arteriole; 12. capillary; 13. smooth muscle, elastic fibers; 14. valve; 15. capillaries; 16. slowly, need; 17. T; 18. the blood-brain barrier; 19. T; 20. muscles in veins contract; 21. ultrafiltration; 22. T; 23. Blood pressure depends on total blood volume and how much is pumped out of the heart in a given time period; 24. At 125, ventricles are contracting. At 86, ventricles are relaxed and atria are contracting. 25. I; 26. F; 27. E; 28. G; 29. A; 30. H; 31. D; 32. D; 33. C; 34. B

37.10. INTERACTIONS WITH THE LYMPHATIC SYSTEM [pp.654-655]
1. C; 2. B; 3. D; 4. A; 5. tonsils; 6. thymus; 7. thoracic; 8. spleen; 9. lymph node(s); 10. lymphocytes; 11. bone marrow; 12. 1. Drainage channels for water and plasma proteins that leaked out of capillaries; 2. Delivers fats absorbed from intestine to blood; 3. Transports cell debris, pathogens, foreign cells to lymph nodes.

SELF-TEST
1. d; 2. e; 3. e; 4. a; 5. b; 6. a; 7. a; 8. a; 9. G; 10. K; 11. F; 12. B; 13. M; 14. E; 15. L; 16. D; 17. J; 18. A; 19. H; 20. C; 21. I

Chapter 38   Immunity

Impacts, Issues: Frankie's Last Wish [p.658]

38.1. INTEGRATED RESPONSE TO THREATS
  [pp.660-661]
38.2. SURFACE BARRIERS [pp.662-663]
38.3. REMEMBER TO FLOSS [p.663]
38.4. INNATE IMMUNE RESPONSES [pp.664-665]
1. antigens; 2. 1000; 3. pathogen-associated molecular patterns; 4. complement; 5. innate; 6. adaptive; 7. skin; 8. mucus; 9. Lysozyme; 10. Gastric; 11. bacterial; 12. pathogens; 13. mucus membranes; 14. T; 15. T; 16. cytokines; 17. C; 18. D; 19. E; 20. G; 21. H; 22. A; 23. F; 24. B; 25. E; 26. A; 27. D; 28. B; 29. F; 30. C; 31. A fever increases the rate of enzyme activity resulting in more rapid metabolism, tissue repair, and activity of phagocytes. It may slow pathogen cell division.

38.5. OVERVIEW OF ADAPTIVE IMMUNITY
  [pp.666-667]
38.6. ANTIBODIES AND OTHER ANTIGEN
  RECEPTORS [pp.668-669]
38.7. THE ANTIBODY-MEDIATED IMMUNE
  RESPONSE [pp.670-671]
38.8. THE CELL-MEDIATED RESPONSE
  [pp.672-673]
1. innate; 2. adaptive; 3. MHC marker (surface antigens); 4. non-self; 5. lymphocytes; 6. B cell

(B lymphocyte); 7. T cell (T lymphocyte); 8. viruses (bacteria or fungus or protest); 9. cancer (tumor); 10. identity; 11. antigen; 12. antigen-presenting; 13. effector; 14. memory; 15. B; 16. T; 17. recognition of antigens; 18. T; 19. cytokines; 20. proteins; 21. T cells; 22. antigen presenting cells; 23. D; 24. C; 25. B; 26. A; 27. E; 28. C; 29. E; 30. F; 31. I; 32. B; 33. J; 34. D; 35. H; 36. G; 37. A; 38. G; 39. E; 40. A; 41. F; 42. B; 43. C; 44. D

38.9. ALLERGIES [p.673]
38.10. VACCINES [p.674]
38.11. IMMUNITY GONE WRONG [p.675]
38.12. AIDS REVISITED—IMMUNITY LOST
  [pp.676-677]
1. E; 2. H; 3. A; 4. G; 5. C; 6. J; 7. I; 8. F; 9. B; 10. D; 11. primary; 12. antigens; 13. T; 14. secondary; 15. reverse transcriptase; 16. too much; 17. T; 18. Edward Jenner

SELF-TEST
1. e; 2. b; 3. b; 4. a; 5. a; 6. e; 7. e; 8. d; 9. H; 10; D; 11. E; 12. J; 13. B; 14. G; 15. C; 16. I; 17. F; 18. A

Chapter 39   Respiration

Impacts, Issues: Up in Smoke [p.680]

39.1. THE NATURE OF RESPIRATION
  [pp.682-683]
39.2. GASPING FOR OXYGEN [p.683]
39.3. INVERTEBRATE RESPIRATION
  [pp. 684-685]
39.4. VERTEBRATE RESPIRATION [pp.686-687]
1. oxygen; 2. carbon dioxide; 3. carbon dioxide; 4. oxygen; 5. circulatory system; 6. water, solutes; 7. urinary system; 8. G; 9. D; 10. B; 11. G; 12. E; 13. A; 14. F; 15. Cold, fast flowing; 16. T; 17. first; 18. four; 19. T; 20. b; 21. c; 22. a; 23. a; 24. c; 25. d; 26. a; 27. c; 28. b; 29. a; 30. b

39.5. HUMAN RESPIRATORY SYSTEM
  [pp.688-689]
39.6. CYCLIC REVERSALS IN AIR PRESSURE
  GRADIENTS [pp.690-691]
1. alveoli (B); 2. bronchiole (J); 3. alveolar sac (L); 4. oral cavity (A); 5. pleural membrane (K);

6. intercostal muscles (H); 7. diaphragm (C); 8. nasal cavity (G); 9. pharynx (F); 10. epiglottis (M); 11. larynx (E); 12. trachea (I); 13. lung (D); 14. a; 15. a; 16. b; 17. b; 18. a; 19. b; 20. b; 21. a; 22. b; 23. C; 24. G; 25. A; 26. F; 27. D; 28. B; 29. E

39.7. GAS EXCHANGE AND TRANSPORT
  [pp.692-693]
39.8. RESPIRATORY DISEASES AND DISORDERS
  [pp.694-695]
39.9. HIGH CLIMBERS AND DEEP DIVERS
  [pp.696-697]
1. b; 2. a; 3. a (or c); 4. b; 5. a; 6. b; 7. c; 8. a; 9. G; 10. E; 11. B; 12. H; 13. C; 14. F; 15. A; 16. D

SELF-TEST
1. I; 2. F; 3. K; 4. L; 5. E; 6. H; 7. O; 8. G; 9. M; 10. B; 11. D; 12. A; 13. C; 14. N; 15. J; 16. a; 17. e; 18. a; 19. c; 20. a; 21. a; 22. a; 23. a; 24. d; 25. e

Chapter 40   Digestion and Human Nutrition

Impacts, Issues: Hormones and Hunger [p.700]

40.1. THE NATURE OF DIGESTIVE SYSTEMS [pp.702-703]
40.2. OVERVIEW OF THE HUMAN DIGESTIVE SYSTEM [pp.704-705]
40.3. FOOD IN THE MOUTH [p.705]
1. Leptin; 2. ghrelin; 3. obesity; 4. gastric bypass surgery; 5. nutrition; 6. mechanically (chemically); 7. chemically (mechanically); 8. absorption; 9. circulatory; 10. incomplete; 11. circulatory; 12. complete; 13. opening; 14. Movements; 15. secretion; 16. ruminants; 17. crown; 18. stomach; 19. small intestine; 20. anus; 21. accessory; 22. pancreas; 23. salivary amylase; 24. epiglottis; 25. esophagus; 26. salivary glands; 27. liver; 28. gallbladder; 29. pancreas; 30. anus; 31. large intestine (colon); 32. small intestine; 33. stomach; 34. esophagus; 35. pharynx; 36. mouth (oral cavity); 37. E; 38. G; 39. C; 40. J; 41. I; 42. D; 43. A; 44. F; 45. H; 46. B

40.4. FOOD BREAKDOWN IN THE STOMACH AND SMALL INTESTINE [pp.706-707]
40.6. ABSORPTION FROM THE SMALL INTESTINE [pp.708-709]
40.7. THE LARGE INTESTINE [p.710]
40.7. METABOLISM OF ABSORBED ORGANIC COMPOUNDS [p.711]
1. C; 2. G; 3. F; 4. A; 5. D; 6. B; 7. H; 8. E; 9. a. mouth, small intestine (duodenum); small intestine (jejunum/ileum); monosaccharide (sugar); b. stomach, small intestine (duodenum); small intestine (jejunum/ileum); amino acids; c. small intestine (duodenum); small intestine (jejunum/ileum); fatty acids, monoglycerides; d.

small intestine (duodenum); small intestine (jejunum/ileum); nucleotides (nucleotide base and monosaccharide); 10. acidic; 11. stomach; 12. T; 13. gastrin; 14. T; 15. osmosis; 16. lipids; 17. gastrin; 18. T

40.8. HUMAN NUTRITIONAL REQUIREMENTS [pp.712-713]
40.9. VITAMINS, MINERALS, AND PHYTOCHEMICALS [pp.714-715]
40.10. WEIGHTY QUESTIONS, TANTALIZING ANSWERS [pp.716-717]
1. vegetables and fruits; 2. milk (dairy); 3. whole grains; 4. refined; 5. saturated; 6. trans; 7. sugar; 8. salt; 9. high glycemic; 10. fat; 11. Fructose; 12. Mediterranean; 13. olive; 14. Low-carb; 15. vitamins; 16. minerals; 17. coenzymes; 18. E; 19. I; 20. H; 21. J; 22. F; 23. G; 24. C; 25. D; 26. A; 27. B; 28. a. 2070; b. 2900; c. 1230; 29. Consulting text Figure 40.14 (men's column, 6 feet tall) yields 178 pounds as his ideal weight. 195-178=17 pounds overweight. Maintaining 178 pounds at a low activity level would require an intake of 1,780 kilocalories/day.

SELF-TEST
1. b; 2. b; 3. a; 4. e; 5. d; 6. c; 7. c; 8. C; 9. H; 10. B; 11. A; 12. G; 13. E; 14. D; 15. F

Chapter 41   Maintaining the Internal Environment

Impacts, Issues: Truth in a Test Tube [p.720]

41.1. MAINTENANCE OF EXTRACELLULAR FLUID [p.722]
41.2. HOW DO INVERTEBRATES MAINTAIN FLUID BALANCE [pp.722-723]
41.3. FLUID REGULATION IN VERTEBRATES [pp.724-725]
1. Diabetes mellitus; 2. bacterial infection; 3. protein; 4. LH (luteinizing hormone); 5. pregnant; 6. menopause; 7. ten; 8. interstitial; 9. blood; 10. extracellular; 11. urinary; 12. volume; 13. homeostasis; 14. D; 15. J; 16. F; 17. B; 18. I; 19. A; 20. E; 21. C; 22. H; 23. G; 24. less; 25. losing; 26. T; 27. uric acid, urea; 28. T; 29. larger; 30. gain; 31. skin

41.4. THE HUMAN URINARY SYSTEM [pp.726-727]
41.5. HOW URINE FORMS [pp.728-729]
41.6 REGULATION OF WATER INTAKE AND URINE FORMATION [pp.730-731]
41.7. ACID-BASE BALANCE [p.731]
1. kidney; 2. ureter; 3. urinary bladder; 4. urethra; 5. kidney cortex; 6. kidney medulla; 7. ureter; 8. kidneys; 9. nephrons; 10. Bowman's capsule; 11. glomerular; 12. renal corpuscle; 13. proximal; 14. loop of Henle; 15. collecting duct; 16. ureter; 17. urine; 18. ureter; 19. urinary bladder; 20. urethra; 21. reflex; 22. skeletal (sphincter); 23. voluntary; 24. Glomerular capillaries; 25. proximal (convoluted)

tubule; 26. Bowman's capsule; 27. distal (convoluted) tubule; 28. peritubular capillaries; 29. collecting duct; 30. loop of Henle; 31. c; 32. a; 33. b; 34. c; 35. a; 36. B; 37. E; 38. D; 39. C; 40. A; 41. pH; 42. bicarbonate (HCO_3); 43. T; 44. little; 45. too little; 46. high

41.8. WHEN KIDNEYS FAIL [p.732]
41.9. HEAT GAINS AND LOSSES [p.733]
41.10. TEMPERATURE REGULATION IN MAMMALS [pp.734-735]
1. C; 2. E; 3. A; 4. B; 5. D; 6. F; 7. E; 8. D; 9. B; 10. A; 11. C; 12. G; 13. F; 14. E; 15. a; 16. c; 17. c; 18. b; 19. a; 20. c; 21. c; 22. b; 23. b; 24. d

SELF-TEST
1. c; 2. e; 3. d; 4. b; 5. d; 6. b; 7. d; 8. e; 9. b; 10. a; 11. d

---

Chapter 42   animal Reproductive Systems

Impacts, Issues: Male or Female? Body or Genes? [p.738]

42.1. MODES OF REPRODUCTION [pp.740-741]
42.2. REPRODUCTIVE SYSTEM OF HUMAN MALES [pp.742-743]
42.3. SPERM FORMATION [pp.744-745]
1. C; 2. F; 3. B; 4. I; 5. G; 6. H; 7. J; 8. A; 9. E; 10. D; 11. prostate gland; 12. urinary bladder; 13. urethra; 14. penis; 15. testis; 16. ejaculatory duct; 17. seminal vesicle; 18. bulbourethral gland; 19. vas deferens; 20. epididymis; 21. scrotum; 22. I; 23. E; 24. G; 25. B; 26. H; 27. J; 28. C; 29. D; 30. A; 31. F

42.4. REPRODUCTIVE SYSTEM OF HUMAN FEMALE [pp.746-747]
42.5. FEMALE TROUBLES [p.747]
    42.6. PREPARATIONS FOR PREGNANCY [pp.748-749]
42.7. FSH AND TWINS [p.750]
1. ovary; 2. oviduct (fallopian tube); 3. clitoris; 4. labium minora; 5. labium majora; 6. uterus; 7. myometrium; 8. endometrium; 9. vagina; 10. E; 11. J; 12. G; 13. C; 14. A; 15. B; 16. I; 17. F; 18. D;

19. H; 20. F; 21. J; 22. A; 23. H; 24. G; 25. C; 26. I; 27. E; 28. D; 29. B; 30. Higher levels of follicle stimulating hormone (FSH) are associated with an increased probability of twin formation. This relationship only applies to fraternal twins, not identical twins.

42.8. WHEN GAMETES MEET [pp.750-751]
42.9. PREVENTING OR SEEKING PREGNANCY [pp.752-753]
42.10. SEXUALLY TRANSMITTED DISEASES [pp.754-755]
1. parasympathetic; 2. Oxytocin; 3. T; 4. one; 5. *zona pellucidum*; 6. diploid zygote; 7. after fertilization; 8. million; 9. F; 10. E; 11. I; 12. J; 13. C; 14. H; 15. A; 16. D; 17. B; 18. G; 19. E; 20. H; 21. D; 22. I; 23. B; 24. G; 25. A; 26. F; 27. C

SELF-TEST
1. b; 2. c; 3. e; 4. a; 5. a; 6. b; 7. d; 8. c; 9. d; 10. e

Chapter 43   Animal Development

Impacts, Issues: Mind Boggling Births [p. 758]

43.1. STAGES OF REPRODUCTION AND DEVELOPMENT [pp.760-761]
43.2. EARLY MARCHING ORDERS [pp.762-763]
1. Gamete formation; 2. fertilization; 3. zygote; 4. Cleavage; 5. blastomeres; 6. Gastrulation; 7. Ectoderm; 8. endoderm; 9. mesoderm; 10. F; 11. B; 12. C; 13. A; 14. E; 15. D; 16. D; 17. H; 18. E; 19. A; 20. I; 21. G; 22. J; 23. B; 24. F; 25. C; 26. Cytoplasmic localization means that cytoplasmic contents of the egg are not evenly distributed. As a result, as cleavage occurs, blastomeres have different contents. This leads to differences in cell differentiation.

43.3. FROM BLASTULA TO GASTRULA [pp.764-765]
43.4. SPECIALIZED TISSUES AND ORGANS FORM [p.765]
43.5. AN EVOLUTIONARY VIEW OF DEVELOPMENT [p.766]
1. induction; 2. morphogens; 3. gradient; 4. cytoplasmic localization; 5. homeotic; 6. position; 7. physical; 8. surface-to-volume; 9. architectural; 10. phyletic; 11. changes the developmental pattern; 12. T; 13. induction; 14. cell differentiation; 15. morphogens; 16. concentration gradient; 17. bones and skeletal muscles; 18. T; 19. Physical; 20. phyletic; 21. One group of cells has an effect on adjacent cells

and regulates their developmental pattern; 22. Apoptosis is when cells self-destruct at an appropriate time to create a particular physical shape. During development, there are times when cells are needed initially but then need to disappear at a later time. They do so by self-destruction. One example is webbing between digits on the hand that disappears later so that fingers are separated.; 23. (1) Molecules in different areas of unfertilized egg induce local expression of master genes creating chemical gradients. (2) Depending on location in gradient, other master genes are activated or inhibited, forming additional gradients. (3) Depending on position within gradient, homeotic genes are expressed regulating development of specific body parts.

43.6. OVERVIEW OF HUMAN DEVELOPMENT [p.767]
43.7. EARLY HUMAN DEVELOPMENT [pp.768-769]
43.8. EMERGENCE OF THE VERTEBRATE BODY PLAN [p.770]
43.9. THE FUNCTION OF THE PLACENTA [p.771]
43.10. EMERGENCE OF DISTINCTLY HUMAN FEATURES [pp.772-773]

1. implantation; 2. chorionic villi; 3. placenta; 4. eighth; 5. fetus; 6. b; 7. a; 8. b; 9. b; 10. a; 11. b; 12 a; 13. b; 14. b; 15. b; 16. a; 17. K; 18. I; 19. F; 20. A; 21. G; 22. B; 23. E; 24. C; 25. L; 26. J; 27. D; 28. H;

29. D; 30. B; 31. A; 32. C; 33. c; 34. a; 35. b; 36. c; 37. a; 38. c; 39. b; 40. c; 41. c; 42. The placenta exchanges all materials between the blood of the mother and the blood of the fetus. Materials include nutrients and oxygen to the fetus and wastes and CO_2 to the mother's blood.; 43. Since the blood of the mother and the fetus never mix, the mother's system is not exposed to the different proteins of the fetus that could be treated as antigens. Therefore, no antibodies are made.; 44. embryonic disk; 45. amniotic cavity; 46. pharyngeal arches; 47. somites; 48. limb bud

43.11. MOTHER AS PROVIDER AND PROTECTOR [pp.774-775]
43.12. BIRTH AND LACTATION [p.776]

1. folate; 2. iodine; 3. morning sickness; 4. organs; 5. teratogens; 6. toxoplasmosis; 7. developmental problems; 8. stillborn; 9. Rubella; 10. alcohol; 11. caffeine; 12. carbon monoxide; 13. oxygen; 14. thalidomide; 15. acne; 16. cranial or facial deformities; 17. G; 18. E; 19. J; 20. A; 21. H; 22. C; 23. B; 24. D; 25. I; 26. F; 27. C; 28. D; 29. A; 30. E; 31. B

SELF-TEST
1. a; 2. b; 3. c; 4. a; 5. c; 6. e; 7. d; 8. b; 9. e; 10. a; 11. c; 12. b; 13. d; 14. e

---

Chapter 44   Animal Behavior

Impacts, Issues: My Pheromones Made Me Do It [p.780]

44.1. BEHAVIORAL GENETICS [pp.782-783]
44.2. INSTINCT AND LEARNING [pp.784-785]
1. c; 2. b; 3. e; 4. f; 5. a; 6. d; 7. g

44.3. ADAPTIVE BEHAVIOR [p.786]
44.4. COMMUNICATION SIGNALS [pp.786-787]
44.5. MATES, OFFSPRING, AND REPRODUCTIVE SUCCESS [pp.788-789]
1. b; 2. i; 3. j; 4. f; 5. g; 6. e; 7. h; 8. a; 9. c; 10. d; 11. k; 12a. midwife toad; 12b. hanging flies; 12c. bison, lion, elk, elephant seals; 12d. sage grouse

44.6. LIVING IN GROUPS [pp.790-791]
44.7. WHY SACRIFICE YOURSELF? [pp.792-793]
44.8. HUMAN BEHAVIOR [p.793]
1. a; 2. d; 3. c; 4. b; 5. e; 6. diploid; 7. genetic; 8. ½; 9. ancestors; 10. genetically; 11. offspring; 12. ¼; 13. Sterile; 14. self-sacrifice; 15. relatives; 16. family; 17. queens; 18. genes

SELF-TEST
1. c; 2. d; 3. b; 4. d; 5. a; 6. c; 7. c; 8. a; 9. d; 10. d

Chapter 45   Population Ecology

Impacts, Issues: The Numbers Game [p.796]

45.1. POPULATION DEMOGRAPHICS [p.798]
45.2. ELUSIVE HEADS TO COUNT [p.799]
45.3. POPULATION SIZE AND EXPONENTIAL GROWTH [pp.800-801]
1. k; 2. h; 3. d; 4. i; 5. b; 6. f; 7. a; 8. g; 9. j; 10. l; 11. c; 12. e; 13. 100; 14. capture was not random, some of the individuals die during the study period, no migration, animals become trap shy and avoid the traps; 15. f; 16. g; 17. b; 18. c; 19. a; 20. h; 21. d; 22. e; 23a. it increases; 23b. it decreases; 23c. it must increase; 24. 100,000; 25a. 100,000; 25b. 300,000

45.4. LIMITS ON POPULATION GROWTH [pp.802-803]
45.5. LIFE HISTORY PATTERNS [pp.804-805]
45.6. NATURAL SELECTION AND LIFE HISTORIES [pp.806-807]
1. a; 2. c; 3. b; 4. e; 5. d; 6. f; 7. g; 8. b; 9. c; 10. a; 11. a; 12. c; 13. b

45.7. HUMAN POPULATION GROWTH [pp.808-809]
45.8. FERTILITY RATES AND AGE STRUCTURE [pp.810-811]
45.9. POPULATION GROWTH AND ECONOMIC EFFECTS [pp.812-813]
45.10. RISE OF THE SENIORS [p.813]
1. woodlands; 2. vegetarians; 3. two; 4. agricultural; 5. migratory; 6. grains; 7. grasses; 8. wheat; 9. rice; 10. animals; 11. Irrigation; 12. nutrition; 13. diseases; 14. birth; 15. dependent; 16. cholera; 17. clean water; 18. antibiotics; 19. death; 20. b; 21. a; 22. c; 23. a; 24. c; 25. a; 26. b; 27. d; 28. a; 29. c

SELF-TEST
1. d; 2. a; 3. b; 4. d; 5. a; 6. a; 7. a; 8. c; 9. b; 10. a; 11. c; 12. a; 13. a

---

Chapter 46   Community Structure and Biodiversity

Impacts, Issues: Fire Ants in the Pants [p.816]

46.1. WHICH FACTORS SHAPE COMMUNITY STRUCTURE? [p.818]
46.2. MUTUALISM [p.819]
1. community; 2. climate, topography; 3. temperature, moisture; 4. food; 5. species; 6. traits; 7. adapt; 8. interactions; 9. shift; 10. natural; 11. disturbance; 12. h; 13. d; 14. g; 15. b; 16. c; 17. a; 18. f; 19. e

46.3. COMPETITIVE INTERACTIONS [pp.820-821]
46.4. PREDATOR-PREY INTERACTIONS [pp.822-823]
46.5. AN EVOLUTIONARY ARMS RACE [pp.824-825]
1. b; 2. e; 3. f; 4. c; 5. a; 6. d; 7. b; 8. c; 9. a; 10. c; 11. d

46.6. PARASITE- HOST INTERACTIONS [pp. 826-827]
46.7. STRANGERS IN THE NEST [pp. 827]
46.8. ECOLOGICAL SUCCESSION [pp. 828-829]
46.9. SPECIES INTERACTIONS AND COMMUNITY INSTABILITY [pp. 830 – 831]
1. Parasites; 2. nutrients; 3. habitat; 4. predation; 5. mates; 6. sterility; 7. sex ratio; 8. birth; 9. death; 10. competition; 11. death; 12. infection; 13. bad; 14. reproductive; 15. offspring; 16. favor; 17. less; 18. b; 19. a; 20. b; 21. b; 22. a; 23. a; 24. d; 25. b; 26. e; 27. c; 28. f; 29. a; 30. g

46.10. EXOTIC INVADERS [pp. 832-833]
46.11. BIOGEOGRAPHIC PATTERNS IN COMMUNITY STRUCTURE [pp. 834 - 835]
1a. tropical latitudes intercept more intense sunlight, receive more rainfall and have a longer growing season, 1b. tropical communities have been evolving longer than temperate ones; 1c. species richness may be self-reinforcing; 2. island C

SELF-TEST
1. b; 2. b; 3. c; 4. a; 5. b; 6. d; 7. d; 8. d; 9. a; 10. d; 11. a

Chapter 47   Ecosystems

Impacts, Issues: Bye-Bye, Blue Bayou [pp. 838]

47.1. THE NATURE OF ECOSYSTEMS
   [pp.840- 841]
47.2. THE NATURE OF FOOD WEBS [pp. 842- 843]
47.3. ENERGY FLOW THROUGH ECOSYSTEMS
   [pp. 844- 845]
1. H; 2. C; 3. K; 4. F; 5. M; 6. O; 7. B; 8. I; 9. N;
10. A; 11. P; 12. G; 13. L; 14. D; 15. J; 16. E;
17. Producer; 18. Herbivore; 19. Carnivore;
20. Decomposer; 21. Energy transfers in aquatic
ecosystems are due to a variety of factors; the cell
walls of most alga is composed of lignin instead of
cellulose. This makes the alga much more digestible
than its plant counterparts. Aquatic ecosystems
usually have a higher percentage of cold blooded
organisms than land ecosystems. Cold blooded
organisms don't need as much energy in order to
maintain a constant internal temperature.

47.4 BIOLOGICAL MAGNIFICATION [p. 846]
47.5 BIOGEOCHEMICAL CYCLES [p. 847]
1. The concentration of a substance that can be
found in the tissues of an organisms will increase as it
moves up the trophic levels; 2. DDT disrupts the
metabolic activities of many animals, it is also a very
toxic substance to many species. Because it is not
selective about which species it will harm it often
harms beneficial species as well as harmful ones.;
3. biogeochemical; 4. essential; 5. nutrients;

6. nitrogen; 7. reservoirs; 8. water; 9. water;
10. carbon; 11. ecosystems; 12. sedimentary;
13. seafloor; 14. crust

47.6 WATER CYCLE [pp.848- 849]
47.7 CARBON CYCLE [pp. 850- 851]
47.8 GREENHOUSE GASES, GLOBAL WARMING
   [pp.852- 853]
1. A; 2. F; 3. E; 4. D; 5. C; 6. B; 7. H; 8. E; 9. C;
10. A; 11. F; 12. D; 13. B; 14. G; 15. The gas layer
helps keep the Earth warm enough to support life,
but rising temperatures may cause glaciers to melt, a
rise in seal levels and the disruption of global weather
patterns.;16. the burning of fossil fuels by humans as
a source of energy

47.9 NITROGEN CYCLE [pp.854- 855]
47.10 PHOSPHORUS CYCLE [pp. 856-857]
1. F; 2. E; 3. C; 4. H; 5. A; 6. B; 7. D; 8. G; 9. The
benefits include the ability to survive in harsh/
nitrogen poor environments. The costs are the loss of
the by products of photosynthesis to the bacteria.

SELF- TEST
1. c; 2. c; 3. d; 4. c; 5. a; 6. a; 7. b; 8. a; 9. a; 10. a;
11. b; 12. d; 13. a; 14. a

Chapter 48   The Biosphere

Impacts, Issues: Surfers, Seals, and the Sea [p.860]

48.1. GLOBAL AIR CIRCULATION PATTERNS
   [pp.862-863]
48.2. SOMETHING IN THE AIR [pp.864-865]
48.3. THE OCEANS, LANDFORMS, AND
   CLIMATES [pp.866-867]
1. Climate; 2. Wind; 3. ocean currents; 4. sunlight;
5. warm; 6. equator; 7. global air; 8. moisture;
9. warms; 10. ascends; 11. moisture; 12. descends;
13. moisture; 14. ascends; 15. moisture; 16. descends;
17. easterlies; 18. westerlies; 19. C; 20. A; 21. D; 22. B;
23. E; 24. A; 25. D; 26. B; 27. C; 28. E

48.4. BIOGEOGRAPHIC REALMS AND BIOMES
   [pp.868-869]
48.5. SOILS OF MAJOR BIOMES [p.870]
48.6. DESERTS [p.871]
1. H; 2. F; 3. A; 4. G; 5. I; 6. B; 7. J; 8. E; 9. D; 10. C;
11. A; 12. E; 13. C; 14. B; 15. D

48.7. GRASSLANDS, SHRUBLANDS, AND
   WOODLANDS [pp.872-873]
48.8. MORE RAIN, BROADLEAF FORESTS [p.874]
48.9. YOU AND THE TROPICAL FORESTS [p.875]
48.10. CONIFEROUS FORESTS [p.876]
48.11. TUNDRA [p.877]
1. D; 2. B; 3. C; 4. A; 5. E; 6. F; 7. G; 8. J; 9. H; 10. I;
11. K

48.12. FRESHWATER ECOSYSTEMS [pp.878-879]
48.13. "FRESH" WATER? [p.880]
48.14. COASTAL ZONES [pp.880-881]
1. D; 2. B; 3. H; 4. C; 5. A; 6. G; 7. E; 8. F;
9. Oligotrophic lakes: deep, steeply banked; large deep-water volume relative to surface-water volume; highly transparent; water blue or green; low nutrient content; oxygen abundant through all levels throughout the year; not much phytoplankton; green algae and diatoms dominant; aerobic decomposers favored in profundal zone; low biomass in profundal zone. Eutrophic lakes: shallow with broad littoral; small deep-water volume relative to surface-water volume; limited transparency; water green to yellow or brownish-green; high nutrient content; oxygen depleted in deep water during summer; abundant, thick masses of phytoplankton; cyanobacteria dominant; anaerobic decomposers in profundal zone; high biomass in profundal zone.; 10. Spring overturn: winds cause vertical currents that forces oxygen-rich water in the surface layers to move down and nutrient-rich water from the lake's depth to move upwards.Fall overturn: the upper layer cools and sinks; this causes the oxygen-rich water to move downward while the nutrient-rich water moves upward.

48.15. THE ONCE AND FUTURE REEFS
   [pp.882-883]
48.16. THE OPEN OCEAN [pp.884-885]
48.17. CLIMATE, COPEPODS, AND CHOLERA
   [pp.886-887]
1. G; 2. F; 3. E; 4. J; 5. C; 6. I; 7. D; 8. B; 9. A; 10. H;
11. K; 12. Benthic; 13. Pelagic; 14. Neritic;
15. Oceanic

SELF-TEST
1. a; 2. d; 3. b; 4. d; 5. e; 6. b; 7. c; 8. b; 9. d; 10. c;
11. c; 12. a; 13. a; 14. a

Chapter 49   Human Impacts on the Biosphere

Impacts, Issues: A Long Reach [pp. 890]

49.1. THE EXTINCTION CRISIS [pp.892-893]
49.2. CURRENT THREATS TO SPECIES
   [pp.894-895]
49.3. THE UNKNOWN LOSSES [p.896]
1. extinction; 2. replaced; 3. 99; 4. biodiversity;
5. Five; 6. geologic; 7. 10; 8. 6th; 9. humans
10. B, E; 11. F; 12. C, I; 13. H; 14. A, D, G;
15. China is trying to protect existing habitat, create corridors that connect suitable habitat and existing populations, as well as expand captive breeding programs.;16. Acid rain, pesticide residues, fertilizer runoff, and emissions of greenhouse gasses degrade habitats.

49.4. ASSESSING BIODIVERSITY [pp.896-897]
49.5. EFFECTS OF DEVELOPMENT AND
   CONSUMPTION [pp.898-899]
1. C; 2. F; 3. B; 4. E; 5. A; 6. D; 7. Survey the range of biodiversity, investigate the evolutionary and ecological origins of biodiversity, and identify ways to maintain and use biodiversity in ways that benefit human populations.; 8. C; 9. D; 10. F; 11. B; 12. E;
13. A; 14. 39.8; 15. 22.4; 16. 6.8; 17. 8.2; 18. 28.3;
19. 10.3; 20. 39.7; 21. 21.6

49.6. THE THREAT OF DESERTIFICATION
   [p.900]
49.7. THE TROUBLE WITH TRASH [p.901]
49.8. MAINTAINING BIODIVERSITY & HUMAN
   POPULATIONS [pp.902-903]
1. 4.6; 2. 1/3; 3. burned; 4. landfill; 5. 100; 6. 50;
7. 10; 8. Recycling; 9. disposal; 10. C; 11. A; 12. D;
13. F; 14. E; 15. B

SELT-TEST
1. a; 2. c; 3. d; 4. d; 5. b; 6. a; 7. b; 8. d; 9. d